Sources and Composition of Ambient Particulate Matter

Sources and Composition of Ambient Particulate Matter

Editor

Manousos-Ioannis Manousakas

MDPI • Basel • Beijing • Wuhan • Barcelona • Belgrade • Manchester • Tokyo • Cluj • Tianjin

Editor
Manousos-Ioannis Manousakas
Paul Scherrer Institut (PSI)
Villigen
Switzerland

Editorial Office
MDPI
St. Alban-Anlage 66
4052 Basel, Switzerland

This is a reprint of articles from the Special Issue published online in the open access journal *Atmosphere* (ISSN 2073-4433) (available at: www.mdpi.com/journal/atmosphere/special_issues/Ambient_PM).

For citation purposes, cite each article independently as indicated on the article page online and as indicated below:

LastName, A.A.; LastName, B.B.; LastName, C.C. Article Title. *Journal Name* **Year**, *Volume Number*, Page Range.

ISBN 978-3-0365-0995-2 (Hbk)
ISBN 978-3-0365-0994-5 (PDF)

© 2021 by the authors. Articles in this book are Open Access and distributed under the Creative Commons Attribution (CC BY) license, which allows users to download, copy and build upon published articles, as long as the author and publisher are properly credited, which ensures maximum dissemination and a wider impact of our publications.

The book as a whole is distributed by MDPI under the terms and conditions of the Creative Commons license CC BY-NC-ND.

Contents

About the Editor vii

Manousos-Ioannis Manousakas
Special Issue Sources and Composition of Ambient Particulate Matter
Reprinted from: *Atmosphere* **2021**, *12*, 462, doi:10.3390/atmos12040462 1

Sacha Moretti, Apostolos Salmatonidis, Xavier Querol, Antonella Tassone, Virginia Andreoli, Mariantonia Bencardino, Nicola Pirrone, Francesca Sprovieri and Attilio Naccarato
Contribution of Volcanic and Fumarolic Emission to the Aerosol in Marine Atmosphere in the Central Mediterranean Sea: Results from Med-Oceanor 2017 Cruise Campaign
Reprinted from: *Atmosphere* **2020**, *11*, 149, doi:10.3390/atmos11020149 5

Shijie Cui, Ruoyuan Lei, Yangzhou Wu, Dandan Huang, Fuzhen Shen, Junfeng Wang, Liping Qiao, Min Zhou, Shuhui Zhu, Yingge Ma and Xinlei Ge
Characteristics of Black Carbon Particle-Bound Polycyclic Aromatic Hydrocarbons in Two Sites of Nanjing and Shanghai, China
Reprinted from: *Atmosphere* **2020**, *11*, 202, doi:10.3390/atmos11020202 25

Olga Popovicheva, Alexey Ivanov and Michal Vojtisek
Functional Factors of Biomass Burning Contribution to Spring Aerosol Composition in a Megacity: Combined FTIR-PCA Analyses
Reprinted from: *Atmosphere* **2020**, *11*, 319, doi:10.3390/atmos11040319 39

Manousos Ioannis Manousakas, Kalliopi Florou and Spyros N. Pandis
Source Apportionment of Fine Organic and Inorganic Atmospheric Aerosol in an Urban Background Area in Greece
Reprinted from: *Atmosphere* **2020**, *11*, 330, doi:10.3390/atmos11040330 59

César Marina-Montes, Luis Vicente Pérez-Arribas, Jesús Anzano and Jorge O. Cáceres
Local and Remote Sources of Airborne Suspended Particulate Matter in the Antarctic Region
Reprinted from: *Atmosphere* **2020**, *11*, 373, doi:10.3390/atmos11040373 77

Subin Yoon, Sascha Usenko and Rebecca J. Sheesley
Fine and Coarse Carbonaceous Aerosol in Houston, TX, during DISCOVER-AQ
Reprinted from: *Atmosphere* **2020**, *11*, 482, doi:10.3390/atmos11050482 89

Sofia Caumo, Roy E. Bruns and Pérola C. Vasconcellos
Variation of the Distribution of Atmospheric *n*-Alkanes Emitted by Different Fuels' Combustion
Reprinted from: *Atmosphere* **2020**, *11*, 643, doi:10.3390/atmos11060643 103

Mathieu Lachatre, Gilles Foret, Benoit Laurent, Guillaume Siour, Juan Cuesta, Gaëlle Dufour, Fan Meng, Wei Tang, Qijie Zhang and Matthias Beekmann
Air Quality Degradation by Mineral Dust over Beijing, Chengdu and Shanghai Chinese Megacities
Reprinted from: *Atmosphere* **2020**, *11*, 708, doi:10.3390/atmos11070708 123

Elena Hristova, Blagorodka Veleva, Emilia Georgieva and Hristomir Branzov
Application of Positive Matrix Factorization Receptor Model for Source Identification of PM10 in the City of Sofia, Bulgaria
Reprinted from: *Atmosphere* **2020**, *11*, 890, doi:10.3390/atmos11090890 145

Giovanni Lonati, Nicola Pepe, Guido Pirovano, Alessandra Balzarini, Anna Toppetti and Giuseppe Maurizio Riva
Combined Eulerian-Lagrangian Hybrid Modelling System for PM2.5 and Elemental Carbon Source Apportionment at the Urban Scale in Milan
Reprinted from: *Atmosphere* **2020**, *11*, 1078, doi:10.3390/atmos11101078 161

Alex Vinson, Allie Sidwell, Oscar Black and Courtney Roper
Seasonal Variation in the Chemical Composition and Oxidative Potential of $PM_{2.5}$
Reprinted from: *Atmosphere* **2020**, *11*, 1086, doi:10.3390/atmos11101086 181

Styliani Pateraki, Kyriaki-Maria Fameli, Vasiliki Assimakopoulos, Kyriaki Bairachtari, Alexandros Zagkos, Theodora Stavraka, Aikaterini Bougiatioti, Thomas Maggos and Nikolaos Mihalopoulos
Differentiation of the Athens Fine PM Profile during Economic Recession (March of 2008 Versus March of 2013): Impact of Changes in Anthropogenic Emissions and the Associated Health Effect
Reprinted from: *Atmosphere* **2020**, *11*, 1121, doi:10.3390/atmos11101121 195

Moustapha Kebe, Alassane Traore, Manousos Ioannis Manousakas, Vasiliki Vasilatou, Ababacar Sadikhe Ndao, Ahmadou Wague and Konstantinos Eleftheriadis
Source Apportionment and Assessment of Air Quality Index of $PM_{2.5-10}$ and $PM_{2.5}$ in at Two Different Sites in Urban Background Area in Senegal
Reprinted from: *Atmosphere* **2021**, *12*, 182, doi:10.3390/atmos12020182 207

Chrysoula Betsou, Evangelia Diapouli, Evdoxia Tsakiri, Lambrini Papadopoulou, Marina Frontasyeva, Konstantinos Eleftheriadis and Alexandra Ioannidou
First-Time Source Apportionment Analysis of Deposited Particulate Matter from a Moss Biomonitoring Study in Northern Greece
Reprinted from: *Atmosphere* **2021**, *12*, 208, doi:10.3390/atmos12020208 223

About the Editor

Manousos-Ioannis Manousakas

Manousos-Ioannis Manousakas received his PhD in environmental monitoring and analysis in 2014 from the Chemistry Department at the University of Patras, Greece. After his studies, he worked as a collaborating researcher at NCSR Demokritos and the Institute of Chemical Engineering Sciences in Greece, and at Paul Scherrer Institute in Switzerland. His key research interests include atmospheric pollution, specifically particulate-related pollution; chemical analysis; and source apportionment. He has a strong international network of collaborators, evidenced by the diverse colleagues with whom he has collaboratively published research papers in scientific journals, and has participated in several scientific projects from different funding programs. He serves as a source apportionment expert for many projects worldwide organized by the International Atomic Energy Agency (IAEA). He is a reviewer board member for five journals, and up until 2021, he has served as a reviewer for over seventeen journals and as a guest editor for three journals.

atmosphere

Editorial

Special Issue Sources and Composition of Ambient Particulate Matter

Manousos-Ioannis Manousakas

Laboratory of Atmospheric Chemistry, Paul Scherrer Institute (PSI), 5232 Villigen, Switzerland; manousos.manousakas@psi.ch

Citation: Manousakas, M.-I. Special Issue Sources and Composition of Ambient Particulate Matter. *Atmosphere* **2021**, *12*, 462. https://doi.org/10.3390/atmos12040462

Received: 26 March 2021
Accepted: 5 April 2021
Published: 7 April 2021

Publisher's Note: MDPI stays neutral with regard to jurisdictional claims in published maps and institutional affiliations.

Copyright: © 2021 by the author. Licensee MDPI, Basel, Switzerland. This article is an open access article distributed under the terms and conditions of the Creative Commons Attribution (CC BY) license (https://creativecommons.org/licenses/by/4.0/).

1. Introduction

Research related to ambient particulate matter (PM) remains very relative today due to the adverse effects PM have on human health. PM are pollutants with varying chemical compositions and may originate from many different emission sources, which directly affects their toxicity. To formulate effective control and mitigation strategies, it is necessary to identify PM sources and estimate their influence on ambient PM concentration, a process that is known as source apportionment (SA). Depending on the geographical location and characteristics of an area, many anthropogenic and natural sources may contribute to PM concentration levels, such as dust resuspension, sea salt, traffic, secondary aerosol formation (both organic and inorganic), industrial emissions, ship emissions, biomass burning, power plant emissions, etc.

Different methodological approaches have been used over recent years to study the aforementioned topics, but some scientific challenges remain, mainly related to the subjects of real-time chemical analysis and SA, uncertainty estimation of SA results, and analytical optimization for PM samples. Additionally, there are areas in the world for which results regarding the composition and sources of PM are still scarce.

This Special Issue's target was to include studies on all aspects of PM chemical characterization and source apportionment regarding the inorganic and/or organic fraction of PM.

2. Results

This special issue includes 14 published studies referring to different regions around the world: Europe (seven), Asia (two), N. America (two), S. America (one), Africa (one), and Antarctica (one). The wide variety of areas included in the issue provide a good overview of particulate-related pollution worldwide. Even though the classification is not always easy as the studies discuss more than one subject, the publications of the issue can be divided into two sub-groups: source apportionment or contributions of specific sources to PM (nine papers), and composition, characterization, and characteristics of PM (five papers).

The first sub-group includes nine papers. In the study entitled "Contribution of Volcanic and Fumarolic Emission to the Aerosol in Marine Atmosphere in the Central Mediterranean Sea: Results from Med-Oceanor 2017 Cruise Campaign", the authors Moretti et al. [1] studied the contribution of the geogenic sources volcanoes and fumaroles to the aerosol in marine atmosphere in the central Mediterranean basin. The study was carried out in the framework of the Med-Oceanor measurement program and aimed in assessing the impact to the aerosol of the most important Mediterranean volcanoes (Mount Etna, Stromboli Island, and Marsili Seamount) and solfatara areas (Phlegraean Fields complex, Volcano Islands, Ischia Island, and Panarea submarine fumarole). Using factor analysis and SEM/EDX (Scanning Electron Microscopy with Energy Dispersive X-Ray Analysis) analysis for the source apportionment, anthropogenic and natural sources including shipping emissions, volcanic and fumarolic load, as well as sea spray were identified as the main factors affecting aerosol levels in the study area.

In another study that took place in Patra, Greece with the title "Source Apportionment of Fine Organic and Inorganic Atmospheric Aerosol in an Urban Background Area in Greece", the authors Manousakas et al. [2] identified and quantified the contributions to both the inorganic and the organic fractions of PM in the area. To meet that goal, both on- and off-line techniques were used, including elemental composition, organic and elemental carbon (OC and EC) measurements, and high-resolution Aerosol Mass Spectrometry (AMS) from different techniques. The results of the two methods were synthesized, showcasing the complementarity of the two methodologies for fine PM source identification. The synthesis suggests that the contribution of biomass burning is quite robust, but that the exhaust traffic emissions are not due to local sources and may also include secondary OA from other sources.

In a publication regarding the sources of total suspended particulate (TSP) in Antarctic, "Local and Remote Sources of Airborne Suspended Particulate Matter in the Antarctic Region", the authors Marina-Montes et al. [3] identified a potentially significant role of terrestrial inputs, marine environments, and biological inputs. Air mass back trajectories were used to confirm the elemental source. These trajectories revealed that both crustal and marine inputs occurred following different pathways and were influenced by the Antarctic Circumpolar pattern.

Yoon et al. [4], in their publication "Fine and Coarse Carbonaceous Aerosol in Houston, TX, during DISCOVER-AQ", investigated the major sources and trends of particulate pollution in Houston. Total suspended particulate (TSP) and fine particulate matter ($PM_{2.5}$) samples were collected and analyzed. Characterization of organic carbon (OC) and elemental carbon (EC) combined with real-time black carbon (BC) concentration provided insight into the temporal trends of $PM_{2.5}$ and coarse PM in Houston in 2013. Lachatre et al. [5], "Air Quality Degradation by Mineral Dust over Beijing, Chengdu and Shanghai Chinese Megacities", quantified the degradation of air quality by dust over Beijing, Chengdu and Shanghai megacities using the three dimensions (3D) chemistry transport model CHIMERE, which simulates dust emission and transport online. According to their findings, annual dust contributions to the PM_{10} budget over Beijing, Chengdu and Shanghai were evaluated respectively as 6.6%, 9.5%, and 9.3%, while they estimated that dust outbreaks largely contribute to poor air quality events during springtime.

In a study conducted in Sofia, Bulgaria, Hristova et al. [6] identified the source contributions to PM_{10} during the period January 2019–January 2020. The results from the source apportionment study showed that the resuspension factor is the main contributor to the total PM_{10} mass (25%), followed by biomass burning (23%), mixed SO_4^{2-} (19%), secondary (16%), traffic (9%), industry (4%), nitrate rich (4%), and fuel oil burning (0.4%) in Sofia.

Lonati et al. [7] performed air quality modeling at the very local scale, within the urban area of the Milan city center in Italy, is performed through a hybrid modeling system (HMS) that combines the CAMx Eulerian model with AUSTAL2000 Lagrangian model. Results show that the outcome of the Eulerian model at the local scale is only representative of a background level, similar to the Lagrangian model's outcome for the green area receptor, but fails to reproduce concentration gradients and hot-spots, driven by local-source emissions.

In another interesting study that took place in Dakar, Senegal, "Source Apportionment and Assessment of Air Quality Index of $PM_{2.5-10}$ and $PM_{2.5}$ in at Two Different Sites in Urban Background Area in Senegal", Kebe et al. [8] have quantified $PM_{2.5}$ and PM_{coarse} sources in the area using Positive Matrix Factorization (PMF), and specifically the version of the Enviromental Protection Agency of USA (US EPA), the PMF 5. Four PM sources were identified: industrial emissions, mineral dust, traffic emissions, and sea salt/secondary sulfates. The study showcased the importance of natural sources such as dust resuspension in countries located near to the arid regions of Africa.

In the final study of this sub-group, "First-Time Source Apportionment Analysis of Deposited Particulate Matter from a Moss Biomonitoring Study in Northern Greece", Betsou et al. [9] have moss samples as biomonitors of deposited PM. A total of one hundred and five samples, mainly of the Hypnum cupressiforme Hedw moss species, were collected from the Northern Greece during 2015–2016, which also included samples from the metalliferous area of Skouries. Using the PMF model, five sources were identified in the region: soil dust, aged sea salt, road dust, lignite power plants, and an Mn-rich source.

The second sub-group includes five papers. In the study of Cui et al. [10], "Characteristics of black carbon particle-bound polycyclic aromatic hydrocarbons in two sites of Nanjing and Shanghai, China", the sources of PAHs (Polycyclic aromatic hydrocarbons) and refractory black carbon (rBC) were explored. This work, for the first time, investigated exclusively the rBC-bound PAH properties by using a laser-only Aerodyne soot-particle aerosol mass spectrometer (SP-AMS). Two datasets were used from urban Shanghai during the fall of 2018 and in suburban Nanjing during the winter of 2017, respectively. A multi-linear regression algorithm combined with PMF analyses on sources of PAHs revealed that the industry emissions contributed the majority of PAHs in Nanjing (~80%), while traffic emissions dominated PAHs in Shanghai (~70%).

In the study of Popovicheva et al. [11], "Functional factors of biomass burning contribution to spring aerosol composition in a megacity: Combined FTIR-PCA analyses", the authors used Principal Component Analysis (PCA) on infrared Fourier transmission (FTIR) spectroscopy data to estimate sources of aerosols in Moscow megacity in the spring of 2017. Principal component loadings of 58%, 21%, and 11% of variability reveal the functional factors of transport, biomass burning, biogenic, dust, and secondary aerosol spring source impacts. Caumo et al. [12], in their study "Variation of the distribution of atmospheric n-alkanes emitted by different fuels' combustion", presented the emission profiles of n-alkanes for different vehicular sources in two Brazilian São Paulo and Salvador using PCA. According to the analysis, the principal factors were attributed to mixed sources and to bus emissions. Pateraki et al. [13], "Differentiation of the Athens Fine PM Profile during Economic Recession (March of 2008 Versus March of 2013): Impact of Changes in Anthropogenic Emissions and the Associated Health Effect" evaluated the impact of the anthropogenic contribution to the fine PM chemical profile in Athens, Greece. They concluded that although the monitoring location was traffic-impacted, the heating sector, from both wood-burning and fossil fuel, proved to be the driving force for the configuration of the obtained PM picture.

In the last paper of the collection by Vinson et al. [14], "Seasonal Variation in the Chemical Composition and Oxidative Potential of $PM_{2.5}$", the purpose of this study was to analyze and compare the oxidative potential and elemental composition of $PM_{2.5}$ collected along two highways in central Oregon, USA in the winter (January) and summer (July–August). The oxidative potential (nM DTT consumed/µg $PM_{2.5}$/min) differed between seasons with summer samples having nearly a two-fold increase when compared to the winter. Significant negative correlations were observed between DTT consumption and several elements as well as with $PM_{2.5}$ mass, but the findings were dependent on if the data was normalized by $PM_{2.5}$ mass.

3. Conclusions

The studies in the collection can be divided into two sub-groups: source apportionment or contributions of specific sources to PM (nine papers), and composition, characterization, and characteristics of PM (five papers). All of the published studies in the first sub-group provide valuable information about PM sources in different regions worldwide and they showcase the importance and great relevance of source apportionment studies even today. Even though specific tools are preferred for such studies, such as PMF, various other tools that can be used that are described in the collection.

The studies from the second sub-group focus on certain chemical PM species and report PM's chemical characteristics in different areas and PM originating from specific sources.

The entire collection of studies provided valuable insights on PM chemical characterization and source apportionment regarding the inorganic and/or organic fraction of PM.

Funding: This research received no external funding.

Institutional Review Board Statement: Not applicable.

Informed Consent Statement: Not applicable.

Data Availability Statement: Data sharing not applicable.

Acknowledgments: The editor would like to thank the authors for their contributions to this Special Issue, and the reviewers for their constructive and helpful comments to improve the manuscripts. The editor is grateful to Assistant Editor Xue Liang for her kind support in processing and publishing this Special Issue.

Conflicts of Interest: The author declares no conflict of interest.

References

1. Moretti, S.; Salmatonidis, A.; Querol, X.; Tassone, A.; Andreoli, V.; Bencardino, M.; Pirrone, N.; Sprovieri, F.; Naccarato, A. Contribution of volcanic and fumarolic emission to the aerosol in marine atmosphere in the central mediterranean sea: Results from med-oceanor 2017 cruise campaign. *Atmosphere* **2020**, *11*, 149. [CrossRef]
2. Manousakas, M.I.; Florou, K.; Pandis, S.N. Source Apportionment of Fine Organic and Inorganic Atmospheric Aerosol in an Urban Background Area in Greece. *Atmosphere* **2020**, *11*, 330. [CrossRef]
3. Marina-Montes, C.; Pérez-Arribas, L.V.; Anzano, J.; Cáceres, J.O. Local and remote sources of airborne suspended particulate matter in the antarctic region. *Atmosphere* **2020**, *11*, 373. [CrossRef]
4. Yoon, S.; Usenko, S.; Sheesley, R.J. Fine and coarse carbonaceous aerosol in Houston, TX, during DISCOVER-AQ. *Atmosphere* **2020**, *11*, 482. [CrossRef]
5. Lachatre, M.; Foret, G.; Laurent, B.; Siour, G.; Cuesta, J.; Dufour, G.; Meng, F.; Tang, W.; Zhang, Q.; Beekmann, M. Air quality degradation by mineral dust over Beijing, Chengdu and Shanghai Chinese megacities. *Atmosphere* **2020**, *11*, 708. [CrossRef]
6. Hristova, E.; Veleva, B.; Georgieva, E.; Branzov, H. Application of positive matrix factorization receptor model for source identification of PM10 in the City of Sofia, Bulgaria. *Atmosphere* **2020**, *11*, 890. [CrossRef]
7. Lonati, G.; Pepe, N.; Pirovano, G.; Balzarini, A.; Toppetti, A.; Riva, G.M. Combined eulerian-lagrangian hybrid modelling system for pm2.5 and elemental carbon source apportionment at the urban scale in milan. *Atmosphere* **2020**, *11*, 1078. [CrossRef]
8. Kebe, M.; Traore, A.; Manousakas, M.I.; Vasilatou, V.; Ndao, A.S.; Wague, A.; Eleftheriadis, K. Source Apportionment and Assessment of Air Quality Index of PM2.5–10 and PM2.5 in at Two Different Sites in Urban Background Area in Senegal. *Atmosphere* **2021**, *12*, 182. [CrossRef]
9. Betsou, C.; Diapouli, E.; Tsakiri, E.; Papadopoulou, L.; Frontasyeva, M.; Eleftheriadis, K.; Ioannidou, A. First-Time Source Apportionment Analysis of Deposited Particulate Matter from a Moss Biomonitoring Study in Northern Greece. *Atmosphere* **2021**, *12*, 208. [CrossRef]
10. Cui, S.; Lei, R.; Wu, Y.; Huang, D.; Shen, F.; Wang, J.; Qiao, L.; Zhou, M.; Zhu, S.; Ma, Y.; et al. Characteristics of black carbon particle-bound polycyclic aromatic hydrocarbons in two sites of Nanjing and Shanghai, China. *Atmosphere* **2020**, *11*, 202. [CrossRef]
11. Popovicheva, O.; Ivanov, A.; Vojtisek, M. Functional factors of biomass burning contribution to spring aerosol composition in a megacity: Combined FTIR-PCA analyses. *Atmosphere* **2020**, *11*, 319. [CrossRef]
12. Caumo, S.; Bruns, R.E.; Vasconcellos, P.C. Variation of the distribution of atmospheric n-alkanes emitted by different fuels' combustion. *Atmosphere* **2020**, *11*, 643. [CrossRef]
13. Pateraki, S.; Fameli, K.M.; Assimakopoulos, V.; Bairachtari, K.; Zagkos, A.; Stavraka, T.; Bougiatioti, A.; Maggos, T.; Mihalopoulos, N. Differentiation of the Athens Fine PM Profile during Economic Recession (March of 2008 Versus March of 2013): Impact of Changes in Anthropogenic Emissions and the Associated Health Effect. *Atmosphere* **2020**, *11*, 1121. [CrossRef]
14. Vinson, A.; Sidwell, A.; Black, O.; Roper, C. Seasonal Variation in the Chemical Composition and Oxidative Potential of PM2.5. *Atmosphere* **2020**, *11*, 1086. [CrossRef]

Article

Contribution of Volcanic and Fumarolic Emission to the Aerosol in Marine Atmosphere in the Central Mediterranean Sea: Results from Med-Oceanor 2017 Cruise Campaign

Sacha Moretti [1], Apostolos Salmatonidis [2], Xavier Querol [2], Antonella Tassone [1], Virginia Andreoli [1], Mariantonia Bencardino [1], Nicola Pirrone [1], Francesca Sprovieri [1] and Attilio Naccarato [1,*]

[1] CNR-Institute of Atmospheric Pollution Research, Division of Rende, UNICAL-Polifunzionale, I-87036 Arcavacata di Rende, CS, Italy; sacha.moretti@iia.cnr.it (S.M.); antonella.tassone@iia.cnr.it (A.T.); virginia.andreoli@iia.cnr.it (V.A.); bencardino@iia.cnr.it (M.B.); pirrone@iia.cnr.it (N.P.); f.sprovieri@iia.cnr.it (F.S.)
[2] Institute of Environmental Assessment and Water Research (IDÆA-CSIC), C/Jordi Girona 18, 08034 Barcelona, Spain; apsgeo@cid.csic.es (A.S.); xavier.querol@idaea.csic.es (X.Q.)
* Correspondence: attilio.naccarato@iia.cnr.it

Received: 8 January 2020; Accepted: 27 January 2020; Published: 30 January 2020

Abstract: This work studied the contribution of the geogenic sources volcanoes and fumaroles to the aerosol in marine atmosphere in the central Mediterranean basin. For this purpose, in the framework of the Med-Oceanor measurement program, we carried out a cruise campaign in the summer of 2017 to investigate the impact to the aerosol of the most important Mediterranean volcanoes (Mount Etna, Stromboli Island, and Marsili Seamount) and solfatara areas (Phlegraean Fields complex, Volcano Islands, Ischia Island, and Panarea submarine fumarole). We collected PM_{10} and $PM_{2.5}$ samples in 12 sites and performed chemical characterization to gather information about the concentration of major and trace elements, elemental carbon (EC), organic carbon (OC), and ionic species. The use of triangular plots and the calculation of enrichment factors confirmed the interception of volcanic plume. We integrated the outcomes from chemical characterization with the use of factor analysis and SEM/EDX analysis for the source apportionment. Anthropogenic and natural sources including shipping emissions, volcanic and fumarolic load, as well as sea spray were identified as the main factors affecting aerosol levels in the study area. Furthermore, we performed pattern recognition analysis by stepwise linear discriminant analysis to seek differences in the composition of PM_{10} and $PM_{2.5}$ samples according to their volcanic or solfatara origin.

Keywords: Mediterranean Sea; particulate matter; volcanic area; air quality; carbonaceous compounds; element analysis

1. Introduction

Advancements in the current state of knowledge of atmospheric aerosols are an important research topic because of their implications in environmental and health issues [1]. Nowadays, much of the air pollution results from human activities, but natural sources also contribute to the increase of the exposure level of the population and ecosystem to the polluted air. Typically, the major natural emissions in terms of mass include sea spray (84%) and mineral dust (13%), along with other sources such as biological primary organic aerosols (POA), volcanic emissions, biogenic secondary organic aerosols (SOA), and biogenic sulfate particles [2]. Anthropogenic aerosols contribute only 2% to global emissions, mainly in the form of anthropogenic sulfate (49%) and industrial dust (40%), with additional emissions of anthropogenic nitrate, SOA, and fossil fuel-derived POA. On a global-scale, primary aerosols are dominant over secondary species (98% vs. 2%), whereas on the local-scale, this scenario is reversed [2]. Indeed, the source contributions to ambient aerosols may differ from one region to another since they are strongly affected by a broad spatial variability. Owing to the importance of aerosol investigation on both global and local scales, several international programs aim to assess its chemistry and interaction with the climate, e.g., Charmex (Chemistry-Aerosol Mediterranean Experiment) and IGAC (International Global Atmospheric Chemistry). The study and identification of events affecting air quality are complex tasks in which data handling strategies including aerosol maps, back-trajectory analysis, and receptor modeling support the laboratory analysis on the composition of particulate matter (PM). In this respect, the use of mass spectrometry technologies has been giving a considerable impulse allowing in-depth analyses and characterization of the PM [3] as well as other environmental matrices [4,5]. Natural contributions to the levels of ambient air particulate matter (PM) and their speciation in Europe have been characterized using different approaches in many studies [2,6–13]. The air quality of Mediterranean countries can be affected by four major natural sources of atmospheric aerosol, i.e., African dust, sea spray, wildfire, and volcanic emission [2]. The wind-blow African dust can increase ambient PM levels even above the PM thresholds [12–15], and its contribution that originates from arid and semiarid areas leads to an increase of mineral dust in the atmosphere. Another important pollution source is derived from the dispersed PM emitted from the sea surface, generally referred to as sea spray. Sea spray is produced by the bubble-bursting processes or wind-induced wave breaking, resulting in a spray of particles in the size range from sub-micrometers up to a few micrometers [16]. Carbonaceous aerosol derived from wildfires is another natural source of aerosol constituents due to the burning of forests and other vegetation through natural processes [10]. In Mediterranean countries, fire emissions during summer have special relevance because the weather is dry and hot [17], and fire outbreaks are commonly fanned by strong winds. A further contribution to PM is given by volcanic emissions, which release a considerable variety of compounds, such as water vapor, ash, CO_2, SO_2, and HCl [18] through the summit of volcanic edifices as well as through fumarolic vents, which are generated by surface fractures usually occurring in the neighborhood of volcanoes [19]. Volcanic activity has not only a local impact on the troposphere but also a global influence since emissions may reach the stratosphere [20].

The lack of information about the contribution of volcanic activity to the aerosol in the atmosphere significantly restricts the assessments of environmental and human health risks, which are of considerable concern in the central sector of the Mediterranean Sea because of the presence of numerous active volcanoes including Etna and Stromboli and fumaroles such as those on the Volcano Island and in the area of Phlegraean Fields, or submarine fumaroles near Panarea Island. Several studies focused on the contribution of aerosol emissions from volcanic activities to the ambient air pollutants around the volcanoes or fumaroles using different approaches [21–24]. However, these works studied a single area at a time and provided for an on-land sampling of the aerosol, thereby assessing the impact near the sources. In previous work, our group dealt with the influence of volcanic sources on the atmospheric PM during a cruise campaign in the Mediterranean basin [25]; however, the local volcanic impact in the Mediterranean area has not been adequately investigated yet. In the effort to fill these observation gaps, the Institute of Atmospheric Pollution Research of the National Research Council (CNR-IIA) within the ongoing Med-Oceanor program [26] has started, since 2000, regular cruise measurement campaigns across the Mediterranean Sea. In this context, our work reports about the input from volcanoes and fumaroles to the levels and composition of atmospheric aerosols in the central Mediterranean Sea. We performed this study during the 2017 research cruise campaign covering the study of trace metals, ionic species, organic carbon (OC), and elemental carbon (EC). Aerosol particulate in its PM_{10} and $PM_{2.5}$ fraction was collected and analyzed by different techniques for metal and ions determination. We merged the results of the chemical characterization with the PM investigation using scanning electron microscopy with energy-dispersive X-ray analysis (SEM/EDX) and strategies of data analysis including receptor modeling tools to identify the potential contribution of different sources to the PM in the central Mediterranean area.

2. Studied Area and Measurement Scheme

We performed aerosol sampling from 19 August to 4 September 2017 on-board of the research vessel "Minerva Uno" of the Italian National Research Council (CNR) during the oceanographic campaign Med-Oceanor 2017. Except for the last two days of the operation, the local weather conditions were characterized by low wind speed (between 0.3 and 13.5 m s^{-1}) and by stable average daily values of temperature and pressure, ranging between 23.7 and 29.2 °C and 1009 and 1019 hPa, respectively. The route involved 12 stops (Figure 1), each belonging to volcanic or fumarole areas. The volcanic area comprised Mount Etna, Stromboli Island, and Marsili Seamount, while the fumarole area comprised the Phlegraean Fields complex, Vulcano Island, Ischia Island, and Panarea submarine fumarole. To simplify the identification of the sampling zones, we labeled each stop using the name of the neighboring site.

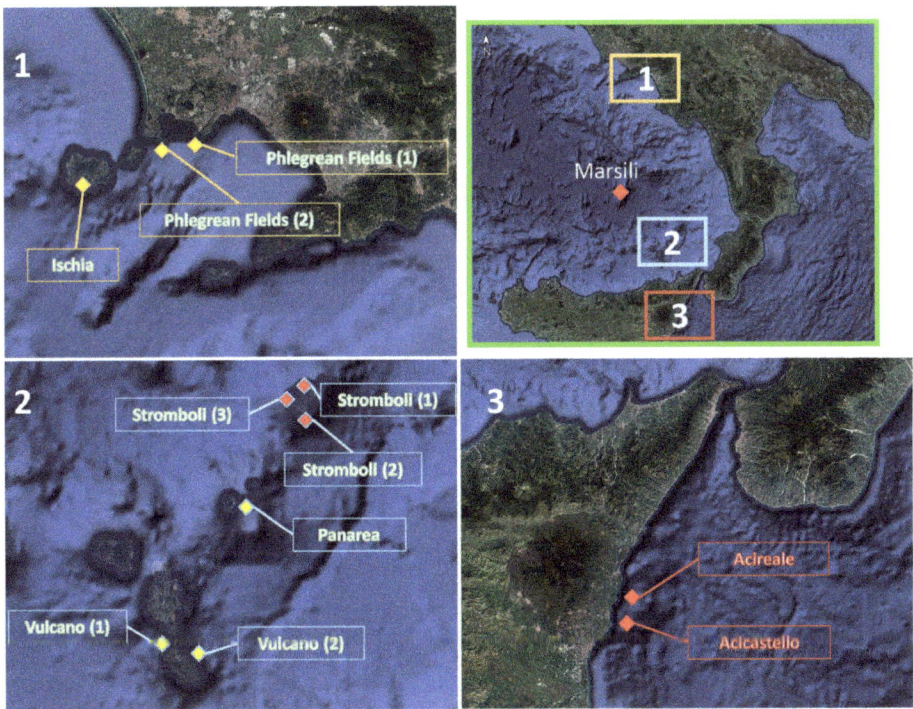

Figure 1. Map of sampling sites; red diamonds represent the volcanic sites while the yellow diamonds represent fumarolic sites: (**1**) Campania area; (**2**) Eolian archipelago; (**3**) Etnean area.

2.1. The Mount Etna

Mount Etna, one of the most active volcanoes in the world, covers an area of about 1250 km^2 and reaches an altitude of about 3340 m a.s.l. (37.754193° N, 14.978929° E) (Figure 1). It has been active during historical time, with frequent paroxysmal episodes separated by passive degassing periods. The edifice consists of a lower shield unit overlain by a stratovolcano [27] that is truncated at 2800–2900 m of altitude by the Ellittico caldera and Cratere del Piano, while the upper part of the volcano culminates with a large summit cone endowed with four active craters (Bocca Nuova, Voragine, Northeast, and Southeast Craters). The general atmospheric circulation in the Etnean area is predominantly westerly to northwesterly. According to these considerations, sampling sites in this zone were located in front of the Acireale and Acicastello coasts.

2.2. Stromboli

Stromboli is an island volcano belonging to the Eolian volcanic arc (926 m a.s.l; 38.7891° N, 15.2131° E) (Figure 1). It is an isolated volcano characterized by permanent activity and continuous gas emissions from potassium-rich basalt. Although Stromboli has a significant hydrothermal system [28], its magma also degasses directly into the atmosphere (i.e., at very high temperature, >1000 °C) from lava-filled vents. Such systems are often described as "open-conduit" volcanoes and can degas very large quantities of unerupted magma. Three different sampling sites around the island of Stromboli were monitored labeled as Stromboli (1), Stromboli (2), and Stromboli (3).

2.3. Marsili

Marsili Seamount is about 60 km long and 20 km wide, rising about 3000 m from the seafloor at over 3500 m water depth (39.236276° N, 14.369875° E) (Figure 1). The possible existence of a very large underwater explosive volcano, together with some encouraging clues that point to Marsili as an important and potential long-lasting-renewable energy resource [29], has reinforced research and exploitation efforts in this area. The existence of an active magmatic chamber at about 2.5 km below the summit was proposed on the basis of petrologic studies of basalts [30], and it is compatible with gravimetric and magnetic data modeling [31], as well as geochemical observations on the summit [32]. We monitored this area by a sampling stop above the crater of the volcano.

2.4. Phlegraean Fields

Solfatara, a crater located within the Phlegraean Fields (458 m a.s.l.; 40.8271° N, 14.1391° E), a quaternary volcanic complex near Naples (Figure 1), is renowned for its vigorous, low-temperature degassing. Indeed, the term Solfatara is applied to many similar manifestations found in other volcanic regions of the world. The hottest fumaroles at Solfatara reach temperatures of 140–160 °C. Previous studies indicate that gas emissions are dominated by H_2O, followed by CO_2 and H_2S [33]. In this area, sampling was performed at two different sites (labeled as Phlegraean Fields (1) and Phlegraean Fields (2)) along the direction of the inland Solfatara crater at a distance of about 1 km.

2.5. Vulcano

The volcanic island of Vulcano is the southernmost of the seven islands forming the Eolian Archipelago in the Tyrrhenian Sea north of Sicily (500 m a.s.l.; 38.4041° N, 14.9621° E). Since the last eruption of Vulcano in 1888–1890, a vigorous fumarole field has been developed in the summit crater, known as La Fossa. Numerous geochemical investigations have focused on Vulcano, especially since increases in gas fluxes and temperatures were observed in the 1970s. Similarly to the Solfatara site, Vulcano does not have magma on the surface but its fumaroles reach high temperatures, well over 300 °C at some vents. In Vulcano, we sampled at two sites labeled as Vulcano (1) and Vulcano (2) (Figure 1).

2.6. Ischia

Ischia (40.7352° N, 13.8517° E) is the westernmost active volcanic complex of the Campania area and belongs to the Phlegraean volcanic district, which also includes Phlegraean Fields (Figure 1). In Ischia, the last eruption took place in 1301, and since that time, several earthquakes affected the island. Thermal manifestations characterize Ischia including thermal waters and fumaroles, which have been well known since Roman times and are the main economic resource of the island at present. The fumarolic activity is mainly concentrated along the faults affecting Mt. Epomeo (789 m a.s.l.), a resurgent block in the central part of the island. In this area, we performed aerosol sampling in a site labeled Ischia (Figure 1).

2.7. Panarea

Panarea Island (Eolian Archipelago) lies in the western sector of a submarine 120 m deep platform constituting the 56 km^2 wide summit of a 2000 m high seamount (38.6369° N, 15.1026° E) (Figure 1). Close to Panarea, a group of islets (Dattilo, Panarelli, Lisca Bianca, Bottaro, Lisca Nera, and Le Formiche) forms an archipelago that surrounds a submerged fumarolic area, which is characterized by relatively shallow depth (<35 m), hosting gas, and thermal water discharges. Geochemical studies of fluids from the submarine fumarolic field of this area started in the early 1990s [34], and were then fostered by the gas burst event that occurred on 3 November 2002 [35]. One sampling site, labeled Panarea, was selected offshore as a monitoring point.

3. Experiments

3.1. Sampling and Laboratory Analysis

The aerosol sampling equipment was fixed on the deck at about 30 m above the sea level on the front of the ship to avoid contamination from the ship exhaust. Daily 24-h ambient PM$_{10}$ and PM$_{2.5}$ samples were collected on quartz fiber filters of 47 mm in diameter (2500 QAT-UP, Pallflex®, Putnam, CT, USA) using a low volume sampler Echo PM (, TCR Tecora, Cogliate, Italy) operating at a flow of 2.3 m^3 h^{-1}. In accordance with the European Standard UNI EN 12341:2014, filters were conditioned for 48 h before and after sampling, at a temperature and relative humidity of 20 ± 1 °C and 50 ± 5%, respectively. The conditioned filters were weighed using a microbalance Crystal Micro (Gibertini, Novate, Italy) in order to determine the mass of both the PM fractions by standard gravimetric procedures, whereupon the filters were analyzed for the determination of the chemical composition. The analysis of the carbonaceous fraction of aerosol (TC = OC + EC) was performed on a 1 cm^2 portion of each filter by a thermal-optical transmission (TOT) technique using an OC/EC Analyzer (Sunset Laboratory Company, Tigard, OR, USA) and following the EUSAAR-II temperature protocol [36]. Briefly, the small punch of the filter was subjected to two stepwise heating cycles, which favored the volatilization and combustion processes of the carbonaceous material. The evolved carbon was converted to methane and subsequently detected by a flame ionization detector (FID) during both temperature cycles, which occurred in a non-oxidizing and oxidizing atmosphere, respectively. Laser light of 670 nm continuously monitored the transmittance through the sample filter in order to correct for pyrolyzed OC conversion to EC.

Chemical characterization of PM samples was performed by Inductively Coupled Plasma Atomic Emission Spectrometry (iCAP 6500, Thermo Scientific) for major elements (Al, Ca, Na, Mg, K, Fe, P, S) and Inductively Coupled Plasma Mass Spectrometry (iCAP-RQ, Thermo Scientific) for trace elements (Li, Ti, V, Cr, Mn, Co, Ni, Cu, Zn, As, Se, Rb, Sr, Cd, Sn, Sb, Ba, La, Ce, Hf, Pb, Bi, Th, U). Half of each filter was acid digested with a mixture of HF:HNO$_3$:HClO$_4$ (2.5:1.25:1.25 mL) in a Teflon reactor for 9 h at a temperature of 90 °C. Each digested sample was driven to dryness, re-dissolved with 1.25 mL HNO$_3$, and made up to a final volume of 25 mL with ultrapure water [37]. Relative analytical errors were estimated using a small amount (15 mg) of the NIST-1633b (fly ash) reference material loaded on a $\frac{1}{4}$ quartz fiber filter. Individual uncertainties were calculated taking into account the analytical uncertainty and the standard deviations of the element concentrations in the blank filters [38]. Since the portions of the PM$_{2.5}$ filters for the sites Acireale, Panarea, and Vulcano (2) were contaminated during the storage, we discarded these samples before the chemical analysis for major and trace elements. For the quantification of ammonium by ion-selective electrode (Orion 9512HPBNWP of Thermo Scientific) and soluble ions, such as sulfate, nitrite, nitrate, fluoride, bromide, chloride, and phosphate, by ion chromatography (Aquion, Thermo Scientific), the leftover filter was water leached in 20 mL sealed PVC bottles (for 14 h at 60 °C), preceded by an ultrasonic bath of 15 min. Sulfur dioxide (SO$_2$) measurements were performed using a Teledyne UV Fluorescence SO$_2$ Analyzer (API Model 100EU) with a sampling flow rate of 0.6 L min^{-1} and a time resolution of 1 min. A particle sampling cassette system (SKC, inlet diameter 1/8 inch), which housed transmission electron microscope (TEM) grids attached to filter support pads (25 mm in diameter), was used to collect particles for offline characterization following a similar procedure described elsewhere [39,40]. The system utilized an SKC Leland pump that was operated at a flow of 5 L min^{-1} for a duration of 4 h. The grids (Agar Scientific Quantifoil 200 Mesh Au) were analyzed offline for morphological and physicochemical particle characterization using a Field Emission Scanning Electron Microscope (JEDL J-7100) coupled with an energy-dispersive X-ray spectroscopy (EDS; retroscattered electron detector) spectrometer.

3.2. Chemical Mass Closure

For the purpose of mass closure, the chemical components were grouped as sea salt (ss), mineral dust, organic matter and elemental carbon (OM + EC), secondary inorganic aerosols (SIA), and trace elements. Each of these components can originate from a variety of sources, some of which play a significant role in their production.

As regards to the sea salt component, six major ions represent more than 99% of the mass of salts dissolved in seawater: four cations, i.e., sodium (ss-Na$^+$), magnesium (ss-Mg^{2+}), calcium (ss-Ca^{2+}), and potassium (ss-K$^+$), and two anions, chloride (ss-Cl$^-$) and sulfate (ss-SO$_4^{2-}$). Thus, sea salt concentration was calculated from Equation (1) [41].

$$[\text{Sea salt}] = [\text{ss-Na}^+] + [\text{ss-Cl}^-] + [\text{ss-Mg}^{2+}] + [\text{ss-K}^+] + [\text{ss-Ca}^{2+}] + [\text{ss-SO}_4^{2-}] \quad (1)$$

To calculate the sea salt sodium [ss-Na$^+$], the value of non-sea salt sodium (nss-Na) was subtracted from the total Na value ([ss-Na$^+$] = [Na] − [nss-Na]) resulting from the chemical analysis. The [nss-Na] value was obtained from the Al concentration multiplied by the Na/Al ratio in the average crustal composition [42]. Based on the seawater composition, the sea salt sulfate [ss–SO$_4^{2-}$] is calculated as 0.2509 times [ss-Na$^+$], sea salt calcium [ss-Ca^{2+}] calculated as 0.038 times [ss-Na$^+$], sea salt potassium [ss-K$^+$] as 0.037 times [ss-Na$^+$], and sea salt magnesium [ss-Mg^{2+}] calculated as 0.1187 times [ss-Na$^+$]. For [ss-Cl$^-$], although the unavoidable reaction of NaCl particles with the atmospheric traces of acidic species including H$_2$SO$_4$ and HNO$_3$ can lead to the volatilization of Cl$^-$ as HCl [37]; thus, determining the underestimation of SS component, it can be reasonably assumed that [ss-Cl$^-$] is approximately equal to the total [Cl$^-$] in a marine environment. As a consequence, the [Cl$^-$] was used in the estimation of [sea salt].

Mineral dust is generated from the suspension of the minerals constituting the soil and can be expressed as the sum of the most common oxide forms [37]. In our study, dust content was estimated using Equation (2) as the sum of the concentrations of Al, Ca, Mg, K, Fe, P, and Ti, each multiplied by factors to convert them to their common oxides and SiO$_2$, calculated multiplying [Al$_2$O$_3$] by 2.5.

$$[\text{Dust}] = [\text{Al}_2\text{O}_3] + [\text{CaO}] + [\text{MgO}] + [\text{K}_2\text{O}] + [\text{Fe}_2\text{O}_3] + [\text{P}_2\text{O}_5] + [\text{TiO}_2] + [\text{SiO}_2] \quad (2)$$

EC was obtained from thermal-optical analysis (as described in Section 3.1), while the OM contribution, which accounts for the unmeasured H, O, N, and S in organic compounds, was calculated according to Equation (3), which considers also the influence of the water-soluble organic carbon (WSOC) on the mass, that for non-urban aerosol is about 40% of the OC mass.

$$\text{OM} = f \times \text{OC} \quad (3)$$

In this equation, OC was experimentally determined by thermal-optical analysis while the f multiplier is a coefficient that is not site- or time-specific. Depending on the extent of OM oxidation and secondary organic aerosol (SOA) formation, values for f vary from 1.2 for fresh aerosol in urban areas [43] to 2.6 for aged aerosol [44–46]. The conversion factor f is usually taken in the range from 1.4 for urban aerosols to 1.8 for remote aerosols. Turpin et al. revisited these conversion factors and proposed values of 1.6 ± 0.2 and 2.1 ± 0.2 for urban and non-urban aerosols, respectively [46]. Based on the previous consideration, the f value was set at 2.1 for the calculations in our work.

The secondary inorganic aerosol was calculated as the sum of non-sea-salt [SO_4^{2-}], [NO_3^-], and [NH_4^+]. Sulfate, in turn, is made by a component in both sea salt and non-sea-salt SO_4^{2-} (ss-SO_4^{2-} and nss-SO_4^{2-}) derived from the atmosphere. With our investigation being in the marine environment, the non-sea-salt sulfate was calculated as [nss-SO_4^{2-}] = [SO_4^{2-}] − 0.25094 [Na^+], based on [SO_4^{2-}]/[Na^+] molar ratio in seawater [37,47,48].

The last contribution to the PM mass closure, which derives from trace elements, was obtained as the sum of the trace element concentrations. The sum of all the above determinations accounted for 81.9% and 83.0% of the PM_{10} and $PM_{2.5}$ average mass, respectively. The remaining undetermined mass can be attributed to the structural and adsorbed water that is not removed during the sample conditioning.

3.3. Data Analysis and Receptor Modeling

In order to characterize the daily atmospheric scenarios influencing PM levels, we used several complementary tools to interpret the different sources of air masses in the studied area. We calculated air mass backward trajectories using the US NOAA HYSPLIT model [49] with GDAS meteorological data. The occurrence of African dust outbreaks was also confirmed by the same tools coupled with the information of the NAAPS aerosol maps from the Naval Research Laboratory, USA (NRL) (Navy Aerosol Analysis and Prediction System) and the BSC-DREAM dust maps [50]. Enrichment factors (EFs) and factor analysis (FA) were used for source identification and apportionment of particulate air pollutants at receptor sites, whereas linear discriminant analysis (LDA) was exploited to perform supervised pattern recognition analysis using the chemical descriptors of PM samples according to their volcanic (Vo) or solfatara (So) origin.

The calculation of EFs is useful to rank the enrichment degree of an element compared to a known source; in this case, the Al concentration was considered as the basis for comparison. EF is defined as the ratio between the concentration of an element i (E^i_{sample}) and that of a reference element (Al_{sample}) in the sample, normalized to the same ratio in the Earth's crust [42]. Specifically, for each element i, the EF_i was calculated according to Equation (4).

$$EF_i = \frac{E^i_{sample}/Al_{sample}}{E^i_{volc}/Al_{volc}} \qquad (4)$$

where *volc* refers to the volcanic rocks (basalt and lava in this case), which were considered as the geogenic source. In our study, the EF_i in Etna and Stromboli plumes was calculated using the aluminum as normalizing element, while the average Etnean lava [23] and the Stromboli basalt [51] were used, respectively, as reference material.

Factor analysis (FA) is widely used to determine the number of independent sources that contribute to the system and elemental source profiles. The association of the factors with the source type is possible by comparing the elements with the highest loading within each component with the elements commonly emitted by recognized source types. In our work, the whole dataset was subjected to factor analysis using the software Statistica (StatSoft,) and Varimax rotation was applied to assure the maximum differentiation between the factors. The number of potential factors was limited to eight, but only six factors were obtained.

Linear discriminant analysis (LDA) is a chemometric tool that defines a set of delimiters that divide the multivariate space of the samples into as many subspaces as the number of the classes. Discriminant functions are computed as a linear combination of variables that maximizes the ratio of variance between categories to variance within categories. An essential restriction in the application of LDA concerns the ratio between the number of samples and the number of variables. Indeed, it is recommended that the number of samples is at least three times the number of variables to obtain a robust model. Therefore, in our study, a stepwise approach (S-LDA) was used for the selection of the variables with a major discriminant power. The S-LDA was used to differentiate between the PM samples collected in the volcanic (Vo) and solfatara (So) sites on the basis of their chemical composition.

The dataset was slightly different from that used for FA because, for LDA, the variables that can be associated exclusively with an anthropogenic origin (Mo, EC, and OC) were not included in the dataset. Furthermore, the Acireale sample was excluded because it was affected by the Saharan Dust Outbreak (SDO), as pointed out from chemical characterization (see Section 4.2). Partial Wilks' λ values were used to figure out the variables that actually contributed to differentiation, i.e., the closer to zero its value is, the higher is the discriminatory importance of this variable in the model.

4. Results and Discussions

4.1. Meteorological Conditions

The trends of temperature, pressure, and wind speed recorded during the cruise by the on-board instrumentation are illustrated in Figure S1. During all the sampling days, the absence of precipitations, low wind intensity (with the maximum speed being 13.5 m s^{-1}), and a mean atmospheric temperature of 26.1 °C characterized the local weather. Information on wind direction and intensity are reported for each sampling site by the wind rose diagrams (Figure S2). Based on the average daily wind speed values and the Beaufort wind scale, we encountered light breeze for eight sampling days and gentle breeze for four sampling days. The atmospheric pressure exhibited the lowest average values and the largest variability between 1 September and 4 September 2017, with a minimum on 2 September, associated with episodes of strong wind and rough sea that forced us to stop in the Harbor of Lipari Island to shelter from the bad weather. The NCEP/NCAR Reanalysis [52] tool shows the composite means and anomalies of the sea level pressure over the Mediterranean basin during the campaign, supporting the worsening of the weather (Figure S3). Indeed, while in August the expansion of the Azores Anticyclone characterized the synoptic conditions in line with seasonal climatology in the Mediterranean, on 2 September the anticyclonic system was confined over the Atlantic Ocean and North of Europe, thus favoring the development of low-pressure systems across Eastern Europe and the Mediterranean basin.

4.2. Percentage Chemical Composition of PM

The mass concentration values for PM_{10} and $PM_{2.5}$ recorded during the campaign ranged between 12.3 and 30.7 µg m^{-3} and 9.4 and 18.3 µg m^{-3}, respectively (Figure 2). The PM_{10} concentration in each site was lower than the daily limit value of 50 µg m^{-3}, as set by the Directive 2015/1480/EC on ambient air quality. It is known that natural episodes, although they occur less frequently than anthropic ones, can influence air quality especially in the surroundings of marine areas and zones with tectonic activity [2]. The percentage chemical composition of the PM samples collected during the cruise is summed up in Table S1 and illustrated in Figure 2.

For the PM_{10}, the main components were carbonaceous (OM + EC) and mineral dust aerosols, with average contributions of 27.2% and 26.6%, respectively. The dust load reached a peak of 44.7% at Acireale, owing to a North African dust outbreak as confirmed by NAAPS, BSC-DREAM, and backward trajectories maps presented in Figure S4. The carbonaceous fraction was above the 20% for the whole campaign with the maximum amount of 38.3%, which was found in Phlegraean Fields (1) and may be due to the interception of air masses moving from the urbanized coast near the sampling area, which is characterized by high vehicular traffic. This assumption is supported by the levels of EC (0.95 µg m^{-3}) and OC (3.64 µg m^{-3}), which were found at their highest concentrations at this stop. The SIA component had an average mass contribution to PM_{10} of 19.9% with a maximum of 26.3% at Stromboli (2). Trace elements and sea salt contributed marginally to the mass, with tracers that accounted for less than 1% and sea salt for about 10%. We detected the maximum value of trace elements in the Ischia sample, where it represented 1.6% of PM_{10} mass. The relatively low contribution of sea salt compared to what can be expected in a marine atmosphere may be attributed to the stable weather encountered for most of the campaign, except for rough sea and strong blowing wind arose at Vulcano (2), where we recorded a higher percentage of the sea salt component (23.7%).

Indeed, the occurrence of substantial sea spray production needs wind with an intensity and duration strong enough to allow its formation [53]. For $PM_{2.5}$, the component prevailing on average was the carbonaceous aerosols (35.3%), which reached the maximum value in the Phlegraean Fields (1) (56.2%), thus reinforcing the preceding hypothesis involving the vehicular traffic contribution in this area. For SIA, the average loading reached 24.7%, thereby showing that its components (i.e., SO_4^{2-}, NH_4^+, and NO_3^-) prevailed in the $PM_{2.5}$ [37,54]. In both fractions, the maximum value of SIA was obtained in the sample Stromboli (2), where the upwind sampling (wind roses presented in Figure S2) led to a better interception of volcanic emissions characterized by high SO_2 values, which affected the increase of SO_4^{2-} concentration because of its reaction with the atmospheric water and the condensation on particles. In support of this, as showed in the time series plots of SO_2 (Figure S5), we recorded a peak of SO_2 at this site, which lasted for about 5 h during that sampling day. As expected, the average mineral dust contribution was smaller in $PM_{2.5}$ than in PM_{10} (17.6% and 25.6%, respectively), with the highest value reached in Acireale because of the SDO (Figure S4). Average sea salt contribution to $PM_{2.5}$ was 4.3%, with the maximum value in Marsili (10.7%), while the trace elements were averagely higher in $PM_{2.5}$ (1.2%) than PM_{10} (0.9%), with the maximum values recorded in Phlegraean Fields (1) and Phlegraean Fields (2) (1.4% and 1.6%, respectively).

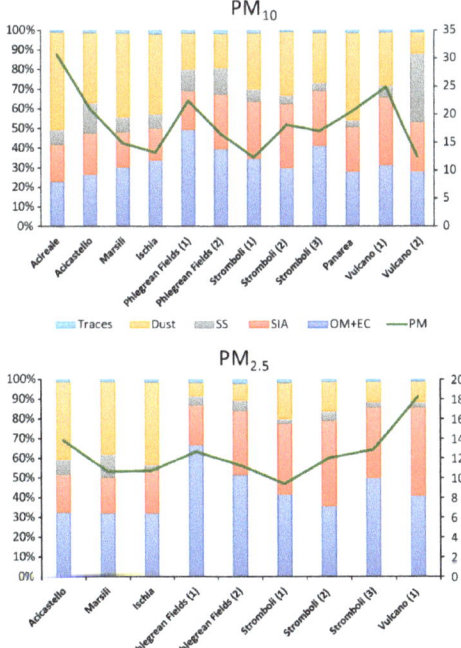

Figure 2. Chemical composition (as a percentage) and mass concentration of particles (expressed in µg m^{-3}) of the PM_{10} and $PM_{2.5}$ samples.

4.3. Volcanic Areas

There are several studies reporting the relevance of volcanic emissions as sources of metal-bearing gases and PM into the atmosphere [51,54–57]. A previous study revealed that volcanogenic trace metals are injected in the atmosphere as volatile elements, emitted from the silicate melt during magma degassing, and later dispersed in atmospheric plumes as sub-micron sized volcanic aerosols and low volatile (refractory or lithophile) elements, transported via coarse volcanic ash fragments [23]. The volcanic area monitored during the campaign, as previously discussed, included Etna and

Stromboli, which are sub-aerial volcanoes with a significant part of their structure under the sea, while the Marsili Seamount is submerged and its activity is still open to debate. In this regard, at the Marsili site, we observed low values for each component constitutive of PM, except for sea salt in $PM_{2.5}$, which reached the highest percentage in this site (10.7%) due to the low contributions of the other pollution sources because the sampling site was over 100 km from inland.

In our study, we used Al, Ca, and Fe, as tracers of crustal emissions to find out the volcanic-derived contribution to PM by comparing their average concentrations in the samples with the literature reference data. In particular, for information about the chemical composition of Etna's and Stromboli's lava and ash, we referred to the work of Calabrese et al. [23] and Allard et al. [51], respectively, while for the data on the chemical composition of Sahara and Sahel dust, we used the values provided by Moreno et al. [58]. The relative proportions of Al/Ca/Fe measured at Acireale and Acicastello are similar to the typical Etna's values in both ash and lava, confirming that samples, though collected at a considerable distance from the source (almost 40 km), are affected by the volcanic plume that spreads throughout the area (Figure 3a). The value of Acireale shown in Figure 3a is shifted toward the values of Sahara and Sahel, thus confirming the intrusion of air masses coming from the African desert region during the sampling. The ternary plot for the samples collected around Stromboli shows a similar composition with the reference values of magma reported in the literature [51]. In particular, the samples identified as Stromboli (2) and Stromboli (3) have a greater similitude with the composition of the basaltic lava, whereas the composition for the sample Stromboli (1) is closer to that of the volcanic ash (Figure 3b).

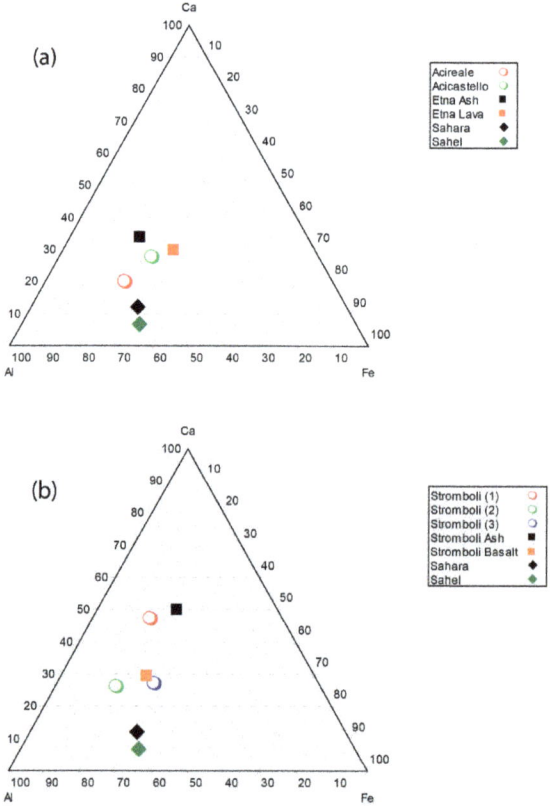

Figure 3. Ternary plot for Etna's sampling area (**a**) and Stromboli's sampling area (**b**).

Besides, the load of the volcanic plume aerosol in the sampled PM was investigated through the calculation of the enrichment factors, which is a useful tool to highlight the enrichment or depletion degree of an element in a geochemical medium relative to reference material. Although the use of EFs could hide some flaws, such as the variable composition of the reference material, or inadequate choice of the element used to normalize the data [59], it provides a first-order insight on the geogenic or anthropogenic origin of the elements. Based on the results of the ternary plot, we calculated the EFs for Acicastello and Stromboli (3), which are the samples identified as those with the greatest volcanic contribution. EFs were computed according to Equation (4) using Al as a reference element because it is a low volatile element at magmatic conditions, and also one of the most immobile elements during the weathering of basalts [60]. Values of EF close to one indicate that there is no enrichment occurring for an element in sampled PM compared to the expected abundance in volcanic basalt. EFs far exceeding one characterize volatile elements enriched in the plume because of their affinity for the magmatic vapor phase or might indicate an anthropogenic origin. Despite some peculiarities, the EFs for the samples Stromboli (3) and Acicastello are overall consistent. The EFs illustrated in Figure S6 suggest that, for both volcanoes, the elements can be grouped into four categories according to the EF values. The strongly enriched elements in PM (EFs > 100) were those with a marked volcanic-volatile character or anthropogenic origin species (i.e., Sn, Sb, Cr, Se). The enriched elements such as Ni, Zn, U, As, Pb, and Tl were those with EFs ranging from 10 to 100, while the elements with EF ≈ 1–10 such as Cu, K, Ba, Rb, and some rare earth elements including Th, Sm, Ce, and La were associated with a geogenic origin. The elements with EFs < 1, such as Co, Ti, Mg, and Sr, can be considered depleted elements compared to geogenic elements and indicate a local geological reduction relative to volcanic rocks. The volatile-anthropogenic species, including Sb, Cr, and Sn, were enriched by two to three orders of magnitude in the PM of both volcanic sites. Furthermore, another enriched element was Ni which may be related to anthropogenic sources and is commonly associated, together with V, to shipping emissions. Among the enriched elements, the presence of Tl only in the Etna sample pointed out the volcanic contribution to PM [23].

4.4. Fumarole Areas

Sulfur dioxide is a valuable tracer of degassing processes by geothermal sources such as fumaroles and volcanoes. In this work, we averaged on a daily basis the 1 min recorded data obtained from the SO_2 analyzer (Figure S5) in order to allow a direct comparison with the nss-SO_4^{2-} concentration values, as obtained from the analysis of the 24 h PM samples. As a result, we found a strong correlation in both PM fractions between the nss-SO_4^{2-} levels in the fumarole area with the corresponding average daily levels of SO_2 concentration, thus supporting the hypothesis that local SO_2 emissions are important for the formation of nss-SO_4^{2-} as reported in Section 4.2 about the SIA. In particular, Figures S7a and S5b illustrate the correlation between the concentrations of SO_2 and nss-SO_4^{2-} for each site in both fractions, with $R^2 = 0.76$ for PM_{10} and $R^2 = 0.74$ for $PM_{2.5}$. This correlation is further improved by neglecting the Vulcano (2) sample (Figure S7c,d), thus achieving $R^2 = 0.95$ and $R^2 = 0.92$ for $PM_{2.5}$ and PM_{10}, respectively. The observed increase in correlation may be due to the different atmospheric conditions that occurred during the sampling in Vulcano (2) when the rough sea and wind gusting may have affected the nss-SO_4^{2-} by favoring trapping of SO_2 with NaCl reactions. This result is confirmed by the concentration of nss-SO_4^{2-} and SO_2 reported in Figure 4, showing the larger gap between these two chemical species in Vulcano (2) compared to Vulcano (1).

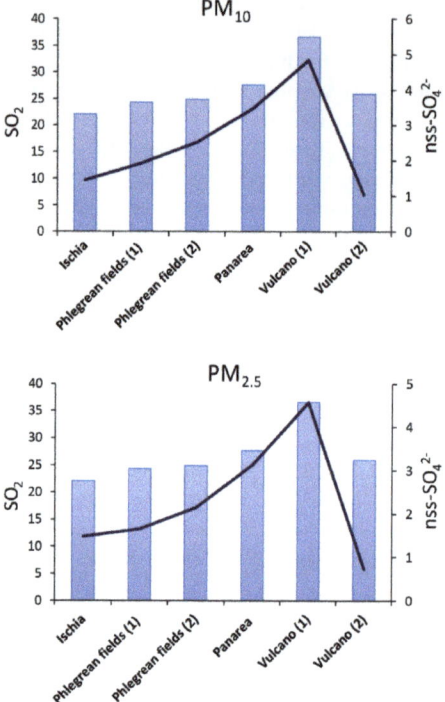

Figure 4. Concentration of sulfur dioxide and sulfate ions (expressed as µg m^{-3}) in the fumarolic area. The histogram bars represent SO_2 average daily concentrations whereas solid line represents nss-SO_4^{2-} collected in particulate matter (PM) samples.

As concerns Panarea, although it is a submarine fumarole source, a noteworthy value of both SO_2 and nss-SO_4^{2-} was detected, confirming the interception of fumarole degassing process. On the contrary, the analysis of trace metals did not present a significant contribution in PM of the elements typically emitted by natural sources, since they possibly dissolved in the path from the emission site and the seawater surface.

4.5. Factor Analysis and Source Identification

We used factor analysis, a tool for source identification and apportionment of particulate air pollutants, to characterize the different sources of aerosols during the campaign. We subjected the dataset to Varimax-rotated factor analysis identifying six factors with Eigenvalues greater than one (Table 1). The six factors together explain 96.9% of the total variance of our dataset.

Table 1. Varimax-rotated factor loadings for the whole dataset of the Med-Oceanor 2017 campaign. Loadings greater than 0.5 are shown in bold.

Variable	Factor1	Factor2	Factor3	Factor4	Factor5	Factor6
PM_x	0.5	0.5	0.3	0.2	**0.6**	−0.3
OC	**0.8**	0.2	0.1	0.2	0.4	0.2
EC	**0.9**	0.2	0.2	−0.1	0.0	−0.1
NH_4^+	−0.1	−0.1	−0.5	0.2	**0.8**	0.0
Al	0.0	**0.9**	0.0	0.2	0.2	−0.4
Fe	0.4	**0.8**	0.3	0.1	0.2	−0.1
Na	0.1	0.2	**0.9**	−0.2	−0.2	0.0
Cl	−0.1	0.2	**0.7**	−0.1	−0.2	**−0.6**
Ca	0.2	0.2	0.0	**0.9**	0.1	0.0
Mg	0.1	**0.6**	**0.7**	0.0	0.1	−0.2
P	0.0	0.2	−0.1	**1.0**	0.1	0.1
nss-SO_4^{2-}	0.1	0.3	−0.1	0.1	**0.9**	−0.2
NO_3^-	0.3	0.0	**0.9**	−0.1	−0.3	0.0
V	0.4	0.2	−0.4	−0.2	**0.7**	0.0
Cr	0.1	**0.8**	0.5	−0.1	0.0	0.2
Mn	**0.7**	**0.6**	−0.2	0.2	0.0	0.2
Zn	0.1	−0.1	−0.2	**1.0**	0.0	0.1
As	**0.8**	−0.3	0.2	0.2	0.2	0.1
Sb	**1.0**	0.1	0.1	0.0	0.0	0.1
REEs	**0.7**	−0.1	0.0	0.0	−0.1	**0.6**
Pb	**0.9**	0.3	0.0	0.1	0.0	0.3
Zr	0.3	−0.1	−0.2	0.2	−0.3	**0.9**
Eigenvalues	5.7	3.8	3.7	3.1	2.8	2.1
%Var	25.9	17.5	17.0	14.3	12.8	9.4
%Cum	25.9	43.4	60.4	74.7	87.5	96.9

- Factor 1: Anthropogenic (pollution). It can be referred to as anthropogenic sources because it has high loadings for carbonaceous material (EC, OC), As, Sb, and Pb. This result found a match with intensive vehicular traffic observed during the sampling at the Phlegraean Fields area and is in accordance with the findings reported in Section 4.2.
- Factor 2: Mineral (e.g., soil, African dust, etc.). This factor is characterized by high loadings for Al, Fe, Mg, Cr, and Mn, and can be associated at the first geogenic source (mineral sources) since these elements are common in the crust, basalt of volcanoes, and also in African dust.
- Factor 3: Aged sea salt (marine aerosol). NO_3^- was added in the FA to describe that this is aged and not fresh sea salt [37,61], indicating that the sampling was not biased by the source of sea spray (limited freshly emitted sea salt particles).
- Factor 4: Factor with unsure attribution. Due to the presence of Zn, it may be attributed to the fumarolic sources [62,63].
- Factor 5: Shipping emissions. This factor can be attributed to the shipping emissions because it has high loadings of PM_x, nss-SO_4^{2-}, NH_4^+ and V. The vanadium indeed is used as a tracer of shipping emission [64].
- Factor 6: Rare earth elements. The sixth factor represents a second potential geogenic source characterized by high loadings of Zr and REEs coming from volcanic ash. Anti-correlated with Cl.

4.6. SEM/EDX Analysis

SEM/EDX analysis is utilized as a complementary tool for the quantitative chemical analysis (ICP). The elemental characteristics of aerosols are apportioned based on the respective sources. SEM/EDX confirms that particles with the specified compositions were present in the sampling area and their morphology is defined. Aerosols with a composition common for volcanic particles (e.g., SiO_2, Mg, and Fe) have been observed, indicating that they may have originated from the volcanic plume (Figure 5a,d).

The anthropogenic factor has influenced the sampling (Figure 5f,j) and consequently the experimental results both in terms of chemical and microscopy analysis interpretation (i.e., factor and SEM analysis). In agreement with FA, aged sea salt (g) and several minerals were also observed.

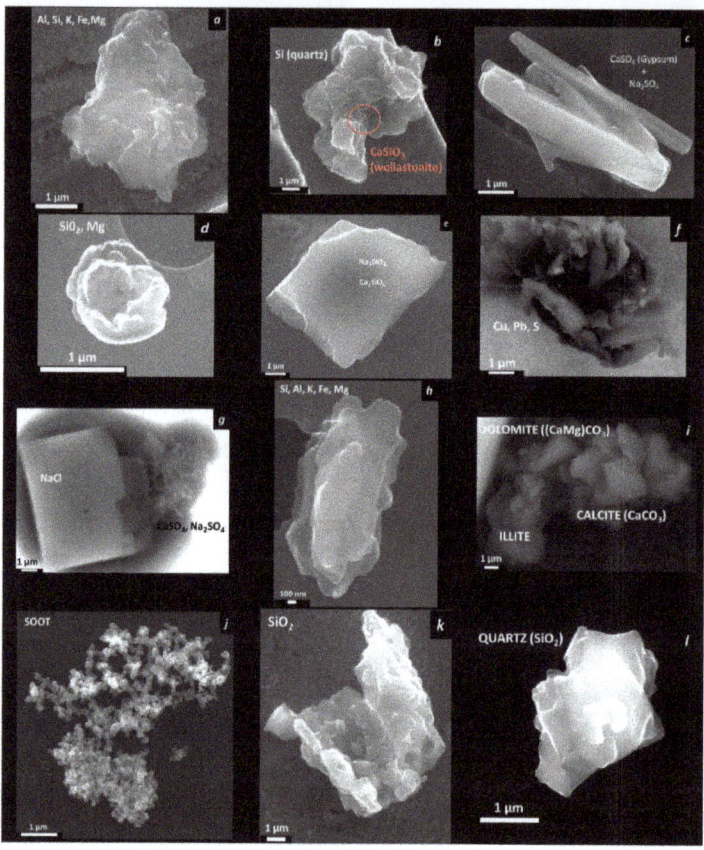

Figure 5. Representative particles sampled from different areas: (**a,b**) Acicastello, (**c,d**) Acireale, (**e–g**) Marsili, (**h–k**), Phlegraean Fields (1), (**l**) Stromboli (1).

4.7. Stepwise Linear Discriminant (S-LDA) Analysis

To seek which variables best discriminate among samples coming from volcanoes (Vo) and solfatara (So), we performed stepwise linear discriminant analysis (S-LDA). The forward S-LDA (F to enter = 5.00 and F to remove = 4.00) retained six elements, namely Zr, As, Hf, Bi, Ca, and Zn, that allowed the construction of a model with satisfactory statistical parameters (Wilks' λ = 0.09560; $F_{(6, 13)}$ = 20.497; $p < 0.00001$). The low Wilks' λ values suggest that the model is highly discriminating, while the low p-level value shows that the classification, which resulted correct for all the samples, occurred with high reliability. Based on the partial Wilks' λ, Zn (0.38) is the independent variable that contributes most to the discrimination, followed by Ca (0.50). The scatterplot of canonical scores (Figure S8) on the discriminant function shows the separation between the groups of the volcanic samples (Vo), with negative root values ranging from −4.38 to −1.56 and the solfatara samples (So) with positive root values in the range between 1.39 and 4.69. We found the predominant presence of Zn in solfatara samples, in accordance to Dekov et al. [32], which detected this element in the aerosol of sulfur-rich sites such as Panarea or other fumarole areas because of its presence as sphalerite (ZnS). On the contrary, the elements Hf and Zr contribute most to characterize volcanic samples. The presence of Hf and Zr agrees with the study of metal emissions from the Stromboli volcano by Allard et al. [51]. Indeed, Hf is one of the elements degassed from magmas as halides, sulfates, sulfides, and/or metals, and is typically found in the aerosol phase of the airborne plume, while the presence of Zr in volcanic samples can be due to its transport through volcanic ash, supporting the evidence previously obtained from factor analysis.

5. Conclusions

In this study, we investigated the impact of volcanic and fumarole sources on PM level and its chemical characteristics in the central Mediterranean Sea. Aerosol samples in marine atmosphere were collected during the research cruise campaign Med-Oceanor 2017 off the coast of the main active volcanic and fumarolic areas in the Mediterranean Sea, which included Mount Etna, the Eolian volcanic arc, Marsili Seamount, and the Phlegraean Fields. The mass concentration of collected PM_{10} resulted, for each site, lower than the daily limit set by the European legislation. The PM_{10} and $PM_{2.5}$ samples were characterized by quantification of major and trace elements, ionic species, as well as elemental and organic carbon content. These chemical data were illustrated by descriptive statistics, histograms, and then analyzed by the computation of mass closure, enrichment factors, factor analysis, and correlation analysis with the support of the daily SO_2 values. The PM_{10} mainly consisted of carbonaceous material and dust, which together made up over 50% of the mass, while carbonaceous fraction and SIA, which reached 60% of the mass, prevalently composed the $PM_{2.5}$. The trace element was the less abundant contributor with an average amount of about 1% in both PM fractions. From the EFs, we found that the particulate from the volcanic areas resulted enriched by Tl, Cu, La, Se, As, and alkali metals, whereas the chemical characterization of PM in fumarole areas showed a good correlation between the sulfur dioxide gas, emitted from fumaroles and sulfate ion, coming from the SO_2 reaction with water and condensation on particles. FA allowed for the identification of six potential sources that contribute to the PM composition, some of which have a natural origin such as volcanic, marine, and geogenic emissions, while other sources are anthropogenic, such as combustion processes, vehicular traffic, and shipping emissions. The SEM/EDX analysis on the sampled PM confirmed the influence of these sources identified with the FA. Stepwise linear discriminant analysis (S-LDA), which was used to seek for a model capable to discriminate between the sample of volcanic and fumarolic origin, pointed out the elements that discriminate the most between these two classes, namely Hf and Zr for volcano samples and Zn and Ca for solfatara samples.

Further investigation on this topic could be addressed in future research for a more thorough assessment of the impact of the volcanic and fumarolic emission on the air quality of the Mediterranean Sea. In this regard, the use of a high-volume sampler would allow sampling the PM with sufficient sensitivity even with a few hours of collection, thus improving the temporal profile of the impact. Similarly, further advancement of the study would be given by the use of cascade impactors, whose data could provide new insight to strengthen the characterization of the marine atmosphere in the neighborhood of volcanoes and fumaroles.

Supplementary Materials: The following are available online at http://www.mdpi.com/2073-4433/11/2/149/s1, Figure S1: Hourly mean meteorological data; Temperature and Pressure (up) and wind speed (down), Figure S2: Wind roses for sampling points, Figure S3: NCEP/NCAR reanalysis maps (mean and anomaly) of the period from the start of the campaign to 31 August (up) and 2 September (down), Figure S4: Acireale maps NAAPS, air mass back trajectory and BSC-Dream, Figure S5: Time series plot of SO_2 for each of the sampling sites; Y-axis: concentration (ppb), X-axis: sampling time (hours), Figure S6: Enrichment Factors for Etna (Acicastello sample) and Stromboli (Stromboli (3) sample). Dashed lines highlight thresholds for EFs = 10 and EFs = 100, Figure S7: Correlation between SO_2 and nss-SO_4^{2-} (as µg m^{-3}) in PM_{10} and $PM_{2.5}$. The plots (a) and (b) comprise all the sampling points, whereas the Vulcano (2) sample is excluded in (c) and (d) plots the plots., Figure S8: LDA scatterplot for aerosol samples collected in volcanic (Vo) or solfatara (So) sites, Table S1: Percentage chemical composition of PM_{10} and $PM_{2.5}$ in each sampling site.

Author Contributions: S.M., Conceptualization, investigation, writing—original draft preparation, writing—review and editing; A.S., investigation, writing—review and editing; X.Q., conceptualization, supervision; A.T., conceptualization, writing—original draft preparation, writing—review and editing; V.A., investigation; M.B., conceptualization; N.P., funding acquisition; F.S., funding acquisition, supervision; A.N., conceptualization, writing—review and editing, supervision, All authors have read and agreed to the published version of the manuscript.

Funding: This research was funded by the European Commission—H2020, the ERA-PLANET programme (www.era-planet.eu; contract no. 689443) within the IGOSP project (www.igosp.eu).

Acknowledgments: The authors are grateful to Valentino Mannarino and Giulio Esposito for the logistic support in the Med-Oceanor campaign.

Conflicts of Interest: The authors declare no conflicts of interest.

References

1. Lelieveld, J.; Evans, J.S.; Fnais, M.; Giannadaki, D.; Pozzer, A. The contribution of outdoor air pollution sources to premature mortality on a global scale. *Nature* **2015**, *525*, 367–371. [CrossRef]
2. Viana, M.; Pey, J.; Querol, X.; Alastuey, A.; de Leeuw, F.; Lükewille, A. Natural sources of atmospheric aerosols influencing air quality across Europe. *Sci. Total Environ.* **2014**, *472*, 825–833. [CrossRef]
3. Naccarato, A.; Tassone, A.; Moretti, S.; Elliani, R.; Sprovieri, F.; Pirrone, N.; Tagarelli, A. A green approach for organophosphate ester determination in airborne particulate matter: Microwave-assisted extraction using hydroalcoholic mixture coupled with solid-phase microextraction gas chromatography-tandem mass spectrometry. *Talanta* **2018**, *189*, 657–665. [CrossRef]
4. Naccarato, A.; Elliani, R.; Sindona, G.; Tagarelli, A. Multivariate optimization of a microextraction by packed sorbent-programmed temperature vaporization-gas chromatography–tandem mass spectrometry method for organophosphate flame retardant analysis in environmental aqueous matrices. *Anal. Bioanal. Chem.* **2017**, *409*, 7105–7120. [CrossRef]
5. Talarico, F.; Brandmayr, P.; Giulianini, P.G.; Ietto, F.; Naccarato, A.; Perrotta, E.; Tagarelli, A.; Giglio, A. Effects of metal pollution on survival and physiological responses in Carabus (Chaetocarabus) lefebvrei (Coleoptera, Carabidae). *Eur. J. Soil Biol.* **2014**, *61*, 80–89. [CrossRef]
6. Aleksandropoulou, V.; Torseth, K.; Lazaridis, M. Contribution of natural sources to PM emissions over the metropolitan areas of athens and Thessaloniki. *Aerosol Air Qual. Res.* **2015**, *15*, 1300–1312. [CrossRef]
7. Bencardino, M.; Andreoli, V.; D'Amore, F.; De Simone, F.; Mannarino, V.; Castagna, J.; Moretti, S.; Naccarato, A.; Sprovieri, F.; Pirrone, N. Carbonaceous Aerosols Collected at the Observatory of Monte Curcio in the Southern Mediterranean Basin. *Atmosphere* **2019**, *10*, 592. [CrossRef]
8. Bencardino, M.M.; Pirrone, N.N.; Sprovieri, F.F. Aerosol and ozone observations during six cruise campaigns across the Mediterranean basin: temporal, spatial, and seasonal variability. *Environ. Sci. Pollut. Res.* **2014**, *21*, 4044–4062. [CrossRef] [PubMed]
9. Beuck, H.; Quass, U.; Klemm, O.; Kuhlbusch, T.A.J. Assessment of sea salt and mineral dust contributions to PM10 in NW Germany using tracer models and positive matrix factorization. *Atmos. Environ.* **2011**, *45*, 5813–5821. [CrossRef]
10. Dinoi, A.; Cesari, D.; Marinoni, A.; Bonasoni, P.; Riccio, A.; Chianese, E.; Tirimberio, G.; Naccarato, A.; Sprovieri, F.; Andreoli, V.; et al. Inter-comparison of carbon content in PM2.5 and PM10 collected at five measurement sites in Southern Italy. *Atmosphere* **2017**, *8*, 243. [CrossRef]
11. Liora, N.; Poupkou, A.; Giannaros, T.M.; Kakosimos, K.E.; Stein, O.; Melas, D. Impacts of natural emission sources on particle pollution levels in Europe. *Atmos. Environ.* **2016**, *137*, 171–185. [CrossRef]
12. Masson, O.; Piga, D.; Gurriaran, R.; D'Amico, D. Impact of an exceptional Saharan dust outbreak in France: PM10 and artificial radionuclides concentrations in air and in dust deposit. *Atmos. Environ.* **2010**, *44*, 2478–2486. [CrossRef]
13. Querol, X.; Pey, J.; Pandolfi, M.; Alastuey, A.; Cusack, M.; Pérez, N.; Moreno, T.; Viana, M.; Mihalopoulos, N.; Kallos, G.; et al. African dust contributions to mean ambient PM10 mass-levels across the Mediterranean Basin. *Atmos. Environ.* **2009**, *43*, 4266–4277. [CrossRef]
14. Bencardino, M.; Sprovieri, F.; Cofone, F.; Pirrone, N. Variability of atmospheric aerosol and ozone concentrations at marine, urban, and high-altitude monitoring stations in southern Italy during the 2007 summer Saharan dust outbreaks and wildfire episodes. *J. Air Waste Manag. Assoc.* **2011**, *61*, 952–967. [CrossRef] [PubMed]

15. Pey, J.; Alastuey, A.; Querol, X. PM10and PM2.5sources at an insular location in the western mediterranean by using source apportionment techniques. *Sci. Total Environ.* **2013**, *456*, 267–277. [CrossRef]
16. O'Dowd, C.D.; de Leeuw, G. Marine aerosol production: a review of the current knowledge. *Philos. Trans. R. Soc. A Math. Phys. Eng. Sci.* **2007**, *365*, 1753–1774. [CrossRef] [PubMed]
17. Faustini, A.; Alessandrini, E.R.; Pey, J.; Perez, N.; Samoli, E.; Querol, X.; Cadum, E.; Perrino, C.; Ostro, B.; Ranzi, A.; et al. Short-term effects of particulate matter on mortality during forest fires in Southern Europe: Results of the MED-PARTICLES project. *Occup. Environ. Med.* **2015**, *72*, 323–329. [CrossRef]
18. von Glasow, R. Atmospheric chemistry in volcanic plumes. *Proc. Natl. Acad. Sci.* **2010**, *107*, 6594–6599. [CrossRef]
19. Helbert, J. Fumarole. In *Encyclopedia of Astrobiology*; Gargaud, M., Amils, R., Quintanilla, J.C., Cleaves, H.J., Irvine, W.M., Pinti, D.L., Viso, M., Eds.; Springer Berlin Heidelberg: Berlin, Germany, 2011; p. 617, ISBN 978-3-642-11274-4.
20. Intergovernmental Panel on Climate Change Summary for Policymakers. In *Climate Change 2013 - The Physical Science Basis*; Intergovernmental Panel on Climate Change (Ed.) Cambridge University Press: Cambridge, UK, 2013; pp. 1–30, ISBN 9788578110796.
21. Arndt, J.; Calabrese, S.; D'Alessandro, W.; Planer-Friedrich, B. Using mosses as biomonitors to study trace element emissions and their distribution in six different volcanic areas. *J. Volcanol. Geotherm. Res.* **2017**, *343*, 220–232. [CrossRef]
22. Bagnato, E.; Aiuppa, A.; Andronico, D.; Cristaldi, A.; Liotta, M.; Brusca, L.; Miraglia, L. Leachate analyses of volcanic ashes from Stromboli volcano: A proxy for the volcanic gas plume composition? *J. Geophys. Res. Atmos.* **2011**, *116*, 1–17. [CrossRef]
23. Calabrese, S.; Aiuppa, A.; Allard, P.; Bagnato, E.; Bellomo, S.; Brusca, L.; D'Alessandro, W.; Parello, F. Atmospheric sources and sinks of volcanogenic elements in a basaltic volcano (Etna, Italy). *Geochim. Cosmochim. Acta* **2011**, *75*, 7401–7425. [CrossRef]
24. Chouet, B.; Dawson, P.; Ohminato, T.; Martini, M.; Saccorotti, G.; Giudicepietro, F.; De Luca, G.; Milana, G.; Scarpa, R. Source mechanisms of explosions at Stromboli Volcano, Italy, determined from moment-tensor inversions of very-long-period data. *J. Geophys. Res. Solid Earth* **2003**, *108*, ESE 7-1–ESE 7-25. [CrossRef]
25. Castagna, J.; Bencardino, M.; D'Amore, F.; Esposito, G.; Pirrone, N.; Sprovieri, F. Atmospheric mercury species measurements across the Western Mediterranean region: Behaviour and variability during a 2015 research cruise campaign. *Atmos. Environ.* **2018**, *173*, 108–126. [CrossRef]
26. Pirrone, N.; Ferrara, R.; Hedgecock, I.M.; Kallos, G.; Mamane, Y.; Munthe, J.; Pacyna, J.M.; Pytharoulis, I.; Sprovieri, F.; Voudouri, A.; et al. Dynamic processes of mercury over the Mediterranean region: Results from the Mediterranean Atmospheric Mercury Cycle System (MAMCS) project. *Atmos. Environ.* **2003**, *37*, 21–39. [CrossRef]
27. Chester, D.K.; Duncan, A.M.; Guest, J.E.; Kilburn, C.R.J. *Mount Etna*; Springer: Dordrecht, The Netherlands, 1986; ISBN 978-94-010-8309-6.
28. Finizola, A.; Sortino, F.; Lénat, J.F.; Aubert, M.; Ripepe, M.; Valenza, M. The summit hydrothermal system of Stromboli. New insights from self-potential, temperature, CO_2 and fumarolic fluid measurements, with structural and monitoring implications. *Bull. Volcanol.* **2003**, *65*, 486–504. [CrossRef]
29. Italiano, F.; De Santis, A.; Favali, P.; Rainone, M.; Rusi, S.; Signanini, P.; Italiano, F.; De Santis, A.; Favali, P.; Rainone, M.L.; et al. The Marsili Volcanic Seamount (Southern Tyrrhenian Sea): A Potential Offshore Geothermal Resource. *Energies* **2014**, *7*, 4068–4086. [CrossRef]
30. Trua, T.; Serri, G.; Marani, M.; Renzulli, A.; Gamberi, F. Volcanological and petrological evolution of Marsili Seamount (southern Tyrrhenian Sea). *J. Volcanol. Geotherm. Res.* **2002**, *114*, 441–464. [CrossRef]
31. Caratori Tontini, F.; Cocchi, L.; Muccini, F.; Carmisciano, C.; Marani, M.; Bonatti, E.; Ligi, M.; Boschi, E. Potential-field modeling of collapse-prone submarine volcanoes in the southern Tyrrhenian Sea (Italy). *Geophys. Res. Lett.* **2010**, *37*, 1–5. [CrossRef]
32. Dekov, V.M.; Kamenov, G.D.; Savelli, C.; Stummeyer, J. Anthropogenic Pb component in hydrothermal ochres from Marsili Seamount (Tyrrhenian Sea). *Mar. Geol.* **2006**, *229*, 199–208. [CrossRef]
33. Chiodini, G.; Avino, R.; Brombach, T.; Caliro, S.; Cardellini, C.; De Vita, S.; Frondini, F.; Granirei, D.; Marotta, E.; Ventura, G. Fumarolic and diffuse soil degassing west of Mount Epomeo, Ischia, Italy. *J. Volcanol. Geotherm. Res.* **2004**, *133*, 291–309. [CrossRef]

34. Italiano, F.; Nuccio, P.M. Geochemical investigations of submarine volcanic exhalations to the east of Panarea, Aeolian Islands, Italy. *J. Volcanol. Geotherm. Res.* **1991**, *46*, 125–141. [CrossRef]
35. Caracausi, A.; Ditta, M.; Italiano, F.; Longo, M.; Nuccio, P.M.; Paonita, A.; Rizzo, A. Changes in fluid geochemistry and physico-chemical conditions of geothermal systems caused by magmatic input: The recent abrupt outgassing off the island of Panarea (Aeolian Islands, Italy). *Geochim. Cosmochim. Acta* **2005**, *69*, 3045–3059. [CrossRef]
36. Cavalli, F.; Viana, M.; Yttri, K.E.; Genberg, J.; Putaud, J.P. Toward a standardised thermal-optical protocol for measuring atmospheric organic and elemental carbon: The EUSAAR protocol. *Atmos. Meas. Tech.* **2010**, *3*, 79–89. [CrossRef]
37. Querol, X.; Alastuey, A.; Rodriguez, S.; Plana, F.; Ruiz, C.R.; Cots, N.; Massagué, G.; Puig, O. PM10 and PM2.5 source apportionment in the Barcelona Metropolitan area, Catalonia, Spain. *Atmos. Environ.* **2001**, *35*, 6407–6419. [CrossRef]
38. Amato, F.; Pandolfi, M.; Escrig, A.; Querol, X.; Alastuey, A.; Pey, J.; Perez, N.; Hopke, P.K.K. Quantifying road dust resuspension in urban environment by Multilinear Engine: A comparison with PMF2. *Atmos. Environ.* **2009**, *43*, 2770–2780. [CrossRef]
39. Salmatonidis, A.; Ribalta, C.; Sanfélix, V.; Bezantakos, S.; Biskos, G.; Vulpoi, A.; Simion, S.; Monfort, E.; Viana, M. Workplace Exposure to Nanoparticles during Thermal Spraying of Ceramic Coatings. *Ann. Work Expo. Health* **2019**, *63*, 91–106. [CrossRef]
40. Salmatonidis, A.; Viana, M.; Pérez, N.; Alastuey, A.; de la Fuente, G.F.; Angurel, L.A.; Sanfélix, V.; Monfort, E. Nanoparticle formation and emission during laser ablation of ceramic tiles. *J. Aerosol Sci.* **2018**, *126*, 152–168. [CrossRef]
41. Seinfeld, J.H.; Pandis, S.N. *Atmospheric Chemistry and Physics; From Air Pollution to Climate Change*; John Wiley & Sons: New York, NY, USA, 2006; ISBN 978-0-471-72018-8.
42. Mason, B. *Principles of geochemistry*, 3rd ed.; JohnWiley&Sons, Inc.: New York, NY, USA, 1966.
43. Chow, J.C.; Watson, J.G.; Edgerton, S.A.; Vega, E. Chemical composition of PM2.5 and PM10 in Mexico City during winter 1997. *Sci. Total. Environ.* **2002**, *287*, 177–201. [CrossRef]
44. Robinson, A.L.; Grieshop, A.P.; Donahue, N.M.; Hunt, S.W. Updating the conceptual model for fine particle mass emissions from combustion systems. *J. Air Waste Manag. Assoc.* **2010**, *60*, 1204–1222. [CrossRef]
45. Roy, A.A.; Wagstrom, K.M.; Adams, P.J.; Pandis, S.N.; Robinson, A.L. Quantification of the effects of molecular marker oxidation on source apportionment estimates for motor vehicles. *Atmos. Environ.* **2011**, *45*, 3132–3140. [CrossRef]
46. Turpin, B.J.; Lim, H.-J. Species Contributions to PM2.5 Mass Concentrations: Revisiting Common Assumptions for Estimating Organic Mass. *Aerosol Sci. Technol.* **2001**, *35*, 602–610. [CrossRef]
47. Cheung, K.; Daher, N.; Kam, W.; Shafer, M.M.; Ning, Z.; Schauer, J.J.; Sioutas, C. Spatial and temporal variation of chemical composition and mass closure of ambient coarse particulate matter (PM10–2.5) in the Los Angeles area. *Atmos. Environ.* **2011**, *45*, 2651–2662. [CrossRef]
48. Mkoma, S.L.; Maenhaut, W.; Chi, X.; Wang, W.; Raes, N. Characterisation of PM10 atmospheric aerosols for the wet season 2005 at two sites in East Africa. *Atmos. Environ.* **2009**, *43*, 631–639. [CrossRef]
49. Stein, A.F.; Draxler, R.R.; Rolph, G.D.; Stunder, B.J.B.; Cohen, M.D.; Ngan, F.; Stein, A.F.; Draxler, R.R.; Rolph, G.D.; Stunder, B.J.B.; et al. NOAA's HYSPLIT Atmospheric Transport and Dispersion Modeling System. *Bull. Am. Meteorol. Soc.* **2015**, *96*, 2059–2077. (Hybrid Single-Particle Lagrangian Integrated Trajectories model). Available online: http://ready.arl.noaa.gov/HYSPLIT.php (accessed on 5 June 2018). [CrossRef]
50. Pérez, C.; Nickovic, S.; Pejanovic, G.; Baldasano, J.M.; Özsoy, E. Interactive dust-radiation modeling: A step to improve weather forecasts. *J. Geophys. Res. Atmos.* **2006**, *111*. (BSC Dust Regional Atmospheric Model). Available online: www.bsc.es/projects/earthscience/DREAM/ (accessed on 5 June 2018). [CrossRef]
51. Allard, P.; Aiuppa, A.; Loyer, H.; Carrot, F.; Gaudry, A.; Pinte, G.; Michel, A.; Dongarrà, G. Acid gas and metal emission rates during long-lived basalt degassing at Stromboli volcano. *Geophys. Res. Lett.* **2000**, *27*, 1207–1210. [CrossRef]
52. Kalnay, E.; Kanamitsu, M.; Kistler, R.; Collins, W.; Deaven, D.; Gandin, L.; Iredell, M.; Saha, S.; White, G.; Woollen, J.; et al. The NCEP/NCAR 40-Year Reanalysis Project. *Bull. Am. Meteorol. Soc.* **1996**, *77*, 437–471. [CrossRef]

53. Laussac, S.; Piazzola, J.; Tedeschi, G.; Yohia, C.; Canepa, E.; Rizza, U.; Van Eijk, A.M.J. Development of a fetch dependent sea-spray source function using aerosol concentration measurements in the North-Western Mediterranean. *Atmos. Environ.* **2018**, *193*, 177–189. [CrossRef]
54. Mather, T.A.; Oppenheimer, C.; Allen, A.G.; McGonigle, A.J.S. Aerosol chemistry of emissions from three contrasting volcanoes in Italy. *Atmos. Environ.* **2004**, *38*, 5637–5649. [CrossRef]
55. Aiuppa, A.; Dongarrà, G.; Valenza, M.; Federico, C.; Pecoraino, G. Degassing of Trace Volatile Metals During The 2001 Eruption of Etna. In *Geophysical Monograph Series*; American Geophysical Union: Washington, DC, USA, 2003; Volume 139, pp. 41–54, ISBN 9781118668542.
56. Allen, A.G.; Mather, T.A.; McGonigle, A.J.S.; Aiuppa, A.; Delmelle, P.; Davison, B.; Bobrowski, N.; Oppenheimer, C.; Pyle, D.M.; Inguaggiato, S. Sources, size distribution, and downwind grounding of aerosols from Mount Etna. *J. Geophys. Res. Atmos.* **2006**, *111*, 1–10. [CrossRef]
57. Bagnato, E.; Aiuppa, A.; Parello, F.; Calabrese, S.; D'Alessandro, W.; Mather, T.A.; McGonigle, A.J.S.; Pyle, D.M.; Wängberg, I. Degassing of gaseous (elemental and reactive) and particulate mercury from Mount Etna volcano (Southern Italy). *Atmos. Environ.* **2007**, *41*, 7377–7388. [CrossRef]
58. Moreno, T.; Querol, X.; Castillo, S.; Alastuey, A.; Cuevas, E.; Herrmann, L.; Mounkaila, M.; Elvira, J.; Gibbons, W. Geochemical variations in aeolian mineral particles from the Sahara-Sahel Dust Corridor. *Chemosphere* **2006**, *65*, 261–270. [CrossRef] [PubMed]
59. Reimann, C.; De Caritat, P. Intrinsic flaws of element enrichment factors (EFs) in environmental geochemistry. *Environ. Sci. Technol.* **2000**, *34*, 5084–5091. [CrossRef]
60. Aiuppa, A.; Allard, P.; D'Alessandro, W.; Michel, A.; Parello, F.; Treuil, M.; Valenza, M. Mobility and fluxes of major, minor and trace metals during basalt weathering and groundwater transport at Mt. Etna volcano (Sicily). *Geochim. Cosmochim. Acta* **2000**, *64*, 1827–1841. [CrossRef]
61. Piazzola, J.; Mihalopoulos, N.; Canepa, E.; Tedeschi, G.; Prati, P.; Zarmpas, P.; Bastianini, M.; Missamou, T.; Cavaleri, L. Characterization of aerosols above the Northern Adriatic Sea: Case studies of offshore and onshore wind conditions. *Atmos. Environ.* **2016**, *132*, 153–162. [CrossRef]
62. Marani, M.P.; Gamberi, F.; Savelli, C. Shallow-water polymetallic sulfide deposits in the Aeolian island arc. *Geology* **1997**, *25*, 815–818. [CrossRef]
63. Savelli, C.; Marani, M.; Gamberi, F. Geochemistry of metalliferous, hydrothermal deposits in the Aeolian arc (Tyrrhenian Sea). *J. Volcanol. Geotherm. Res.* **1999**, *88*, 305–323. [CrossRef]
64. Viana, M.; Kuhlbusch, T.A.J.; Querol, X.; Alastuey, A.; Harrison, R.M.; Hopke, P.K.; Winiwarter, W.; Vallius, M.; Szidat, S.; Prévôt, A.S.H.; et al. Source apportionment of particulate matter in Europe: A review of methods and results. *J. Aerosol Sci.* **2008**, *39*, 827–849. [CrossRef]

© 2020 by the authors. Licensee MDPI, Basel, Switzerland. This article is an open access article distributed under the terms and conditions of the Creative Commons Attribution (CC BY) license (http://creativecommons.org/licenses/by/4.0/).

Article

Characteristics of Black Carbon Particle-Bound Polycyclic Aromatic Hydrocarbons in Two Sites of Nanjing and Shanghai, China

Shijie Cui [1], Ruoyuan Lei [1], Yangzhou Wu [2], Dandan Huang [3], Fuzhen Shen [1], Junfeng Wang [1], Liping Qiao [3], Min Zhou [3], Shuhui Zhu [3], Yingge Ma [3] and Xinlei Ge [1,*]

[1] Jiangsu Key Laboratory of Atmospheric Environment Monitoring and Pollution Control, Collaborative Innovation Center of Atmospheric Environment and Equipment Technology, School of Environmental Sciences and Engineering, Nanjing University of Information Science and Technology, Nanjing 210044, China; csj930429@163.com (S.C.); 18851967099@163.com (R.L.); 20161119082@nuist.edu.cn (F.S.); 15295738628@163.com (J.W.)
[2] School of Earth Sciences, Zhejiang University, Hangzhou 310027, China; wu_yz@zju.edu.cn
[3] State Environmental Protection Key Laboratory of Formation and Prevention of Urban Air Pollution Complex, Shanghai Academy of Environmental Sciences, Shanghai 200233, China; huangdd@saes.sh.cn (D.H.); qiaolp@saes.sh.cn (L.Q.); zhoum@saes.sh.cn (M.Z.); zhush@saes.sh.cn (S.Z.); mayg@saes.sh.cn (Y.M.)
* Correspondence: caxinra@163.com; Tel.: +86-25-58731394

Received: 6 February 2020; Accepted: 11 February 2020; Published: 14 February 2020

Abstract: Airborne polycyclic aromatic hydrocarbons (PAHs) are of great concern to human health due to their potential high toxicity. Understanding the characteristics and sources of PAHs, as well as the governing factors, is therefore critical. PAHs and refractory black carbon (rBC) are both from combustion sources. This work, for the first time, investigated exclusively the rBC-bound PAH properties by using a laser-only Aerodyne soot-particle aerosol mass spectrometer (SP-AMS). This technique offers highly time-resolved PAH results that a traditional offline measurement is unable to provide. We analyzed two datasets conducted in urban Shanghai during the fall of 2018 and in suburban Nanjing during the winter of 2017, respectively. Results show that the average concentration of PAHs in Nanjing was much higher than that in Shanghai. Nanjing PAHs contained more low molecular weight components while Shanghai PAHs contained more high molecular weight ones. PAHs in Shanghai presented two peaks in early morning and evening, while Nanjing PAHs had only one significant morning peak, but remained high throughout the nighttime. A multi-linear regression algorithm combined with positive matrix factorization (PMF) analyses on sources of PAHs reveals that the industry emissions contributed the majority of PAHs in Nanjing (~80%), while traffic emissions dominated PAHs in Shanghai (~70%). We further investigated the relationships between PAHs with various factors. PAHs in both sites tended to positively correlate with primary pollutants, including primary organic aerosol (OA) factors, and gaseous pollutants of CO, NO_2 and SO_2, but negatively correlated with secondary OA factors and O_3. This result highlights the enhancement of rBC-bound PAHs level due to primary emissions and their oxidation loss upon atmospheric aging reactions. High concentration of PAHs seemed to frequently appear under low temperature and high relative humidity conditions, especially in Shanghai.

Keywords: Aerosol mass spectrometry; refractory black carbon; Source; Positive matrix factorization; traffic emissions; industry emissions

1. Introduction

Polycyclic aromatic hydrocarbons (PAHs) are a group of organic species that are ubiquitously present in ambient air, derived mainly from anthropogenic sources, including incomplete combustion or pyrolysis of fossil fuels (such as gasoline, diesel, and coal), solid waste, as well as biomass [1–6]. PAHs themselves are known to possess high health hazards [7–10]. When emitted into the air, PAHs can be further oxidized into other PAH derivatives (such as nitrated and oxygenated PAHs), which are likely more toxic and carcinogenic and have longer atmospheric lifetimes than their parent PAHs [11–15]. Studies on concentrations, size distributions, sources, chemical transformations, as well as the spatial distributions and temporal variations of airborne PAHs, are therefore essential.

As is well known, identification and quantification of PAHs are typically conducted by offline analysis of filter extracts by using the gas chromatography mass spectrometry (GC-MS) technique [16,17], which also has the advantage to differentiate PAH isomers with the same molecular weight (MW). Of course, filter samples often require longer collection time, resulting in lower temporal resolution, which sometimes limit detailed investigations on quick changes of PAH behaviors in ambient [18]. Sampling artifacts are also difficult to eliminate during offline analysis. Online measurement techniques are also developed in the past decades, including the photoelectric aerosol sensor (PAS) [17,19] and Aerodyne aerosol mass spectrometry (AMS) [20]. AMS is an instrument that is designed to determine the aerosol concentration, composition, and size distribution in real time with very fine time resolution. Dzepina et al. [20] first proposed the method to determine PAHs via the quadruple-AMS, and such method can be used by other versions of Aerodyne AMS, such as the high resolution-AMS (HR-AMS) [21,22] and soot-particle AMS (SP-AMS) [23]. A number of studies have used this method to quantify particulate PAHs, studied the chemical evolution of PAHs [16,24–26], as well as PAH emissions from specific sources, including biomass burning [27] and vehicle emissions [28]. The AMS technique offers unique highly time-resolved results on PAHs, whereas offline GC-MS measurement cannot.

Note the emission sources of PAHs are similar as those of refractory black carbon (rBC) or soot, and PAHs are in fact intermediates during soot formation [29], and can be converted to soot during combustion [28,30,31]. Different PAH compounds have different gas-to-particle partitioning properties [32–34]. PAHs with relatively large numbers of benzene rings mainly exist in fine particles [35], and those with smaller MWs may partition into gas phase [36]. In addition, some studies have shown that the particle-bound PAHs can react with ozone (O_3), and form a viscous layer to protect the underlying PAHs from ozonolysis, so that prolongs their chemical lifetimes [37]. Therefore, the PAHs co-emitted with rBC particles might evaporate or be oxidized into other species during their atmospheric transport; PAHs that are emitted separately from rBC may coagulate or adsorb on rBC particles as well. As an aerosol component, rBC also exerts adverse effects on human health [38], although not as toxic as PAHs when inhaling the same amount. Studies of the rBC-bound PAH properties, and its correlation with other factors, can advance our understanding on both rBC and PAH properties, as well as the health co-benefits from rBC and PAH reductions.

The Aerodyne SP-AMS is the most advanced version of Aerodyne AMS, which uses an intra-cavity infrared laser to measure rBC, and another thermal vaporizer to measure non-refractory materials [39]. If we remove the thermal vaporizer, the SP-AMS can exclusively measure rBC-containing particles only, acquiring the concentration, size, and composition of both rBC cores and any species that coat on rBC cores [40,41], including, of course, rBC-bound PAHs. In this study, we aim to use the SP-AMS technique to elucidate the characteristics of the PAHs species associated with rBC particles in two representative megacities in the Yangtze River Delta (YRD) region of China, investigate their concentrations, source contributions, as well as their relationships with a series of factors, including different primary and secondary particle-phase components, gas-phase pollutants, and meteorological parameters.

2. Experiments

2.1. Sampling Sites and Instrumentation

In this study, we conducted SP-AMS measurements in two megacities (as shown in Figure 1). The first study was performed during winter in suburban Nanjing (from 23 January to 24 February, 2017). The SP-AMS was deployed inside the campus of Nanjing University of Information Science and Technology (NUIST). The site was mainly affected by nearby industrial emissions (petrochemical plants, chemical plants, power plants, ironmaking, and steelmaking plants, etc.), residential activities, and road traffic (Nanjing Jiangbei expressway), as marked in Figure 1. The dataset is denoted as NUIST hereafter. The second campaign was performed in urban Shanghai from 31 October to 2 December, 2018. The sampling site was located inside the Shanghai Academy of Environmental Sciences (SAES). It was surrounded by business buildings, and influenced mainly by dense residential and traffic activities (Figure 1). The dataset is denoted as SAES hereafter. The two sites both locate in the densely populated YRD region, but under different environments as indicated above. Therefore, comparison of their results may provide useful insights into the ambient PAH properties emitted from different sources. Of course, since we only have one SP-AMS, it should be noted that the results from two sites were not for the same period, but both measurements were performed in seasons with relatively high particular matter (PM) pollution.

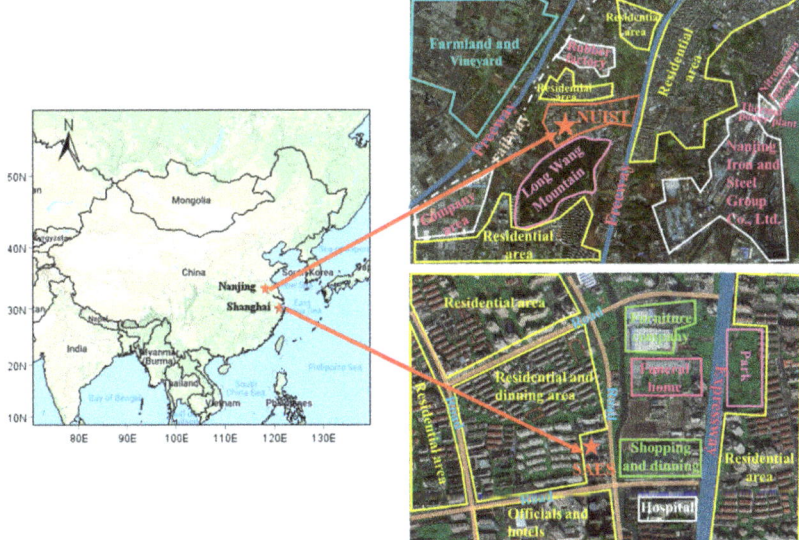

Figure 1. Field measurement sites (Nanjing University of Information Science and Technology—NUIST; Shanghai Academy of Environmental Sciences—SAES) and surrounding areas.

During these two campaigns, the laser-only mode of SP-AMS was operated to measure rBC-containing particles only. More instrument details can be found in our previous work [41]. Concentrations of major gas pollutants (CO, NO_2, SO_2, and O_3) and the meteorological parameters (temperature, relative humidity, wind speed, and direction) were determined simultaneously in parallel with aerosol measurements in both sites.

2.2. PAHs Quantification and Source Apportionment

The SP-AMS data were analyzed by using Igor Pro 6.37 (WaveMetrics, Oregon, USA) with the SQUIRREL (version 1.59D) and PIKA (1.19D) data analysis toolkit (publicly available from: http://cires1.

colorado.edu/jimenez-group/ToFAMSResources/ToFSoftware/index.html#Upgrade). The SP-AMS laser can evaporate the rBC-containing particles, and all species are subjected to 70 eV electron impact (EI) ionization [38], and we can obtain signals of ion fragments from these species. Most PAHs existing in particle phase in ambient air have relatively high molecular weight (>200 g/mol). Based on possible molecular formulas of PAHs and mass-to-charge (m/z) ratios of corresponding fragments, and the fact that PAHs are very resistant to fragmentation (which means that the molecular ions represent most signals of the corresponding parent PAH compounds), Dzepina et al. [20] proposed a well-validated method to identify different PAH compounds and calculate the total PAHs mass concentration. Here, the same methodology was adopted to determine the mass loadings of rBC-bound PAHs in Shanghai and Nanjing. A relative ionization efficiency of 1.4 was used here for PAHs quantification, consistent with previous studies [21,22,26].

Positive matrix factorization (PMF) [42] was employed to deconvolute the rBC-associated organic aerosol (OA) species to a few factors indicative of specific sources or processes. It is a receptor-only model, which requires no prior information about factor profiles or time series. Ulbrich et al. [43] developed an evaluation software tool with a user-friendly interface, which allows us to conduct the PMF analysis on AMS data. Zhang et al. [44] further provide standard procedures to evaluate PMF solutions and choose the optimal one. For NUIST dataset, five OA factors were resolved, as detailed in Wu et al. [45]. Note in Wu et al. [45], the PMF results were used to elucidate the relationship of OA with hygroscopicity of rBC-containing particles, while characteristics of rBC-bound PAHs were not discussed. For the SAES dataset, PMF analysis was conducted on the OA mass spectra including ions with m/zs of 12 to 120. The OA mass spectral matrix was further pre-processed by removing ions with low signal-to-noise ratio (SNR<0.2), down-weighting of ions with SNR of 0.2–2.0, and some time points appearing to be outliers. PMF solutions with a different number of factors, varying f_{peak} values (rotations), and seeds (initial values) were also systematically explored, and finally a six-factor solution was chosen to be the best solution. More details regarding the diagnostic and justification of the PMF results of SAES dataset are presented in Cui et al. [46]; note that work focuses on the chemical properties and aging process of rBC-containing particles, while current work aims to elucidate PAH properties, and compare it with those in Nanjing.

After we obtained the time series of PMF factors, a multi-linear regression method was applied to apportion the PAHs and rBC signals to different sources for NUIST and SAES datasets, respectively (details in Section 3.2). Such method has been used in our previous work to quantify sources of fullerene soot [47], and in some other studies for rBC [48]. Correlation analyses were also performed on PAHs with other aerosol components, gaseous pollutants, as well as meteorological parameters.

3. Results and Discussion

3.1. Overview of PAHs Concentration and Distribution

Figure 2 presents the mass concentrations of rBC and rBC-bound PAHs, normalized mass spectrum of PAHs, and mass contributions of different groups of PAHs. The rBC-bound PAHs concentration in NUIST ranged from 5.4 to 301.9 ng/m^3, with an average value of 46.0(± 38.0) ng/m^3 (mean ± one standard deviation). This average PAHs level is lower than, but in a similar order as that measured by offline GC-MS in the same location (64.3 ng/m^3) during winter of 2014–2015 [49]. PAHs concentration in SAES varied from 0.38 to 42.25 ng/m^3, with a mean of 6.6(± 6.9) ng/m^3. This PAHs level is much lower than that of NUIST, but is in the same order of previous reported values of 7.2 ng/m^3 (suburban area) and 6.5 ng/m^3 (urban area) in Shanghai [50]. Considering that we only measured the portion of PAHs bound with rBC, lower levels therefore could be anticipated. Overall, the results show that PAHs pollution in suburban Nanjing, at least during winter, was more significant than that during fall in Shanghai, likely due to enhanced emissions from nearby industry activities around NUIST; much more heavy-duty diesel trucks around NUIST (no restrictions on those vehicles around this site) than those around SAES might also add the rBC-bound PAH emissions.

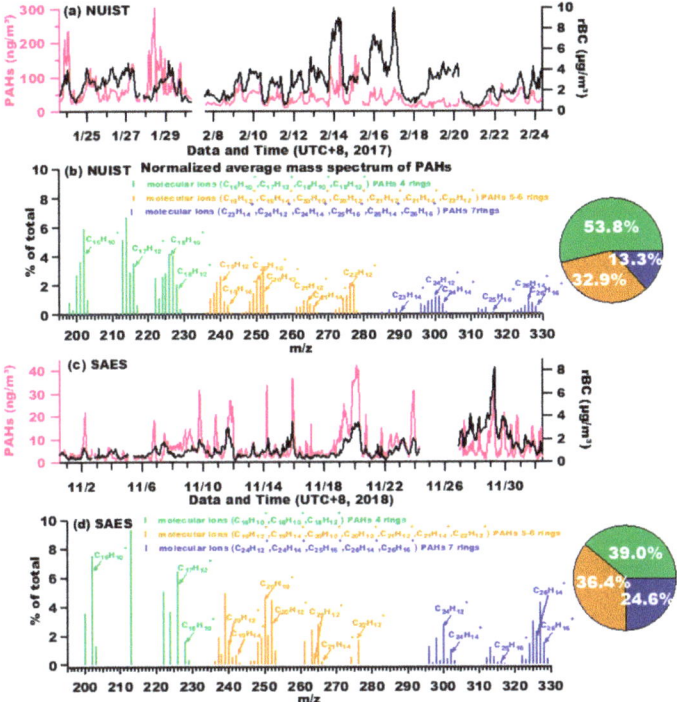

Figure 2. Time series of refractory black carbon (rBC) and rBC-bound polycyclic aromatic hydrocarbons (PAHs) in Nanjing University of Information Science and Technology (NUIST) (**a**) and in Shanghai Academy of Environmental Sciences (SAES) (**c**), and normalized average mass spectrum of PAHs in NUIST (**b**) and in SAES (**d**) (PAH fragments are classified into three groups; pie charts show mass contributions of the three groups of PAHs).

Correspondingly, average rBC concentrations in NUIST and SAES were 2.76 (±1.60) and 1.21 (±0.99) μg/m^3, and the mean mass ratios of rBC-bound PAHs to rBC were 0.0167 and 0.0055, respectively. Moderate correlations were found between time series of PAHs and rBC (0.53 of NUIST and 0.62 of SAES), and there were periods with relatively high PAHs loadings, yet low rBC concentrations (for example, 28 January, 2017 in NUIST), and vice versa (16–18 February, 2017 in NUIST) (Figure 2a). Both results point to different source contributions to and/or influences of atmospheric processes on PAHs and rBC. This is discussed in detail in the following sections.

Figure 2b,d display the normalized mass spectra of PAHs in NUIST and SAES, respectively. The fragments were grouped into PAHs with 4 rings, 5-6 rings, and 7 rings. For NUIST, relatively low MW 4-ring PAHs were dominant (53.8%), followed by 5-6 ring PAHs (32.9%), and 7-ring PAHs (13.3%). In contrast, for SAES, mass contribution of 4-ring PAHs was much less (39.0%), while high MW 7-ring ones became more important (24.6%), and contribution of 5-6 ring PAHs (36.4%) were similar to that of NUIST. Since different groups of PAHs have different gas-to-particle partitioning properties and vapor pressures, different distributions of PAHs in NUIST and SAES might lead to their distinct atmospheric behaviors.

Figure 3 illustrates the diurnal variations of rBC, rBC-bound PAHs and mass ratios of PAHs to rBC. For NUIST, both rBC and PAHs presented morning peaks (3.39 μg/m^3 of rBC, and 68.76 ng/m^3 of PAHs) likely due to elevated traffic or industrial emissions. They had minimums (1.91 μg/m^3 of rBC, and 29.47 ng/m^3 of PAHs) in the afternoon, probably owing to emission decrease and the dilution effect by elevated planetary boundary layer (PBL) height, etc.; their concentrations rose continuously in the early evening and remained at a high level throughout the nighttime. PAHs/rBC ratios also showed a morning peak and an afternoon minimum, reflecting the semi-volatile behavior of PAHs (high evaporation loss due to high temperature in the afternoon) and possible loss due to photochemical reactions of PAHs with atmospheric oxidants. For SAES, beside the morning peaks (1.56 µg/m^3 of rBC, and 10.23 ng/m^3 of PAHs) and afternoon minimums (1.02 µg/m^3 of rBC, and 4.96 ng/m^3 of PAHs), early evening peaks of rBC (1.35 µg/m^3) and PAHs (11.86 ng/m^3) were more apparent than those in NUIST. Such differences between the two datasets may be attributed to diverse source contributions: the NUIST site was affected significantly by industrial emissions, which were relatively stable, and the early evening increases were probably owing to a decrease of PBL height. Meanwhile, the SAES site was predominantly governed by traffic emissions; therefore, in addition to the PBL influence, evening rush hour traffic stood out and led to more obvious enhancements.

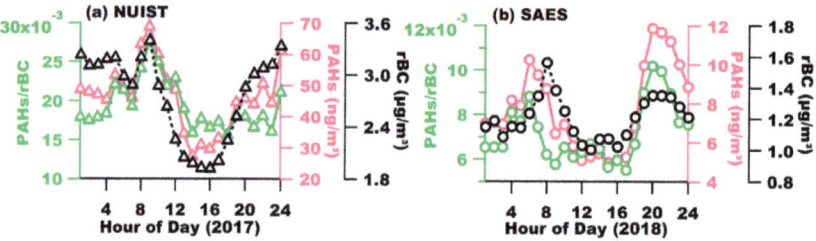

Figure 3. Diurnal patterns of rBC, rBC-bound PAHs, and the mass ratios of PAHs to rBC in NUIST (a) and SAES (b).

3.2. Source Appointments of PAHs and rBC

To further elucidate the PAHs and rBC sources, we employed a multi-linear regression method to apportion the signals of PAHs and rBC to different sources (i.e., different PMF factors). For the NUIST dataset, five factors were resolved, including the traffic-related hydrocarbon-like OA (HOA), industry-related OA (IOA), aged biomass burning OA (ABBOA), semi-volatile oxygenated OA (SV-OOA), and low-volatility oxygenated OA (LV-OOA). Their mass spectral profiles and time series are presented in Wu et al. [45], therefore are not shown here. For SAES dataset, we separated six factors, including the enriched hydrocarbon-like OA (HOA-rich), rBC-enriched OA (rBC-rich), biomass burning OA (BBOA), nitrogen-enriched OA (NOA), less oxidized oxygenated OA (LO-OOA), and more oxidized oxygenated OA (MO-OOA). The high-resolution mass spectra and corresponding time series of those factors are depicted in Figure 4. For NUIST dataset, HOA, IOA, and BBOA were primary OA (POA) factors, SV-OOA and LV-OOA were secondary OA (SOA) factors; for the SAES dataset, HOA-rich, rBC-rich, BBOA and NOA were four POA factors, LO-OOA and MO-OOA were SOA factors.

Note that PAHs and rBC mostly originate from primary sources [1,38] rather than secondary sources; thus, the regressions were only conducted on primary factors. Actually, if we added secondary factors into the fitting, correlations between measured and re-constructed rBC and PAH concentrations would all become weaker; therefore, exclusion of SOA factors was reasonable (SOA factors were, in fact, negatively correlated with PAHs as shown in Section 3.3). Finally, we established the following empirical formulas for PAHs and rBC for NUIST and SAES datasets, respectively.

$$rBC_{\text{re-constructed, NUSIT}} = 1.171 \times [\text{HOA}] + 0.945 \times [\text{IOA}] + 0.543 \times [\text{ABBOA}] + 0.456 \tag{1}$$

$$rBC_{\text{re-constructed, SAES}} = 1.453 \times [rBC\text{-rich}] + 0.240 \times [\text{HOA-rich}] + 0.833 \times [\text{NOA}] + 0.243 \times [\text{BBOA}] + 0.146 \tag{2}$$

where $rBC_{\text{re-constructed, NUSIT}}$ and $rBC_{\text{re-constructed, SAES}}$ (µg/m^3) represent the re-constructed mass concentrations of rBC in NUIST and SAES, respectively. Factors in braces were time series of corresponding factors.

$$\text{PAHs}_{\text{re-constructed, NUSIT}} = 11.44 \times [\text{HOA}] + 44.917 \times [\text{IOA}] + 5.43 \times [\text{ABBOA}] + 26.14 \tag{3}$$

$$\begin{aligned}\text{PAHs}_{\text{re-constructed, SAES}} &= 6.478 \times [r\text{BC-rich}] + 1.474 \times [\text{HOA-rich}] + 0.655 \times [\text{NOA}] \\ &\quad + 2.495 \times [\text{BBOA}] + 1.314\end{aligned} \tag{4}$$

where PAHs$_{\text{re-constructed, NUSIT}}$ and PAHs$_{\text{re-constructed, SAES}}$ (ng/m^3) represent the reconstructed PAHs concentrations in NUIST and SAES, respectively. Meanings of braced factors were the same as those in Equations (1) and (2).

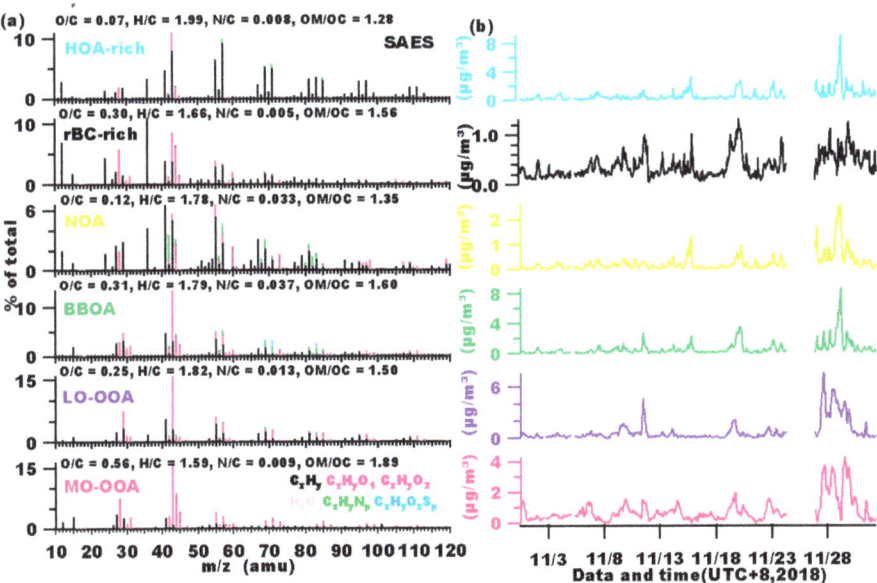

Figure 4. High-resolution mass spectra of the positive matrix factorization (PMF)-resolved factor profiles (classified by six different ion families; elemental ratios were calculated based on Canagaratna et al. [51]) (a), and their corresponding time series in SAES (b).

By using above formulas, re-constructed PAHs and rBC concentrations agreed fairly well with corresponding measured data (all with p <0.01, Pearson's r of 0.93 and 0.97 for rBC, Pearson's r of 0.81 and 0.76 for PAHs of NUIST and SAES, respectively) (as shown in Figures 5 and 6), indicating effectiveness of this method. We then used the regressed coefficients and mean concentrations of factors to calculate their relative contributions to rBC and PAHs, as shown in the pie charts. Results show that source contributions to total POA, rBC, and PAHs were remarkably different for NUIST and SAES datasets. POA in NUIST was comprised of 42.0% IOA, 32.5% ABBOA, and 25.5% HOA; source contributions to rBC was 45.5% IOA, 34.3% HOA, and 20.2% ABBOA; while PAHs appeared to be overwhelmingly contributed by IOA (80.1%), followed by HOA (12.4%), and a minor contribution from ABBOA (7.5%). POA in SAES was composed of 21.6% rBC-rich OA, 25.5% HOA-rich, 30.1% BBOA, and 12.4% NOA. Note rBC-rich OA was likely associated closely with traffic diesel combustion (and possibly some ship emissions), HOA-rich factor was likely linked with traffic gasoline emissions, and NOA was a local POA factor, possibly from industrial activities. Relative mass contribution to rBC was largest from the rBC-rich factor (54.5%), followed by 17.9% NOA, 14.9% HOA-rich, and 12.7% BBOA. For PAHs, direct traffic-related OA contributions were still dominant (50.8% rBC-rich OA and 19.1% HOA-rich); BBOA contribution was a bit more (27.1%) than its contribution to rBC, while NOA contribution (possible industry contribution) was very small (2.9%). Source contributions to PAHs in NUIST and SEAS were found to be contrastingly different.

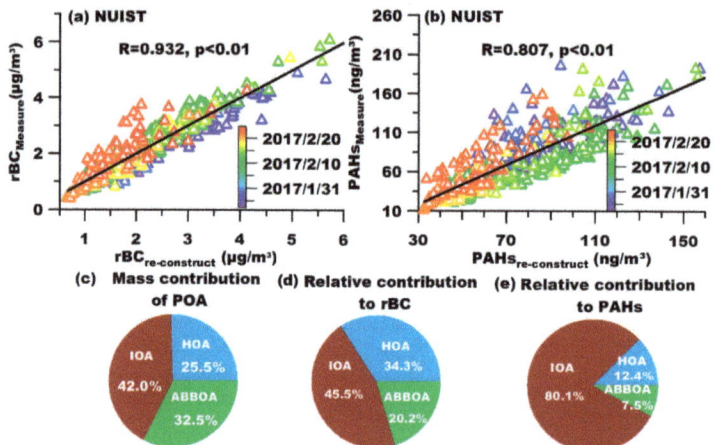

Figure 5. Scatter plots of measured and re-constructed rBC (**a**) and PAHs concentrations (**b**), and pie charts of mass contributions of different factors to the total primary organic aerosol (POA) (**c**), and relative contributions of these factors to rBC (**d**) and PAHs (**e**) in NUIST.

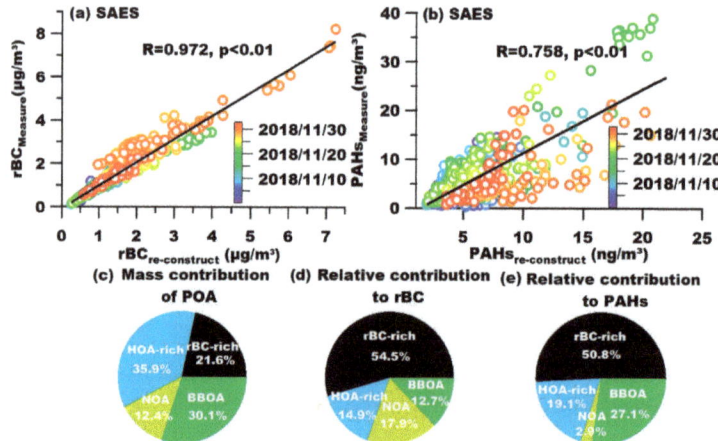

Figure 6. Scatter plots of measured and re-constructed rBC (**a**) and PAHs concentrations (**b**), and pie charts of mass contributions of different factors to the total primary organic aerosol (POA) (**c**), and relative contributions of these factors to rBC (**d**) and PAHs (**e**) in SAES.

3.3. Relationships between PAHs and Other Factors

In general, the mass loading and variation of a certain species in ambient are controlled by an integrated effect from emissions, reactions, and meteorology. Therefore, we explored the relationships of PAH concentrations with primary or secondary species, including both particle-bound and gas-phase species, as well as the meteorological parameters.

Figure 7 presents the variations of PAH concentrations against different OA terms, including the oxidation state (OS_c), mass concentrations of POA and SOA. OS_c is defined as 2 × O/C-H/C [52], here O/C and H/C are oxygen-to-carbon and hydrogen-to-carbon atomic ratios. Generally, PAH concentrations decreased with the increase of OS_c for both NUIST (except a small plateau between −1.15 and −0.95) and SAES. Since OS_c can be treated as a metric of the aging or oxidation degree of OA, this result reflects the loss of PAHs upon atmospheric aging. On the contrary, PAHs had an obvious positive relationship with POA for both datasets (Figure 7b,f), verifying its primary origin. Furthermore, investigations on correlations between PAHs with individual POA factors (not shown here) revealed that PAHs in NUIST correlated the best with IOA, and PAHs in SAES correlated tightest with rBC-rich OA. These results are in line with the pie charts shown in Figures 5 and 6—that IOA and rBC-rich OA were dominant sources of PAHs in NUIST and SAES, respectively.

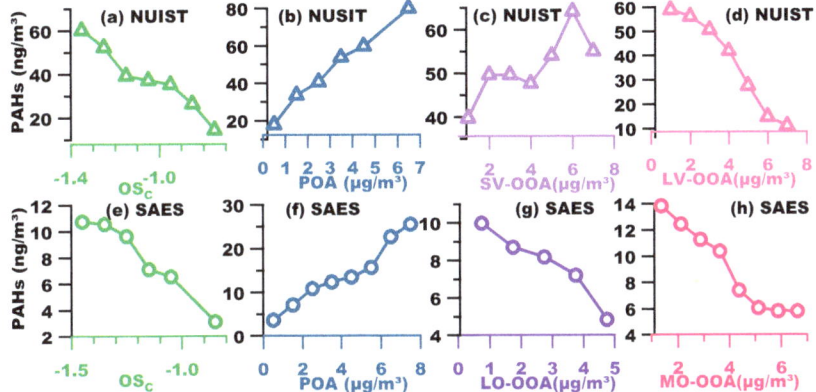

Figure 7. Variations of PAH concentrations versus the oxidation states (OS_c), POA, semi-volatile oxygenated OA (SV-OOA) or less oxidized oxygenated OA (LO-OOA), and low-volatility oxygenated OA (LV-OOA) or more oxidized oxygenated OA (MO-OOA) concentrations: (**a**–**d**) NUIST; (**e**–**h**) SAES.

The relationships between PAHs and SOA factors, however, were a bit complex. In particular, the PAHs in NUIST even presented an increasing trend with SV-OOA (although not highly correlated). We think this positive correlation reflects the semi-volatile behavior of PAHs, similar to that of SV-OOA in NUIST, rather than an indication of secondary production of PAHs. NUIST PAHs did contain more low MW PAHs, which might partially explain this semi-volatile behavior. For SAES, in fact, the factor with relatively low oxidation degree (LO-OOA) had no clear semi-volatile behavior; therefore, the PAHs correlated negatively with it, as expected. For these highly oxygenated SOA factors, LV-OOA in NUIST and MO-OOA in SAES, very clear anti-correlations were observed between them with PAHs, suggesting again the primary not secondary origin of PAHs, as well as oxidation loss of PAHs, along with the formation of large amounts of SOA.

Figure 8 illustrates the relationships between PAHs with those criteria gaseous pollutants, including CO, NO_2, SO_2, and O_3. Note that concentrations of CO, NO_2, and SO_2 in NUIST were all higher than those in SEAS; in particular, NUIST SO_2 concentration was a few times higher than SAES SO_2, indicating a more important role of industrial (likely coal combustion) emissions in NUIST. PAHs in both NUIST and SAES, in general, correlated positively with CO, NO_2, and SO_2, as these gases were also, in large part, from primary emission sources, as PAHs. Only the correlation between PAHs and NO_2 in NUIST was less clear-cut, which was in fact also reasonable, considering that PAHs in NUIST were mainly from industry emissions, yet NO_2 was mainly from traffic. On the other hand, PAH concentrations decreased continuously with the increases of O_3 in both NUIST and SAES. Different from other gases, O_3 is exclusively produced from photochemical reactions; therefore, this anti-relationship may suggest that PAHs were subject to photochemical oxidation, which impacted its atmospheric levels. Results in Figures 6 and 7 together demonstrate the enhancement of the mass concentration of PAHs due to primary emissions, as well as the loss or sink of PAHs due to atmospheric oxidation reactions.

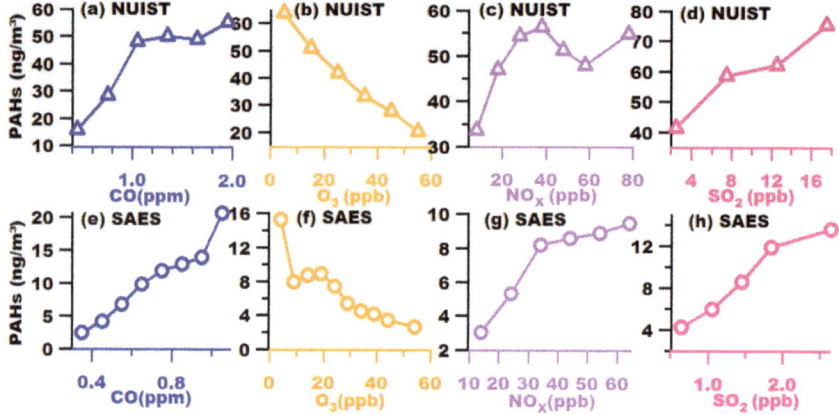

Figure 8. Variations of PAH concentrations versus CO, O_3, NO_2, and SO_2 concentrations: (a–d) NUIST; (e–h) SAES.

At last, we investigated impacts of meteorological parameters (relative humidity (RH), temperature (T), wind speed (WS), and wind direction (WD)) on PAH concentrations (as shown in Figure 9). First, it is interesting to observe that WS had no clear impact on PAH loadings at both sites. The dilution effect by wind speed was not significant unless there were strong winds (WS >3.5 m/s). For NUIST, PAHs concentration from air parcels (east to southeast) was the highest, coincident with the industry zone from that direction (Figure 1). Again, it demonstrates the dominance of industrial emissions to PAHs in NUIST. For SAES, except that PAHs concentration was quite low in air parcels from north to east, other air parcels all brought about relatively high loadings of PAHs, likely due to traffic activities from the freeway and residential area (Figure 1). Regarding the effects of temperature, PAH concentrations in both sites decreased against an increase in temperature, probably owing to two reasons: one is the evaporation loss of PAHs at higher temperatures; another is the oxidation loss (since periods with high temperatures, such as afternoons, often overlapped with periods with strong solar radiation, i.e., strong photochemical oxidation). In both sites, responses of PAH concentrations to RH changes were generally positive (especially in SAES), but there was a drop at very high RH (>90%) probably due to scavenging effects. The positive correlations were more likely owing to that high RH often occurring during a time (i.e., nighttime and early morning) when strong anthropogenic PAH emissions were exacerbated by low PBL heights.

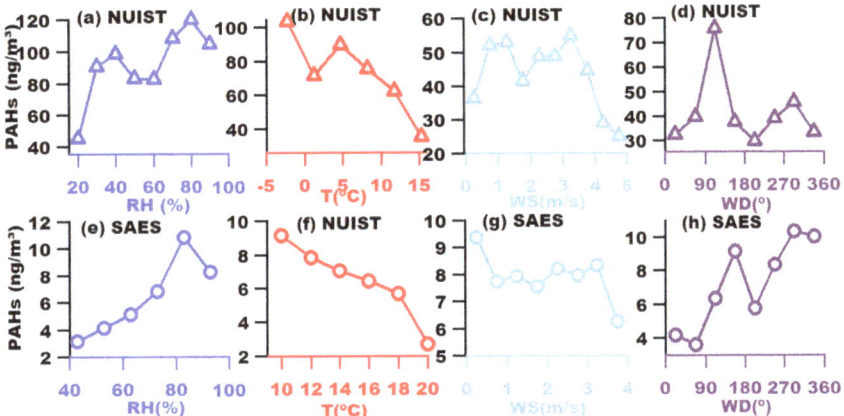

Figure 9. Variations of PAHs concentrations versus relative humidity (RH), temperature (T), wind speed (WS), and wind direction (WD): (**a**–**d**) NUIST; (**e**–**h**) SAES.

4. Conclusions

In this work, for the first time, we applied a laser-only SP-AMS to investigate rBC-bound PAHs during the winter in suburban Nanjing, and during fall in urban Shanghai. Highly time-resolved data allowed us to characterize the mass spectra, diurnal patterns, sources of PAHs, and their correlations with other factors. Average rBC-bound PAHs concentration in Nanjing was 46.0 ng/m^3, much higher than that determined in Shanghai (6.64 ng/m^3). Nanjing PAHs were found to contain more low MW PAHs, while in Shanghai, high MW PAHs contribution was more significant. The diurnal pattern of PAHs concentration in Nanjing showed a clear morning peak and an overall high level throughout the nighttime, while Shanghai PAHs concentration presented bimodal distribution in early morning and evening, indicating a clear impact from traffic activities. Source apportionment on PAHs via a multi-linear regression method using PMF-resolved OA factors as inputs, showed that, indeed, traffic emissions were a major source of PAHs in Shanghai (70%), while industry emissions overwhelmingly dominated PAHs in Nanjing (80%). We further investigated the relationships between PAHs and other factors. PAHs presented overall positive relationships with POA and CO, NO_2, and SO_2, but negatively correlated with OS_c, SOA factors, and O_3, demonstrating enhancement of PAHs level in ambient, due to primary sources and oxidation loss from atmospheric aging. PAHs also decreased with an increase in temperature, again, reflecting possible evaporation and/or oxidation loss of PAHs. High PAHs concentrations often appeared under low T and high RH conditions, especially in Shanghai. Our findings here provide useful insights into the understanding of pollution levels, temporal variability, and the source and lifecycle of PAHs in representative atmospheric environments.

Author Contributions: Conceptualization, X.G.; Data curation, S.C., D.H., F.S., L.Q., M.Z., S.Z. and Y.M.; Formal analysis, S.C., R.L., Y.W., D.H., F.S., and J.W.; Investigation, X.G.; Methodology, R.L., Y.W., F.S. and J.W.; Project administration, X.G.; Resources, L.Q., M.Z., S.Z. and Y.M.; Supervision, X.G.; Writing—original draft, S.C.; Writing—review & editing, X.G. All authors have read and agreed to the published version of the manuscript.

Funding: This research was funded by the National Natural Science Foundation of China (grant number 21777073 and 21976093), the National Key R&D program of China (2018YFC0213802), and the Innovation and Entrepreneurship Training Program for College Students in Jiangsu Province (201810300095X).

Conflicts of Interest: The authors declare no conflict of interest.

References

1. Ravindra, K.; Sokhi, R.; Vangrieken, R. Atmospheric polycyclic aromatic hydrocarbons: Source attribution, emission factors and regulation. *Atmos. Environ.* **2008**, *42*, 2895–2921. [CrossRef]
2. Poulain, L.; Iinuma, Y.; Müller, K.; Birmili, W.; Weinhold, K.; Brüggemann, E.; Gnauk, T.; Hausmann, A.; Löschau, G.; Wiedensohler, A.; et al. Diurnal variations of ambient particulate wood burning emissions and their contribution to the concentration of polycyclic aromatic hydrocarbons (PAHs) in seiffen, germany. *Atmos. Chem. Phys.* **2011**, *11*, 12697–12713. [CrossRef]
3. Abbas, I.; Badran, G.; Verdin, A.; Ledoux, F.; Roumié, M.; Courcot, D.; Garçon, G. Polycyclic aromatic hydrocarbon derivatives in airborne particulate matter: Sources, analysis and toxicity. *Environ. Chem. Lett.* **2018**, *16*, 439–475. [CrossRef]
4. Elzein, A.; Dunmore, R.E.; Ward, M.W.; Hamilton, J.F.; Lewis, A.C. Variability of polycyclic aromatic hydrocarbons and their oxidative derivatives in wintertime beijing, china. *Atmos. Chem. Phys.* **2019**, *19*, 8741–8758. [CrossRef]
5. Wang, Y.; Zhang, Q.; Zhang, Y.; Zhao, H.; Tan, F.; Wu, X.; Chen, J. Source apportionment of polycyclic aromatic hydrocarbons (PAHs) in the air of dalian, china: Correlations with six criteria air pollutants and meteorological conditions. *Chemosphere* **2019**, *216*, 516–523. [CrossRef]
6. Harrison, R.M.; Smith, D.J.T.; Luhana, L. Source apportionment of atmospheric polycyclic aromatic hydrocarbons collected from an urban location in birmingham, U.K. *Environ. Sci. Technol.* **1996**, *30*, 825–832. [CrossRef]
7. Nisbet, I.C.; LaGoy, P.K. Toxic equivalency factors (TEFs) for polycyclic aromatic hydrocarbons (PAHs). *Regul. Toxicol. Pharmacol.* **1992**, *16*, 290–300. [CrossRef]
8. Bostrom, C.-E.; Gerde, P.; Hanberg, A.; Jernstrom, B.; Johansson, C.; Kyrklund, T.; Rannug, A.; Tornqvist, M.; Victorin, K.; Westerholm, R. Cancer risk assessment, indicators, and guidelines for polycyclic aromatic hydrocarbons in the ambient air. *Environ. Health Perspect.* **2002**, *110* (Suppl. 3), 451–488. [CrossRef]
9. Bi, C.; Chen, Y.; Zhao, Z.; Li, Q.; Zhou, Q.; Ye, Z.; Ge, X. Characteristics, sources and health risks of toxic species (pcdd/fs, pahs and heavy metals) in pm2.5 during fall and winter in an industrial area. *Chemosphere* **2020**, *238*, 124620. [CrossRef]
10. Han, F.; Guo, H.; Hu, J.; Zhang, J.; Ying, Q.; Zhang, H. Sources and health risks of ambient polycyclic aromatic hydrocarbons in china. *Sci. Total Environ.* **2020**, *698*, 134229. [CrossRef]
11. Wang, W.; Jariyasopit, N.; Schrlau, J.; Jia, Y.; Tao, S.; Yu, T.W.; Dashwood, R.H.; Zhang, W.; Wang, X.; Simonich, S.L. Concentration and photochemistry of pahs, npahs, and opahs and toxicity of pm2.5 during the beijing olympic games. *Environ. Sci. Technol.* **2011**, *45*, 6887–6895. [CrossRef] [PubMed]
12. Song, M.-K.; Song, M.; Choi, H.-S.; Kim, Y.-J.; Park, Y.-K.; Ryu, J.-C. Identification of molecular signatures predicting the carcinogenicity of polycyclic aromatic hydrocarbons (PAHs). *Toxicol. Lett.* **2012**, *212*, 18–28. [CrossRef] [PubMed]
13. Kim, K.H.; Jahan, S.A.; Kabir, E.; Brown, R.J. A review of airborne polycyclic aromatic hydrocarbons (PAHs) and their human health effects. *Environ. Int.* **2013**, *60*, 71–80. [CrossRef] [PubMed]
14. Garrido, A.; Jiménez-Guerrero, P.; Ratola, N. Levels, trends and health concerns of atmospheric pahs in europe. *Atmos. Environ.* **2014**, *99*, 474–484. [CrossRef]
15. Niu, X.; Ho, S.S.H.; Ho, K.F.; Huang, Y.; Sun, J.; Wang, Q.; Zhou, Y.; Zhao, Z.; Cao, J. Atmospheric levels and cytotoxicity of polycyclic aromatic hydrocarbons and oxygenated-pahs in pm2.5 in the beijing-tianjin-hebei region. *Environ. Pollut.* **2017**, *231*, 1075–1084. [CrossRef]
16. Walgraeve, C.; Chantara, S.; Sopajaree, K.; De Wispelaere, P.; Demeestere, K.; Van Langenhove, H. Quantification of pahs and oxy-pahs on airborne particulate matter in chiang mai, thailand, using gas chromatography high resolution mass spectrometry. *Atmos. Environ.* **2015**, *107*, 262–272. [CrossRef]
17. Polidori, A.; Hu, S.; Biswas, S.; Delfino, R.J.; Sioutas, C. Real-time characterization of particle-bound polycyclic aromatic hydrocarbons in ambient aerosols and from motor-vehicle exhaust. *Atmos. Chem. Phys.* **2008**, *8*, 1277–1291. [CrossRef]
18. Pratt, K.A.; Prather, K.A. Mass spectrometry of atmospheric aerosols–recent developments and applications. Part i: Off-line mass spectrometry techniques. *Mass Spectrom. Rev.* **2012**, *31*, 1–16. [CrossRef]

19. Marr, L.C.; Grogan, L.A.; Wöhrnschimmel, H.; Molina, L.T.; Molina, M.J.; Smith, T.J.; Garshick, E. Vehicle traffic as a source of particulate polycyclic aromatic hydrocarbon exposure in the mexico city metropolitan area. *Environ. Sci. Technol.* **2004**, *38*, 2584–2592. [CrossRef]
20. Dzepina, K.; Arey, J.; Marr, L.C.; Worsnop, D.R.; Salcedo, D.; Zhang, Q.; Onasch, T.B.; Molina, L.T.; Molina, M.J.; Jimenez, J.L. Detection of particle-phase polycyclic aromatic hydrocarbons in mexico city using an aerosol mass spectrometer. *Int. J. Mass Spectrom.* **2007**, *263*, 152–170. [CrossRef]
21. Ge, X.; Setyan, A.; Sun, Y.; Zhang, Q. Primary and secondary organic aerosols in fresno, california during wintertime: Results from high resolution aerosol mass spectrometry. *J. Geophys. Res. Atmos.* **2012**, *117*. [CrossRef]
22. Eriksson, A.C.; Nordin, E.Z.; Nystrom, R.; Pettersson, E.; Swietlicki, E.; Bergvall, C.; Westerholm, R.; Boman, C.; Pagels, J.H. Particulate pah emissions from residential biomass combustion: Time-resolved analysis with aerosol mass spectrometry. *Environ. Sci. Technol.* **2014**, *48*, 7143–7150. [CrossRef] [PubMed]
23. Ye, Z.; Liu, J.; Gu, A.; Feng, F.; Liu, Y.; Bi, C.; Xu, J.; Li, L.; Chen, H.; Chen, Y.; et al. Chemical characterization of fine particular matter in changzhou, china and source apportionment with offline aerosol mass spectrometry. *Atmos. Chem. Phys.* **2017**, *2017*, 2573–2592. [CrossRef]
24. Marr, L.C.; Dzepina, K.; Jimenez, J.L.; Reisen, F.; Bethel, H.L.; Arey, J.; Gaffney, J.S.; Marley, N.A.; Molina, L.T.; Molina, M.J. Sources and transformations of particle-bound polycyclic aromatic hydrocarbons in mexico city. *Atmos. Chem. Phys.* **2006**, *6*, 1733–1745. [CrossRef]
25. Zelenyuk, A.; Imre, D.; Beranek, J.; Abramson, E.; Wilson, J.; Shrivastava, M. Synergy between secondary organic aerosols and long-range transport of polycyclic aromatic hydrocarbons. *Environ. Sci. Technol.* **2012**, *46*, 12459–12466. [CrossRef]
26. Xu, J.; Zhang, Q.; Chen, M.; Ge, X.; Ren, J.; Qin, D. Chemical composition, sources, and processes of urban aerosols during summertime in northwest china: Insights from high-resolution aerosol mass spectrometry. *Atmos. Chem. Phys.* **2014**, *14*, 12593–12611. [CrossRef]
27. Bruns, E.A.; Krapf, M.; Orasche, J.; Huang, Y.; Zimmermann, R.; Drinovec, L.; Močnik, G.; El-Haddad, I.; Slowik, J.G.; Dommen, J.; et al. Characterization of primary and secondary wood combustion products generated under different burner loads. *Atmos. Chem. Phys.* **2015**, *15*, 2825–2841. [CrossRef]
28. Malmborg, V.B.; Eriksson, A.C.; Shen, M.; Nilsson, P.; Gallo, Y.; Waldheim, B.; Martinsson, J.; Andersson, O.; Pagels, J. Evolution of in-cylinder diesel engine soot and emission characteristics investigated with online aerosol mass spectrometry. *Environ. Sci. Technol.* **2017**, *51*, 1876–1885. [CrossRef]
29. Johansson, K.O.; Head-Gordon, M.P.; Schrader, P.E.; Wilson, K.R.; Michelsen, H.A. Resonance-stabilized hydrocarbon-radical chain reactions may explain soot inception and growth. *Science* **2018**, *361*, 997–1000. [CrossRef]
30. Wang, H. Formation of nascent soot and other condensed-phase materials in flames. *Proc. Combust. Inst.* **2011**, *33*, 41–67. [CrossRef]
31. Nielsen, I.E.; Eriksson, A.C.; Lindgren, R.; Martinsson, J.; Nyström, R.; Nordin, E.Z.; Sadiktsis, I.; Boman, C.; Nøjgaard, J.K.; Pagels, J. Time-resolved analysis of particle emissions from residential biomass combustion – emissions of refractory black carbon, pahs and organic tracers. *Atmos. Environ.* **2017**, *165*, 179–190. [CrossRef]
32. Fernández, P.; Grimalt, J.O.; Vilanova, R.M. Atmospheric gas-particle partitioning of polycyclic aromatic hydrocarbons in high mountain regions of europe. *Environ. Sci. Technol.* **2002**, *36*, 1162–1168. [PubMed]
33. Shen, G.; Wang, W.; Yang, Y.; Ding, J.; Xue, M.; Min, Y.; Zhu, C.; Shen, H.; Li, W.; Wang, B.; et al. Emissions of pahs from indoor crop residue burning in a typical rural stove: Emission factors, size distributions, and gas-particle partitioning. *Environ. Sci. Technol.* **2011**, *45*, 1206–1212. [CrossRef] [PubMed]
34. Lv, Y.; Li, X.; Xu, T.T.; Cheng, T.T.; Yang, X.; Chen, J.M.; Iinuma, Y.; Herrmann, H. Size distributions of polycyclic aromatic hydrocarbons in urban atmosphere: Sorption mechanism and source contributions to respiratory deposition. *Atmos. Chem. Phys.* **2016**, *16*, 2971–2983. [CrossRef]
35. Ma, W.-L.; Zhu, F.-J.; Liu, L.-Y.; Jia, H.-L.; Yang, M.; Li, Y.-F. Pahs in chinese atmosphere: Gas/particle partitioning. *Sci. Total Environ.* **2019**, *693*, 133623. [CrossRef]
36. Ye, Z.; Li, Q.; Liu, J.; Luo, S.; Zhou, Q.; Bi, C.; Ma, S.; Chen, Y.; Chen, H.; Li, L.; et al. Investigation of submicron aerosol characteristics in changzhou, china: Composition, source, and comparison with co-collected pm2.5. *Chemosphere* **2017**, *183*, 176–185. [CrossRef]

37. Zhou, S.; Hwang, B.C.H.; Lakey, P.S.J.; Zuend, A.; Abbatt, J.P.D.; Shiraiwa, M. Multiphase reactivity of polycyclic aromatic hydrocarbons is driven by phase separation and diffusion limitations. *Proc. Natl. Acad. Sci. USA* **2019**, *116*, 11658–11663. [CrossRef]
38. Bond, T.C.; Doherty, S.J.; Fahey, D.W.; Forster, P.M.; Berntsen, T.; DeAngelo, B.J.; Flanner, M.G.; Ghan, S.; Kärcher, B.; Koch, D.; et al. Bounding the role of black carbon in the climate system: A scientific assessment. *J. Geophys. Res. Atmos.* **2013**, *118*, 5380–5552. [CrossRef]
39. Onasch, T.B.; Trimborn, A.; Fortner, E.C.; Jayne, J.T.; Kok, G.L.; Williams, L.R.; Davidovits, P.; Worsnop, D.R. Soot particle aerosol mass spectrometer: Development, validation, and initial application. *Aerosol Sci. Technol.* **2012**, *46*, 804–817. [CrossRef]
40. Wang, J.; Zhang, Q.; Chen, M.; Collier, S.; Zhou, S.; Ge, X.; Xu, J.; Shi, J.; Xie, C.; Hu, J.; et al. First chemical characterization of refractory black carbon aerosols and associated coatings over the tibetan plateau (4730 m a.S.L). *Environ. Sci. Technol.* **2017**, *51*, 14072–14082. [CrossRef]
41. Wang, J.; Liu, D.; Ge, X.; Wu, Y.; Shen, F.; Chen, M.; Zhao, J.; Xie, C.; Wang, Q.; Xu, W.; et al. Characterization of black carbon-containing fine particles in beijing during wintertime. *Atmos. Chem. Phys.* **2019**, *19*, 447–458. [CrossRef]
42. Paatero, P.; Tapper, U. Positive matrix factorization: A non-negative factor model with optimal utilization of error estimates of data values. *Environmetrics* **1994**, *5*, 111–126. [CrossRef]
43. Ulbrich, I.M.; Canagaratna, M.R.; Zhang, Q.; Worsnop, D.R.; Jimenez, J.L. Interpretation of organic components from positive matrix factorization of aerosol mass spectrometric data. *Atmos. Chem. Phys.* **2009**, *9*, 2891–2918. [CrossRef]
44. Zhang, Q.; Jimenez, J.; Canagaratna, M.; Ulbrich, I.; Ng, N.; Worsnop, D.; Sun, Y. Understanding atmospheric organic aerosols via factor analysis of aerosol mass spectrometry: A review. *Anal. Bioanal. Chem.* **2011**, *401*, 3045–3067. [CrossRef]
45. Wu, Y.; Liu, D.; Wang, J.; Shen, F.; Chen, Y.; Cui, S.; Ge, S.; Wu, Y.; Chen, M.; Ge, X. Characterization of size-resolved hygroscopicity of black carbon-containing particle in urban environment. *Environ. Sci. Technol.* **2020**. [CrossRef]
46. Cui, S.; Gao, J.; Shen, F.; Wang, J.; Huang, D.; Ge, X. Chemical characteristics of submicron black carbon-containing aerosols in shanghai. *Atmos. Chem. Phys.* (under review).
47. Wang, J.; Onasch, T.B.; Ge, X.; Collier, S.; Zhang, Q.; Sun, Y.; Yu, H.; Chen, M.; Prévôt, A.S.H.; Worsnop, D.R. Observation of fullerene soot in eastern china. *Environ. Sci. Technol. Lett.* **2016**, *3*, 121–126. [CrossRef]
48. Liu, D.; Allan, J.D.; Young, D.E.; Coe, H.; Beddows, D.; Fleming, Z.L.; Flynn, M.J.; Gallagher, M.W.; Harrison, R.M.; Lee, J.; et al. Size distribution, mixing state and source apportionment of black carbon aerosol in london during wintertime. *Atmos. Chem. Phys.* **2014**, *14*, 10061–10084. [CrossRef]
49. Haque, M.M.; Kawamura, K.; Deshmukh, D.K.; Fang, C.; Song, W.; Mengying, B.; Zhang, Y.-L. Characterization of organic aerosols from a chinese megacity during winter: Predominance of fossil fuel combustion. *Atmos. Chem. Phys.* **2019**, *19*, 5147–5164. [CrossRef]
50. Wang, Q.; Liu, M.; Yu, Y.; Li, Y. Characterization and source apportionment of pm2.5-bound polycyclic aromatic hydrocarbons from shanghai city, china. *Environ. Pollut.* **2016**, *218*, 118–128. [CrossRef]
51. Canagaratna, M.R.; Jimenez, J.L.; Kroll, J.H.; Chen, Q.; Kessler, S.H.; Massoli, P.; Hildebrandt Ruiz, L.; Fortner, E.; Williams, L.R.; Wilson, K.R.; et al. Elemental ratio measurements of organic compounds using aerosol mass spectrometry: Characterization, improved calibration, and implications. *Atmos. Chem. Phys.* **2015**, *15*, 253–272. [CrossRef]
52. Kroll, J.H.; Donahue, N.M.; Jimenez, J.L.; Kessler, S.H.; Canagaratna, M.R.; Wilson, K.R.; Altieri, K.E.; Mazzoleni, L.R.; Wozniak, A.S.; Bluhm, H.; et al. Carbon oxidation state as a metric for describing the chemistry of atmospheric organic aerosol. *Nat. Chem.* **2011**, *3*, 133–139. [CrossRef] [PubMed]

© 2020 by the authors. Licensee MDPI, Basel, Switzerland. This article is an open access article distributed under the terms and conditions of the Creative Commons Attribution (CC BY) license (http://creativecommons.org/licenses/by/4.0/).

Article

Functional Factors of Biomass Burning Contribution to Spring Aerosol Composition in a Megacity: Combined FTIR-PCA Analyses

Olga Popovicheva [1,*], Alexey Ivanov [1] and Michal Vojtisek [2]

1. Scobeltsyn Institute of Nuclear Physics, Lomonosov Moscow State University, 119991 Moscow, Russia; aleshaiva255@gmail.com
2. Center for Sustainable Mobility, Czech Technical University in Prague, 160 00 Prague, Czech Republic; michal.vojtisek@fs.cvut.cz
* Correspondence: olga.popovicheva@gmail.com

Received: 2 March 2020; Accepted: 23 March 2020; Published: 25 March 2020

Abstract: Whether the spring season brings additional pollution to the urban environment remains questionable for a megacity. Aerosol sampling and characterization was performed in the urban background of the Moscow megacity in spring 2017, in a period of a significant impact of mass advection from surrounding fire regions. Parametrization of Angstrom absorption exponent (AAE) on low and high values provides periods dominated by fossil fuel (FF) combustion and affected by biomass burning (BB), respectively. The period identification is supported by air mass transportation from the south of Russia through the regions where a number of fires were observed. Functionalities in entire aerosol composition, assigned to classes of organic, ionic compounds, and dust, are inferred by diffusion refection infrared Fourier transmission (FTIR) spectroscopy. Functional markers of urban transport emissions relate to modern engine technology and driving cycles. Regional BB functionalities indicate the fire impacts to the spring aerosol composition. The development of the advanced source apportionment for a megacity is performed by means of combined ambient FTIR data and statistical PCA analysis. PCA of FTIR spectral data differentiate daily aerosol chemistry by low and high AAE values, related to FF- and BB-affected spectral features. PC loadings of 58%, 21%, and 11% of variability reveal the functional factors of transport, biomass burning, biogenic, dust, and secondary aerosol spring source impacts.

Keywords: megacity; aerosols; source; PCA; fossil fuel; biomass burning; functionalities

1. Introduction

Air pollution due to particulate matter (PM) is one of the most important emerging environmental problems. Atmospheric processes and numerous aerosol emissions determine PM chemical properties impacting air quality [1] and human health [2]. The analysis of aerosol chemical composition provides insight into source contributions and strengthens links between particle constituents, health, and environmental impacts. However, the diversity of molecular constituents in the PM organic fraction poses challenges for characterization; detailed chemical analysis is far from being achieved at a molecular level [3].

Fourier transform infrared (FTIR) spectroscopy is a powerful tool for characterizing the aerosol properties, behavior, and origins of the complex organic fraction [4]. It is an analytical technique that captures the signature of a multitude of aerosol constituents and give rise to feature-rich spectral patterns over the mid-IR wavelengths [5]. Aerosol functionalization is highly source-dependent; it varies considerably between urban and rural regions and populated and remote areas, attributing to various classes of oxygen, hydrogen, and nitrogen-containing compounds [6,7].

Particulate emissions from fossil fuel (FF) combustion (motor vehicles, heating plants, energy production) is dominated in urban environment. Health-related properties of aerosols assigned to the biological accessibility and inflammatory potential indicates the importance of the functionalized structure of transport engine-produced particles [8]. For apportionment organic functional groups to sources, recurring FTIR spectral features were associated with available emission tracers [4]. The FF combustion factor had found as a mixture of alkane, hydroxyl, and carboxylic acid groups, with small contributions of amine and carbonyl groups. FF factors have characteristic peak locations common with fuel standards: gasoline, diesel, and oil [9]. Diesel emission analyses based on an evaluation of the relationship between engine, fuel, operating condition, and particle composition highlight the functional markers of the organic structure [10,11].

The main part of the diesel transport is heavy-duty vehicles (e.g., trucks, buses, tractors, etc.), which produce more black carbon than vehicles that run on gasoline [12]. However, the examination of carbonaceous PM emissions and secondary organic aerosol (SOA) formation from modern diesel particle filter (DPF) and catalyst-equipped diesel cars showed that they are markedly lower than from gasoline vehicles [13]. This emphasizes a need for additional quantification of functional markers of gasoline emissions as well.

Biomass burning (BB) affects the composition of particles by a large amount of incompletely oxidized products related to a significant proportion of basic sugars, fatty acids, and aldehydes [14]. Wood combustion releases oxygenated polyromantic hydrocarbons (PAH) and acids with increased temperatures, whereas wood combustion releases oxygenated phenolic compounds and sugars derivatives at lower temperatures [15,16]. Alkane, carboxylic acid, amine, and alcohol functional groups are mainly associated with FF-related sources, while non-acid carbonyl groups are likely from BB events [6].

Multivariate calibration methods were developed to quantify the ambient aerosol organic functional groups and inorganic compounds [17,18]. They are associated to sources of emissions with the highest consistent contributions [4]. Apportionment of organic functional groups has identified FTIR spectral features of emission factors of the respective sources such as FF combustion, biogenic, BB, and ocean sources. Compositional analysis, functional group correlations, and back trajectories were used to identify three types of periods with source signatures: primary biogenic-influenced, urban-influenced, and regional background [7].

However, a complete set of internal standards for calibration of organic components in the atmosphere is not available, in part because the ambient particle composition is not fully known. Quantitative source apportionment developed for FTIR data can be applied only on a given instrument where the calibration was done. This leaves a need for further developing the chemometric techniques to apply to spectroscopy data analyses. In addition, the complexity of ambient mixtures of organic compounds in the atmosphere results in mixtures that cannot be fully resolved by FTIR spectroscopy [3], which requires the application of complementary aerosol characterization. From the other side, optical aerosol characteristics such as spectral absorption can act as complementary for identifying the source impact on the aerosol composition [19–22].

A promising approach for analyzing the FTIR spectral data is a chemometric tool such as principal component analysis (PCA). It had been used for determination of the physico-chemical properties of multi-component mixtures and discrimination between different chemical compositions [14,23,24]. The advantage of combined FTIR-PCA is that it can investigate chemical variations between well-defined species that can provide a clear parametrization for the analyses of spectral differences and identification of characteristic chemical components [25]. In urban areas, the apportionment of multifunctional aerosol compounds remains largely uncertain because of the plurality of emission sources. In the Mexico City Metropolitan Area (the second largest megacity in the world), FTIR-based studies found motor vehicle emission, oil burning, BB, and crustal components to be the main sources [6].

Moscow is the largest city in Europe. Moscow often faces serious traffic congestion problems because of the increased vehicle numbers; the total count of vehicles was registered as much as

4.6 million by the end of 2017 [26]. The uniform interdepartmental information and statistical system (www.fedstat.ru) reported around 4,640,000 vehicles, including 90.4% and 8.5% of light- and heavy-duty cars and trucks, respectively, as well as 1.1% of buses. According to the Department of the Federal State Statistics Service, in the Moscow megacity, transport gaseous emissions compose up to 93% of the gross pollution from mobile sources [26].

Despite Moscow facing air quality problems [27], a lack of aerosol chemical characterization is significant. The pollution data presented for Moscow between global megacities are available only for particulate mass of size less than 10 µm (PM10) and 2.5 µm (PM2.5) masses [28]. The impact of BB is assumed negligible because the central heating system is operated. However, during an extreme smoke event in the Moscow megacity, a waste range of hydroxyl, aliphatic, aromatic, acid and non-acid carbonyl, and nitro compounds had indicated the intensive wildfires around a city as a major source of pollution [29]. The first steps to source the apportionment on the bases of organic and inorganic composition characterization was made in

This paper is devoted to aerosol characterization of the Moscow megacity urban background in spring season, when the megacity is most affected by a plurality of emission sources. FTIR spectroscopy complemented by the functional groups represent the classes of compounds in the entire aerosol composition. Analyzed spectral absorption provides the parametrization of Angstrom Absorption Exponent for identification of the periods dominated by FF and affected by agriculture and residential BB. Functional markers for diesel emissions related to modern engine technology and driving cycles as well as regional BB indicate to which extent the FF and BB emission sources impact the aerosol composition in urban environment. Because the identification of functional markers for gasoline car emissions had not been done before, we have conducted the sampling campaign and characterize the functional structure of gasoline particulate emission, based on European standard protocols. Comparison of the day-to-day functional group composition shows the changes over the range of major functional factors influenced the aerosol composition. We attempted to interpret each PC loading in terms of chemical composition, relating them to functional factors.

2. Experiments

2.1. Ambient Sampling Campaign

The intensive campaign was performed in the southwest of the Moscow city from 17 April to 25 May 2017. Aerosol sampling was conducted at the rooftop of two–story building of the Meteorological Observatory of Moscow State University (MO MSU) (55.07' N, 37.52' E) (Figure 1). MO MSU is located at the territory of the MSU campus, in an area of Vorob'evy Gory hills, which is well-ventilated. There are no industrial plants or commercial areas nearby. At about 800 m to the north and north-west from the MO MSU, there are the residential area and the Lomonosovsky prospect highway, respectively. The closed industrial enterprises are central heating stations at a distance of 3 km from the MSU. Therefore, the MO MSU is considered an urban background station. The recent aerosol and its radiative effects observed at MO MSU during the AEROCITY 2018 experiment was presented in [30].

Sampling of particles with dimeter less than 10 µm (PM10) was performed on 47 mm quartz fiber filters (preheated at 600 °C in advance) at 24 h intervals from 8 p.m., totaling 38 samples. Measurements of meteorological parameters (temperature, precipitation) was performed each 3 h by the MO MSU meteorological service. PM10 mass concentrations were collected by the Mosecomonitoring Agency using the tampered element oscillating microbalance TEOM 1400a (Thermo Environmental Instruments Inc. Franklin, MA, USA).

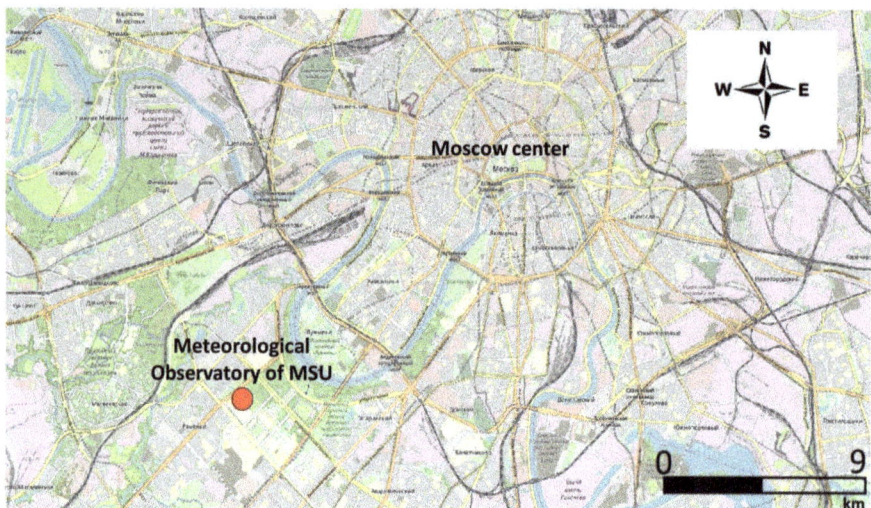

Figure 1. Map of Moscow city. Meteorological station of Moscow State University is indicated.

To evaluate the air mass transportation impact, the backward trajectories (BWT) were generated using the HYbrid Single-Particle Lagrangian Integrated Trajectory (HYSPLIT) model of the Air Resources Laboratory (ARL) [31] with a coordinate resolution equal to 1° × 1° of latitude and longitude. The potential origin areas were investigated using two-day back trajectories for air masses arriving each 12 h to the MO MSU at 500 m heights above sea level (A.S.L.). Fire information was obtained from Resource Management System (FIRMS), operated by the Earth Science Data Information System (ESDIS). Daily maps were related to the computed trajectories, providing a clear picture of the geographical location of fires, with a resolution of several kilometers. The number of fires that could affect air masses transported to the MO MSU was calculated as the sum of fires that occurred at a distance of 0.5° on both latitude and longitude from the BWT.

2.2. Near-Gasoline Source Sampling Campaign

A Ford Focus car with a downsized three-cylinder 1.0-L turbocharged gasoline direct injection (DISI) EcoBoost engine (12-valve, 999 cm^3 displacement, Euro 6) and a Škoda Fabia car with a naturally aspirated multi-point injection (MPI) engine (1394 cm^3 displacement, Euro 5) were used for emission sampling. Ordinary gasoline without oxygenated compounds with a nominal research octane number of 95, meeting CSN EN228 specifications, was obtained from a local gas station (EurOil). The Ford Focus car was tested on a chassis dynamometer using the full length of the Common Artemis Driving Cycles (CADC), including urban, rural, and motorway (130 km/h maximum speed) parts. The exhaust was routed into a full-flow dilution tunnel, from which samples were taken on quartz fiber and Teflon membrane filters during three runs of the CADC. The results of vehicle emissions tests have been described previously elsewhere [32].

2.3. Methods and Techniques

FTIR spectra of filter samples were acquired using an IRPrestige-21 spectrometer (Shimadzu, Kyoto, Japan) in a diffuse reflection mode, described in details elsewhere [10,11]. Spectra were recorded in the 450 to 4000 cm^{-1} range with 4 cm^{-1} resolution and 100 accumulated scans. IR Solution software was applied to subtract the spectrum of blank substrates as well as to perform the Kubelka-Munk (K-M) conversion into a quasi-quantitative spectrum correlating with the sample concentration. Basis line correction was done by a subtraction of basis line function spectra.

To address the possible inhomogeneity of the sample loading, spectra were collected from three–five different spots on each sample. One spectrum was taken as representative of the entire sample, because it shared the biggest number of common absorption bands with others. Spectra in the range 3600–4500 and 1790–2500 cm^{-1} were not considered because dominant absorption of water vapors and CO_2, respectively. Because blank quartz filters exhibit strong absorption bands in the 770–860 and 972–1530 cm^{-1} spectral ranges, causing the majority of uncertainty, their examination was excluded from the following analyses.

The FTIR spectrum was divided into the vibration band regions, which are unique for each chemical compound [33]. Extensive knowledge is required for the identification of the functional groups. The common approach used in FTIR studies for the interpretation of the IR absorbance peaks is based on the referencing to spectroscopic guides and bands observed in previously published data [34,35]. Because of the possible overlapping of vibration bands and in order to avoid the discrepancy obtained in measurements performed by various spectral modes and apparatus, the interpretation of the IR spectra in this work was based on the database purposely built using the measurements performed on the same FTIR setup. We based our methods on the approach for identification of functional groups represented by wavenumbers of the absorption bands typical for ambient aerosols, according to previous comprehensive field observations and calibrations [5,18,34]. Additionally, the analyses of IR spectral features of near-source emissions performed in previous studies on the same apparatus [10,36] were summarized.

The database was completed through the use of a set of authentic chemical standards. Supplementary Materials Table S1 lists classes and compounds included into the database. We address 16 classes of organic compounds (alkanes, alkenes, polyaromatics, carboxylic acids, carbohydrates, amino acids, amines, aldehydes, ketones, esters, lactones, quinones, nitrocompounds, formats, alcohols, humic-like substances (HULIS), and sugars) and inorganic compounds (sulfates, nitrates, carbonates, and sulfuric acid). Absorbance peaks related to organic (hydroxyl O-H, carbonyl C=O, aliphatic C-C-H and C=C-H, polyaromatic C=C and C=C-H, C-O, N-H, C-N, -NO_2, C-O-C) as well as ionic (NO_3^-, CO_3^{2-}, NH_4^+, SO_4^{2-}) functional groups are assigned in Table S2.

Off-line examination of light attenuation of particles deposited on quartz filter samples was performed using the multiple wavelength light transmission instrument (transmissometer) based on the methodology of [19]. The intensity of light attenuation through quartz filters was measured at seven wavelengths from the near-ultraviolet to near-infrared spectral region. Five different areas of the sample filter were analyzed in order to assess the possible heterogeneity of the sample. Then, the averaged attenuation (ATN) was used for the parametrization of the dependence of the attenuation (ATN) on the wavelength λ using a power law relationship:

$$ATN = k\lambda^{-AAE} \tag{1}$$

where the Absorption Angstrom Exponent (AAE) is a measure of a strength of the spectral variation of aerosol light absorption. It was shown that light attenuation is primarily due to particle light absorption [37]. In the optical transmission method employed here, the aerosol particles were collected using reflective quartz fiber filters, which brings uncertainty because light scattering by the filter fibers provides the embedded particles multiple opportunities to absorb light. Black carbon (BC) produced by high-temperature combustion sources fit within the Rayleigh scattering regime for near-visible wavelengths with a theoretical λ^{-1} relationship [38]. Weak spectral dependence of ATN with AAE around 1 was found for diesel soot and urban aerosols produced by fossil fuels combustion [19], and as much as 4.1 for peat burning [39]. Spectral absorption of BB shows the combined impact of both BC absorbing from 670 nm down to 500 nm and brown carbon (BrC), which increases the absorption below 500 nm.

2.4. Functional Markers

Fuel combustion processes yield the basis particulate functionalized structure determined by a high and low combustion temperature [4,16,40]. Characterization of near-source emissions was conducted in order to emphasize the specific atmospheric pollutant functional patterns from the major local sources and then to compare them with those identified in the urban environment [22,29]. Observations of dominant alkane functionalities of particles emitted by an Opel Astra diesel engine [41] and BMW and John Deere engines operating at the stationary and transient conditions [42] are in a good agreement with the quantification of total aliphatic compounds in diesel emission constituting 68% carbon by weight [43]. Aromatic C=C functionalities accompanies aliphatic C-C-H, with less prominent carbonyl C=O and nitro-ONO_2 functionalities, in the particulate emission of the most widely used Iveco Tector heavy-duty diesel engine operated in a World Harmonized Transient Cycle (WHTC) using conventional EN 590 diesel fuel [11]. A similar functional pattern was observed in emissions of off-road diesel engines [10] operated in standard Non-Road Steady State cycle, which is identical to the ISO-8178 non-road engine emission certification procedure.

Therefore, we assumed that diesel transport emissions significantly impact the aerosol composition in Moscow environment. Then, we suggested that the diesel emission functionalized structures of the common transport system operated at the most widely distributed fuels and driving cycles described above as functional markers for transport impact onto Moscow aerosol composition. Since real-world engine operating conditions are best represented by a transient cycle (where both engine speed and fuel consumption vary) [44], we took into account the functional patters of particulates emitted during transient diesel engine operation cycles as markers for the following source apportionment.

In order to address the gasoline transport emission, the functionalized patterns of particulate matter emitted from the DISI engine, using gasoline fuel with an octane number of 95 of CSN EN228 specifications, were analyzed. For this study, spectra obtained at an urban ARTEMIS driving cycle commonly applied in a city were used as a functional marker for diesel transport emission.

Since BB occurs in various phases and for different biomasses, we identified the functional pattern that can represent the regional BB features impacted the aerosol composition in a city. With a purpose to quantify the functional markers for Moscow regional wildfire and residential emission, the small-scale experimental open fires were conducted in the Moscow region at the location where intensive forest fires were observed [29]. Both combustion phases (low-temperature smoldering and open flame (flaming)) represent the BB process [45]. Therefore, for the purpose of this study, we used both spectral data for smoldering and flaming of regional BB as functional markers. To address agriculture fires relating to grass combustion, the inside and above grass burning spectra data were additionally considered, which were collected from the measurement campaign conducted during the peat burning event near Moscow [39].

Carbonates (CO_3^{2-}) were identified in the range 880–860 cm^{-1}; their presence in the PM size fraction was confirmed by thermo-optical measurements of carbonates in form of carbonate carbon [46]. Relative concentrations of carbonates increased from low to high amounts of smoke, thus showing the impact of re-suspended soil particles during intensive agricultural fires on the composition of coarse ambient aerosols. We should also note the prominent similarity of the position of the 880 cm^{-1} vibration band of carbonates in ambient aerosols, similar to spectra of flaming emission. This indicates the wide distribution of dust in the urban atmosphere as well as dust of soil evolved by air convection during fires impacted the city atmosphere [22].

The soil-related particles are due to transportation, construction, agriculture, and wind erosion. Dust functionalities are specific features of coarse particles related to soil; Blanco and McIntyre (1972) reported quartz, silicates, and kaolinite to be the main constituents of coarse PM. We refer to the published data for the bands of dust-related functionalities in silicates, quartz, and kaolinite [47]. The band that appeared at 914 and 950 cm^{-1} is attributed to Al-OH vibrations in the octahedral sheet structure of kaolinite [48] and Si-OH stretching [49], respectively.

2.5. Principal Component Analyses

PCA is a well-known chemometric procedure that allows the rotating the space spanned by the original variables to a new space, spanned by the Principal Components (PCs), in which most of the information contained in the original data is reported in the first (generally two or three) PCs [50]. The PCs are obtained using both the covariance data matrices (scaling by mean-centered data). Visualizing the two-dimensional (2D) plot of PC1 vs. PC2 (and/or PC3) allows studying the behavior of samples (in the scores plot) and variables (in the loadings plot).

For the preparation of the data matrix for PCA analyses, after conversion to K-M mode, the smoothing of the FTIR spectra was performed by a method of smoothing spline, using the minimizing of the mischief function and penalties for irregularities. Then, spectra were subjected to secondary derivatization by an approach described elsewhere [25]; the second derivative is beneficial in highlighting the difference between spectrum and minimization of the noise. A matrix for 25 daily FTIR spectral intensities and variables of 797 wavenumbers was built. Principal component analysis (PCA) decomposes the data matrix and concentrates the source of the variability into the first few PCs.

3. Results and Discussions

3.1. FF and BB—Affected Periods

PM10 mass concentrations over the sampling period show a strong variation from the lowest of 8 µg/m^3 to the highest value of 64 µg/m^3, on average 22 ± 16 µg/m^3 (Figure 2). The longest episode of the highest PM10 was observed at MO MSU on 29 and 30 April, on average 55 ± 19 µg/m^3. On nearly the same days, from 29 April until 2 May, the ambient temperature approached an abnormally high level for this season: +21 °C. We observed relatively good correlation between high PM10 and the local temperature maximums (Figure 2), with correlation coefficient R^2 equal 0.68.

Figure 2. PM10 mass concentrations and temperature over the sampling period at MO MSU.

The spectral dependence of the light attenuation (ATN) for daily samples was found to be well approximated by a power law equation (1). The Absorption Angstrom Exponent (AAE) was obtained as the slope of the linear regression with R^2 around 0.9. Variation of the ATN spectral dependence during the whole sampling period exhibits the range of AAE from 1.03 to 1.95 (Figure 3). The values above 1.0 may indicate that brown carbon (BrC) associated with organic carbon in addition to BC contributes significantly to the measured light absorption of biomass smoke aerosols in ultraviolet and visible spectral regions [51].The authors in [19] obtained a high value of 2.5 for smoke of savanna fires, while peat bog burning near Moscow region was characterized by AAE around 4.2 [39].

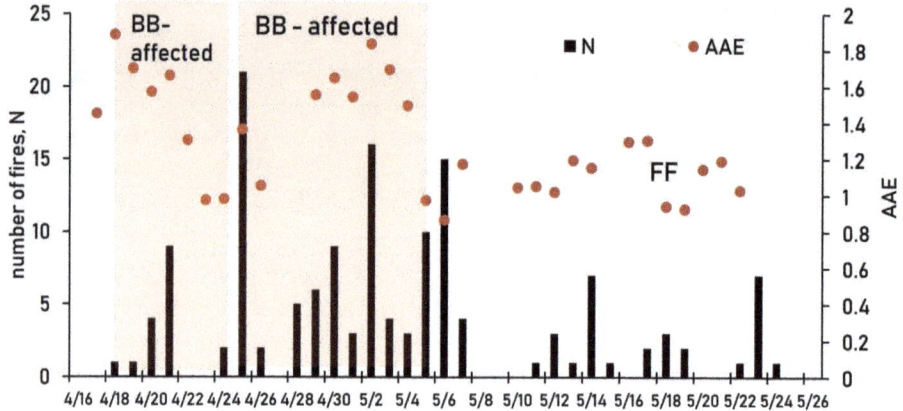

Figure 3. Absorption Angstrom Exponent (AAE) obtained from spectral dependence of aerosol light attenuation and number of fires (N) passed by air masses.

In spring time, fires are usually observed south of Moscow, an agriculture practice with the purpose to remove last year's grass on the fields that is widespread in this season. BB in residential areas around the city can also be pronounced, especially during the May holidays from 1 to 6 May, because of the high temperatures during that time. Therefore, we could assume that the analyses of the aerosol spectral dependence of light absorption during the studied time can indicate the impact of fire-affected air mass on the aerosol chemistry in Moscow.

From 17 to 22 April, AAE values higher than 1.4 were observed (Figure 3). During this period, the air masses were transported from the north and passed the agriculture fires close to Moscow (Figure 4). On 23 April, the direction of air mass transportation changed to the west, while from 25 April to 2 May the direction was consistently from the south. On 23 and 24 April, the AAE dropped to 1.0. Since 25 April, the AAE became higher again, in correlation with a large number of fires observed in the south of Moscow (Figure 4). Moreover, the period from 30 April–5 May coincided with a vacation period in Russia when the warm temperatures (Figure 2) stimulated intensive residential activity around Moscow city, such as garden cleaning, grass burning, and barbecues. After 4 May, the direction of air mass transportation changed to the arctic region (Figure 4), leading to a drop of temperatures of a few degrees. On 11–13 May and 17–20 May, the air mass changed its direction from the west and east, respectively (Figure 4). On 5 May, the AAE dropped to 0.97 and no longer exceeded 1.3, even though the trajectories passed the regions of fires. This can be explained by small amounts of fires and frequent precipitation in this period.

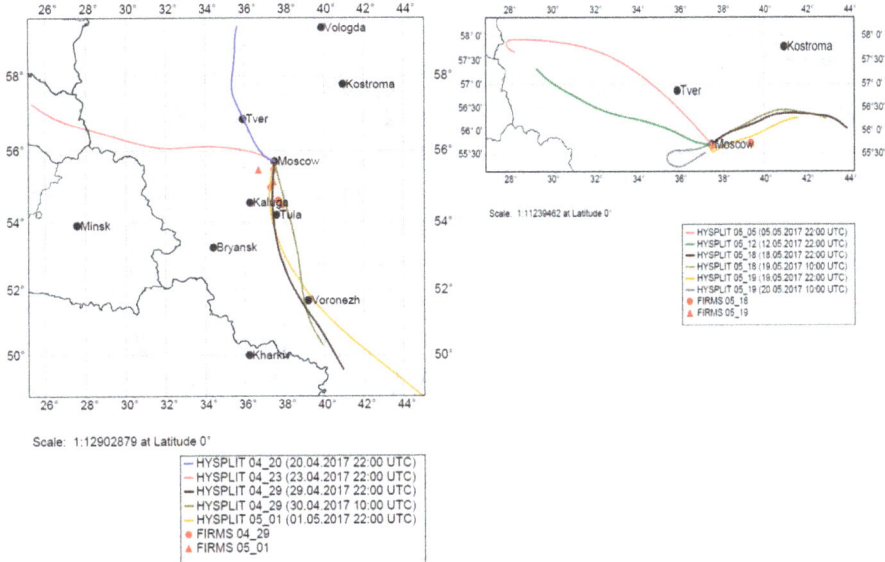

Figure 4. Two-day back air mass transportation from the HYSPLIT model in the days of the BB-affected period (**left**) and FF period (**right**), 500 m A.S.L. Fires observed by FIRMS in indicated days are marked by circles and triangles.

The light spectral absorption can be applied as a source-specific optical marker for impact of BB on urban aerosols. The sampling period of our study can be divided into periods of low and high AAE. The formal criteria for parametrization were chosen to be based on the observations performed in an urban environment, where the separation between BB-affected and FF periods was proposed at an AAE level of around 1.3 [21]. Thus, we addressed the days of the weak spectral dependence with AAE = 1.08 ± 0.02 to the dominant impact of fossil fuel combustion emissions with less impact of biomass burning during the "FF" period (Figure 3). Days of the dominant impact of biomass burning emissions during "BB-affected" period show a high spectral dependence with an averaged value AAE = 1.61 ± 0.02. Similar results for traffic emissions were obtained in the Hanoi megacity with an AAE equal to 1.3 (Popovicheva et al., 2017b), indicating that traffic produces significantly less BrC and more BC than biomass burning. The number of fires passed by air masses relates to BB-influenced days (Figure 3).

3.2. FF-Related FTIR Spectral Features

In the Moscow megacity, anthropogenic PM emissions occurred due to traffic, industry, heating, waste recycling, and construction. Biomass was not used as a fuel for domestic burning because of the central heating system, which is different from many European cities. Various absorption bands relating to different classes of organic and ionic compounds were observed in FTIR spectra of spring aerosols; the most frequently indicated wavenumbers are shown in Figures S1 and S2. Saturated C-C-H, unsaturated C=C-H, and aromatic C=C and C=C-H vibrations suggest the presence of alkanes, alkens, and polyaromatic hydrocarbons (PAH), respectively. Carboxyl C=O groups represent functionalities in carboxylic acids, ketones, esters, anhydrides, and quinones. Hydroxyl -OH groups are associated with alcohols, while aromatic -NO$_2$ and -NH groups indicate the presence of nitrogen-oxy compounds and amines, respectively. Attributes of salts are SO$_4^{2-}$ and CO$_3^{2-}$ bands in sulfates and carbonates, respectively.

FTIR spectra of days assigned to the FF period are shown in Figure S1. The most frequent feature of all spectra are the aliphatic C-C- symmetric/asymmetric stretches (2926–2855 cm^{-1}) of methylene > CH$_2$ groups in alkanes. At higher wavenumbers, at around 3223 and 3405 cm^{-1}, the vibrations of ammonium NH_4^+ and N-H amine groups, respectively, were prominent. In the same range, the O-H groups in alcohols and sugars can be identified. The most frequently observed bands in the range of 1738–1715 cm^{-1} and 1587 cm^{-1} are carbonyl C=O and C=C, respectively. C=C stretching is attributed to either aromatic compounds or microcrystalline structure of soot particles due to its polyaromatic character [52,53]. In both cases, the IR-inactive C=C mode is augmented by either sufficient asymmetry of the carbon material polyaromatic structure (e.g., by defects) or related to carbonyl groups conjugated with aromatic segments [54]. At 880 cm^{-1} carbonate CO_3^{2-} groups dominate.

The representative spectra for two days of the FF period (05.18 and 05.19) are shown in Figure 5, together with representative spectra of transport source emissions. In order to assign the absorption bands of ambient aerosols to the transport impact, the correlation with a pattern of a functional marker for diesel and gasoline particulate emissions was analyzed. Spectra of particulate emissions from heavy-duty diesel, off-road diesel, and gasoline direct injection engine operated in World Harmonized Transient cycle (Diesel_WHTC), Non-Road Steady State cycle ISO-8178 (Diesel_ ISO-8178), and Urban portion of the Artemis cycle (term as MPI gasoline URBAN), respectively, are shown in Figure 5. Both diesel engine emissions demonstrate the high similarity for bands at 1585, 1668, 1719, and 2920–2859 cm^{-1} of aromatic C=C, aldehydes, acids/ketones C=O, and alkane C-C-H functional groups, respectively. Gasoline engine emission coincides by a wide band of C=C group. It is worth to note that a prominent band at 1589 cm^{-1} associated with carbonyls was persistently observed in high-temperature combustion emissions, which was also found in gas flaring particulates [40]. Such findings prove that soot produced at high-temperature combustion can be identified by the similar chemical structure related to the aromatic ring stretching mode enhanced in intensity by O-containing functional groups.

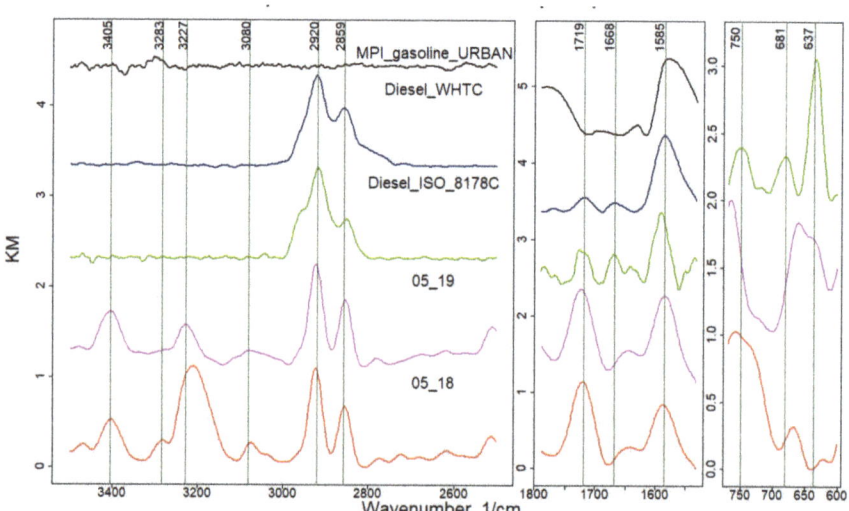

Figure 5. FTIR spectra for 05.18 and 05.19 from the FF period. Spectra of a heavy-duty diesel engine operated in the World Harmonized Transient Cycle and with conventional EN 590 diesel fuel engine (Diesel_WHTC), an off-road diesel engine operated in a Non-Road Steady State cycle ISO-8178 (Diesel_ ISO_8178), and an multi-point injection (MPI) gasoline engine operated in an Urban cycle (MPI_gasoline_URBAN) represent road and off-road diesel and gasoline emissions, respectively.

There is a band in the representative spectra for two days of the FF period which always coincides for all spectra in Figure 5, it is one peaked at 1589 cm^{-1} and other at 1719 cm^{-1} for C=C and C=O, respectively. Bands of alkanes peaked at 2926–2855 cm^{-1}; moreover, alkenes observed at 1645 cm^{-1} are identical between spectra of the FF period and diesel emissions functional markers, but different from the gasoline ones. Polyaromatic C=C=H at 750 cm^{-1} are similar for spectra of the FF period and Diesel_ ISO-8178. Thus, we considered that C-C-H, C=C, and C=O groups in alkane, aromatic, and oxidized compounds act as a functional marker of transport vehicle emissions in the Moscow environment (Table 1).

Table 1. Patterns of functional groups acting as functional markers of transport vehicle and regional BB smoldering and flaming emissions in Moscow megacity.

Wavenumbers, cm^{-1}	Diesel/Gasoline Transport	Wavenumbers, cm^{-1}	Smoldering/Flaming
637	SO_4^{2-}	617–621	SO_4^{2-}
750	C=C-H	762	C=C-H
1585	C=C	1530	-NO_2
1668	C=O	1618	C=C
1719	C=O	1730–1680	C=O
2920–2859	C-C-H	2926–2851	C-C-H
		~3401	O-H, N-H

Additionally, to organic functionalities, in the FF period sulfates, SO_4^{2-} in various salts and sulfuric acids are well observed in the wide range of 661 to 617 cm^{-1}. Sulfates were consistently observed in transport vehicle emissions [10,41,42] due to fuel and lubrication oil contaminations. They show the prominent absorption band on the spectra of off-road diesel engine operated in Non-Road Steady State cycle ISO-8178 (Figure 5).

Secondary particles that are mainly ammonium sulfate and nitrate formed in the air from regional or local gaseous emissions of sulfur dioxide and oxides of nitrogen reacting with ammonia [55]. SO_4^{2-} is well correlated with the NH_4^+ absorption band at 3227 and 638 cm^{-1}, due to the formation of internally mixed particles during long-term transport from urban sources [7] and BB season [46].

The bands observed in the FF period at 3405 cm^{-1} can be associated to N-H in amines and amino acids, and to O-H in alcohols. Functionalities of amino acids are found to be well correlated with hydroxyls and carbohydrates; they are classified as biogenic functional groups because they originate from biogenic sources [7]. Additionally, the band at 3474 cm^{-1}, assigned to O-H in sugars, alcohols, and carbohydrates, may prove the biogenic impact to aerosol composition. Such compounds are not emitted by diesel/gasoline transport; together with sulfates and ammonium, they demonstrate the mixing of secondary and biogenic aerosols in ambient urban environment. Related to sugars, alcohols, and carbohydrates, the band at 3474 cm^{-1} is also prominent during the FF period, indicated by the biogenic emission already observed during spring time in other studies [56].

3.3. BB-Related FTIR Spectral Features

FTIR spectra of days assigned to the BB-affected period are shown in Figure S2. The most frequent feature of all spectra are the aliphatic C-C-symmetric/asymmetric stretches (2926–2851 cm^{-1}) of methylene >CH_2 groups in alkanes. Besides this group, other functionalities are much more variable and difficult to be described for comparison only by observation.

In Figure 6, the representative spectra for two days of the BB-affected period (05.01 and 04.29) are shown, together with spectra for smoldering and flaming of Moscow regional biomass. BB spectra of both biomasses exhibit a very wide unresolved band near 3401 cm^{-1} from various O-H and N-H in carbohydrates (including levoglucosan), alcohols, and amines. Aliphatic C-C-H and C=C stretching of aromatic rings are observed at 2926–2851 and 1614 cm^{-1}, respectively. A wide band of oxidized functionalities in the range from 1730 to 1700 cm^{-1} relates to carbonyl C=O groups. Polyaromatic C=C=H at 754 and 710 cm^{-1}, as well as features of sulfates at 617 cm^{-1}, are prominent for flaming, while strong absorption was observed at 667 cm^{-1} due to sulfate emissions in K_2SO_4, $MgSO_4$, and Na_2SO_4 in the smoldering phase. The presence of SO_4^{2-} absorption bands at 617 m^{-1} and their slightly prominent feature in smoldering smoke indicate the formation of secondary sulfates in the Moscow environment, a phenomenon usually relating to the aging of biomass burning aerosols [57].

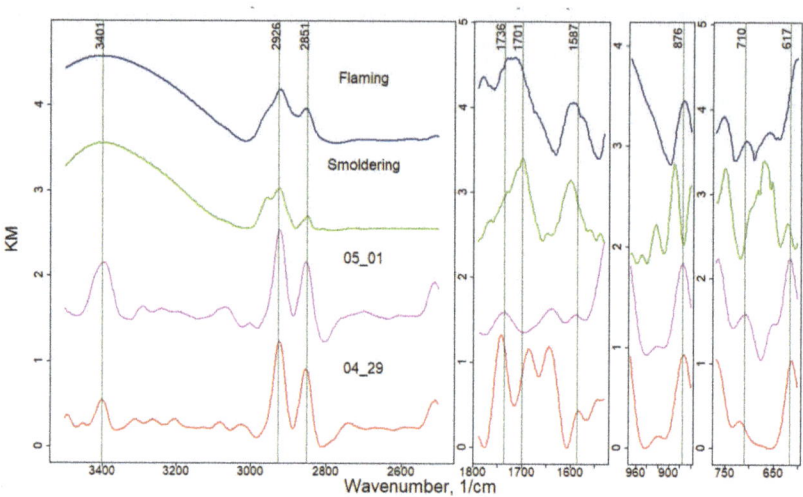

Figure 6. FTIR spectra for 05.01 and 04.29 from the BB-affected period. Spectra of the Moscow regional biomass represented by spruce flaming and mixture smoldering fires.

It should be noted that during an extreme smoke event, observed in the Moscow megacity in August 2010, smoldering and flaming of regional wildfires near a city were a source for very intensive persistent carbonyls found near 1736 cm^{-1} [29]. The photochemical aging of BB smoke could result in the formation of secondary organic aerosols (SOA), which are represented through high dicarboxylic acid concentrations [58]. Additionally, in flaming spectra, a band of organic −NO$_2$ groups are located at 1530 cm^{-1}, similar to the very prominent vibrations there were observed in the peat bog smoke inside and above the grass during long-lasting peat burning near the Moscow region [39].

Analyses of the correlation between absorption bands of aerosols during the days of the BB-affected period and regional BB spectra indicate many similarities. In the range of the highest wavenumbers, ambient aerosols have a prominent peak at the same wavenumber of 3401 cm^{-1}, defining the similar absorption compounds with regional BB; however, it is much narrower and likely dominated by N-H amines. A wide band of oxidized functionalities is localized near 1736 cm^{-1}, while -NO$_2$ is always prominent at 1530 cm^{-1}. Polyaromatic C=C=H at 765 and 710 cm^{-1} are similar to the regional BB spectra; sulfates always dominate at 617 cm^{-1}, which are related to the vibrations in Ca$_2$SO$_4$. The representative spectra in Figure 6 show the days of the highest fire impact when BWT indicates the areas of fires (Figure 3). Therefore, we justify that a BB functional marker pattern of OH, N-H, C-C-H, C=O, C=C, C=C-H, -NO$_2$, and SO$_4^{2-}$ groups in carbohydrates, alcohols, amines, alkanes, oxidized compounds, aromatic, nitrocompounds, and sulfates can act as a functional marker of regional BB emissions in the Moscow environment (Table 1). The presence of small absorption at 3474 cm^{-1} reflects the co-existence of BB and bioaerosols, which was already observed during spring time [56].

Dust functionalities related to soil appeared at 923 cm^{-1}, attributed to Al-(OH) vibrations in kaolinite, and related to Si-OH stretching around 950 cm^{-1}. Carbonates (CO$_3^{2-}$) were identified in the range 880–860 cm^{-1}; their presence in the PM size fraction was confirmed by thermo-optical measurements of carbonates in the form of carbonate carbon [46]. Relative concentrations of carbonates are increasing from low to high smoke, thus showing the impact of re-suspended soil particles during intensive agricultural fires on the composition of coarse ambient aerosols. The IR spectra of the coarse samples exhibited a peak around 870 cm^{-1}, which was due to asymmetric vibrations of calcium carbonate (CaCO$_3$) [59]. During the BB-affected period, we should note the prominent similarity of the position of 880 cm^{-1} vibration band of carbonates CO$_3^{2-}$ in ambient aerosols, similar to spectra of flaming emission. This indicates the wide distribution of dust in the urban atmosphere as well as that evolved by air convection during biomass burning, which impacted the city atmosphere [22].

3.4. Combined FTIR-PCA Analyses

In order to investigate the variations in functionalized structures and detect similarities between daily aerosol chemistry, a combined FTIR spectroscopy and PCA were used. The score PC1 × PC2 plots calculated using the FTIR spectra data matrix for the whole sampling period is shown in Figure 7. The highest principal component (PC) axes 1, 2, and 3 account for 57.7, 21.1, and 11.1% of the total variance in the data set, respectively. There are three days, namely 04.29, 05.01, and 05.19, for which FTIR spectra show very high PC1 variability with respect to the other ones, described by the highest PC1 values. Additionally, these days were when the highest temperature and PM10 mass concentrations were observed (Figure 2).

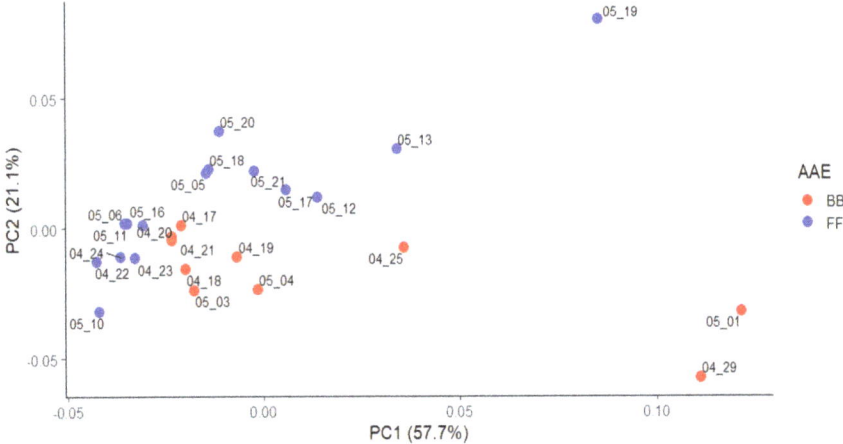

Figure 7. PCA score plots (PC1 × PC2) for FTIR spectra of sampling days marked by red and blue for AAE in the range of 1.61 ± 0.02 and 1.08 ± 0.02, relating to the BB-affected and FF periods, respectively.

The scatter plot of PC1 against PC2 shows the similarities as well as the differentiation between aerosol composition of days separated on two episodes according to the AAE parametrization. A big group of days assigned to the high AAE is clustered at negative PC2, their chemical composition is characterized by a similar pattern of the functional groups. It indicates that there is a BB-related factor that strongly influences aerosol chemistry, separating those days into the negative direction of PC2. On the other hand, the PC2 × PC3 plot does not demonstrate such separation well. The loadings plot in Figure 8 show the spectral variability explained by the first PCs. Marked extremums are the absorption bands identified on FTIR spectra of daily samples.

Table 2 summarizes the results of PCA analysis providing the most important variables as functional factors. It shows the principal components explaining PC1 (Factor 1, 58%), PC2 (Factor 2, 21%), and PC3 (Factor 3, 11%), totaling 89.9% of the data set variance. Factor 1 had the highest absolute value 0.21 for dust carbonates CO_3^{2-} (878 cm^{-1}), showing the impact of re-suspended soil particles, dust generated by the wind erosion, and transportation during spring season. Polyaromatics (762 cm^{-1}) from combustion emissions and sulfates (617 cm^{-1}) of BB features were also described by a high PC1, equal to 0.133. On the other hand, alkanes (2926–2855 m^{-1}), alkenes (1645 cm^{-1}), C=C (1591 cm^{-1}) related to diesel/gasoline transport, and esters/carboxylic acids (1736 cm^{-1}) showed a PC1 value in the range of 0.046–0.023. In Factor 2, sulfates (621 cm^{-1}) from secondary aerosols demonstrated a high variability, with a PC2 value of 0.21. Furthermore, dust-related features (947 and 882 cm^{-1}) with carboxylic acids/ketones (1715 cm^{-1}) from transport and aldehydes (1686 cm^{-1}) in BB emissions were best explained by PC2 in the high absolute value range of 0.13–0.06. Strong features of sugars/alcohols/carbohydrates (3474 cm^{-1}) were observed in biogenic aerosols, characterized by an absolute value of PC2 equal to 0.02. Factor 3 reveals the biggest impact of various sulfates (667, 638cm^{-1}) of long transportation, as well of dust carbonates (883 cm^{-1}) at the highest PC3 near 0.13, supported by C=C (1618 cm^{-1}) agriculture/residential fires with a PC3 of 0.075.

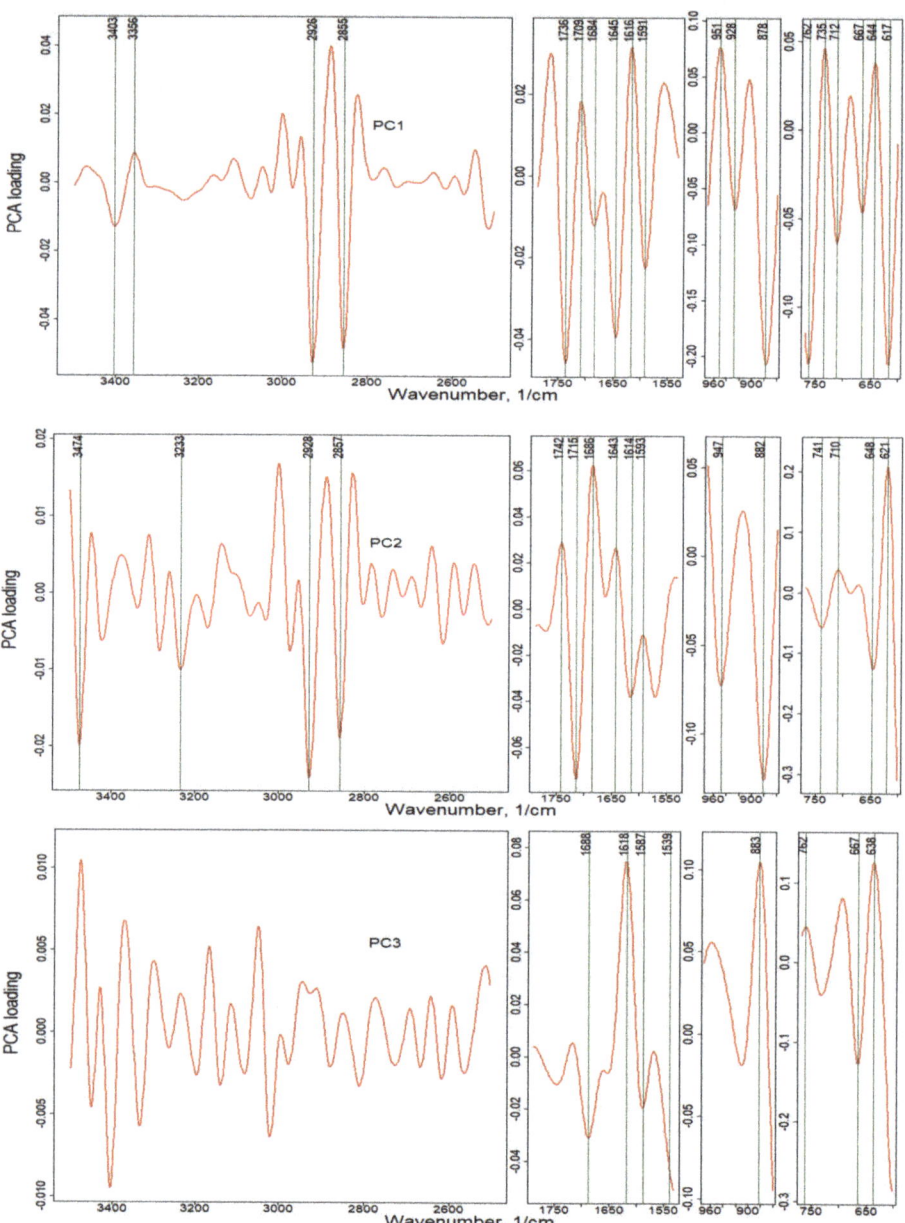

Figure 8. PC1, PC2, and PC3 loadings. Marked extremums are the absorption bands identified on FTIR spectra of daily samples.

Table 2. Functional factors presented by principal components of PCA analysis of FTIR data.

Wavenumber, 1/cm	PC1 Loading (58%)	PC2 Loading (21%)	PC3 Loading (11%)	Functional Groups
617	−0.133			SO_4^{2-} sulphates
621		0.21		SO_4^{2-} sulphates
638			0.126	SO_4^{2-} sulfuric acid, sulphates
644	0.038			SO_4^{2-} sulfuric acid
648		−0.125		SO_4^{2-} sulfuric acid, sulphates
667	−0.046		−0.126	SO_4^{2-} sulphates, sulfuric acid
710		0.038		CO_3^{2-} carbonates; SO_4^{2-} sulphates; C=C-H polyaromatics
712	−0.064			CO_3^{2-} carbonates; C=C-H polyaromatics
735	0.046			C=C-H polyaromatics
741		−0.058		C=C-H polyaromatics
762	−0.132		0.046	C=C-H polyaromatics
878	−0.207			CO_3^{2-} carbonates; C=C-H polyaromatics
882		−0.126		CO_3^{2-} carbonates; C=C-H polyaromatics
883			0.105	CO_3^{2-} carbonates; C=C-H polyaromatics
928	−0.068			O-H organic acids
947		−0.073		O-H organic acids
951	0.076			O-H organic acids
1539			−0.043	-NO_2 nitrocompounds
1587			−0.019	N-H amino acid; -NO_2 nitrocompounds
1591	−0.023			N-H amino acid; C=C polyaromatics
1593		−0.011		N-H amino acid; C=C polyaromatics
1614		−0.038		C=C polyaromatics; N-H amino acid
1616	0.032			C=C polyaromatics; N-H amino acid
1618			0.075	C=C polyaromatics; N-H amino acid
1643		0.027		C=C alkenes
1645	−0.04			C=C alkenes
1684	−0.012			C=O aldehydes
1686		0.063		C=O aldehydes
1688			−0.031	C=O aldehydes, carboxylic acid
1709	0.018			C=O carboxylic acid, ketones, aldehydes
1715		−0.074		C=O carboxylic acid, ketones, aldehydes
1736	−0.046			C=O esters, carboxylic acid
1742		0.029		C=O esters, carboxylic acid
2855	−0.049			C-C-H aliphatic hydrocarbons; O-H organic acids
2857		−0.019		C-C-H aliphatic hydrocarbons; O-H organic acids
2926	−0.053			C-C-H aliphatic hydrocarbons; C=C alkenes
2928		−0.024		C=C alkenes
3233		−0.01		NH_4^+ ammonium
3356	0.009			N-H amino acid; O-H carbohydrates and alcohols
3403	−0.013			N-H amines and amino acid; O-H alcohols
3474		−0.02		O-H sugar, alcohols and carbohydrates

4. Conclusions

Spring aerosols in the Moscow urban background were analyzed for particle-associated organics, ions, and dust functionalities. Classes of organic/inorganic compounds in the aerosol composition were inferred from analyses of FTIR spectral absorbance. Sixteen organic compound classes and two ionic inorganic and dust-related species were identified in spring aerosols. The composition of daily ambient aerosols demonstrated aliphatic, aromatic, carbonyl, and sulfate compound absorbance as features of traffic emissions. The functionalities of diesel/gasoline emission were found in ambient aerosols by their functional markers, showing the dominant impact of transport emissions from the diesel/gasoline vehicles operated in the Moscow megacity at the most common engine cycle conditions. Specific bands of amines, sugars, alcohols, and carbohydrates indicated the biogenic activity, while prominent ammonium and sulfates absorbance was assigned to secondary inorganic formation typical for urban environment. Because the sampling site at the Meteorological Observatory took place in a residential area, far from highways and industrial and agricultural activities, the main source of ammonium sulfates in PM were likely from secondary aerosols. Dust-related functionalities were proved by carbonates and kaolinites.

Complex chemistry and devised organic composition requires a source-specific optical marker for the assessment of a potential impact of biomass burning on urban aerosols. Parametrization of the sampling duration on FF and BB-affected periods by low (below 1.3) and high (above 1.3) AAE supports the relative contribution of agriculture fires/residential BB to urban Moscow aerosol composition, which is dominated by FF combustion. The BB functional marker reveals the pattern of carbohydrates, alcohols, amines, alkanes, oxidized compounds, aromatic, nitrocompounds, and sulfates observed in regional BB emissions. Air mass arriving in the Moscow area due to long-term transportation from the south of Russia impacts the air quality of the city, especially when the direction of transportation correlates well with fire-affected regions.

For the first time for the Moscow megacity environment, a factorial analysis such as principal component analysis (PCA) was used complementary to FTIR. FTIR-PCA allowed the distinguishing between daily aerosol composition according to high AAE, which identified BB spectral features during BB-affected periods. Chemometric techniques discriminated day-to-day changes in a range of major factors influenced by the aerosol composition. PCA analysis for FTIR data provided the most significant variables in main PC loadings showing the functional factors of transport, biomass burning, biogenic, dust, and secondary aerosol spring source impacts. Factor 1, explaining 57.7% of the variability, demonstrated the highest impact of carbonates, polyaromatics from combustion emissions, and BB-related sulfates following by alkanes, alkenes, esters/carboxylic acids, and C=C functional marker related to diesel/gasoline transport. Factor 2 explained 21.1% of the variability; sulfates from secondary aerosols dominated, while dust, carboxylic acids/ketones from transport, aldehydes in BB emission, and sugars/alcohols/carbohydrates of biogenic sources explained the aerosol composition to a large extend. Factor 3 explained 11.1% of the variability, revealing the biggest impact of sulfates and dust carbonates, supported by the C=C marker of agriculture/residential fires. Based on the combined FTIR-PCA analyses, it was suggested that traffic and biogenic emissions affected by biomass burning were the dominating sources of particle-bound organic compounds in spring aerosols in the urban background of the megacity environment.

Supplementary Materials: The following are available online at http://www.mdpi.com/2073-4433/11/4/319/s1, text, title, Table S1: Classes and standards of organic and inorganic compounds used for the identification of functional groups in ambient aerosols, Table S2: Stretch and bend vibrations of the functional groups, assigned to classes of organic and inorganic compounds, Figure S1: FTIR spectra of the FF period, Figure S2: FTIR spectra of the BB-affected period.

Author Contributions: Conceptualization and paper writing: O.P.; field and simulation work: A.I.; methodology and sampling of vehicle emissions: M.V. All authors have read and agreed to the published version of the manuscript.

Funding: Observation data analyses concerning FF impact in the Moscow environment was supported by the Russian Science Foundation (RSF), project N19-773004 titled "Integrated technology for environment assessment of Moscow megacity based on chemical analysis of microparticle composition in the "atmosphere - snow - road dust - soil - surface water" system (Megacity)". Vehicle tests were supported by the Czech Science Foundation project 18-04719S. Financial support was obtained from the Russian Fond for basic Research (RFBR) project N 20-55 12001 for FTIR- PCA studies of biomass burning impact.

Conflicts of Interest: The authors declare no conflict of interest.

References

1. Fuzzi, S.; Baltensperger, U.; Carslaw, K.; Decesari, S.; Denier Van Der Gon, H.; Facchini, M.; Fowler, D.; Koren, I.; Langford, B.; Lohmann, U. Part. matter, air quality and climate: Lessons learned and fure needs. *Atmos. Chem. Phys.* **2015**, *15*, 8217–8299. [CrossRef]

2. Pope III, C.A.; Dockery, D.W. Health effects of fine particulate air pollution: Lines that connect. *J. Air Waste Manag. Assoc.* **2006**, *56*, 709–742. [CrossRef] [PubMed]
3. Russell, L.M.; Bahadur, R.; Hawkins, L.N.; Allan, J.; Baumgardner, D.; Quinn, P.K.; Bates, T.S. Organic aerosol characterization by complementary measurements of chemical bonds and molecular fragments. *Atmos. Environ.* **2009**, *43*, 6100–6105. [CrossRef]
4. Russell, L.M.; Bahadur, R.; Ziemann, P.J. Identifying organic aerosol sources by comparing functional group composition in chamber and atmospheric particles. *Proc. Natl. Acad. Sci. USA* **2011**, *108*, 3516–3521. [CrossRef] [PubMed]
5. Ruggeri, G.; Takahama, S. Development of chemoinformatic tools to enumerate functional groups in molecules for organic aerosol characterization. *Atmos. Chem. Phys.* **2016**, *16*, 4401–4422. [CrossRef]
6. Liu, S.; Takahama, S.; Russell, L.; Gilardoni, S.; Baumgardner, D. Oxygenated organic functional groups and their sources in single and submicron organic particles in MILAGRO 2006 campaign. *Atmos. Chem. Phys. Discuss.* **2009**, *9*, 4567–4607. [CrossRef]
7. Coury, C.; Dillner, A.M. ATR-FTIR characterization of organic functional groups and inorganic ions in ambient aerosols at a rural site. *Atmos. Environ.* **2009**, *43*, 940–948. [CrossRef]
8. Steiner, S.; Czerwinski, J.; Comte, P.; Popovicheva, O.; Kireeva, E.; Müller, L.; Heeb, N.; Mayer, A.; Fink, A.; Rothen-Rutishauser, B. Comparison of the toxicity of diesel exhaust produced by bio-and fossil diesel combustion in human lung cells in vitro. *Atmos. Environ.* **2013**, *81*, 380–388. [CrossRef]
9. Guzman-Morales, J.; Frossard, A.; Corrigan, A.; Russell, L.; Liu, S.; Takahama, S.; Taylor, J.; Allan, J.; Coe, H.; Zhao, Y. Estimated contributions of primary and secondary organic aerosol from fossil fuel combustion during the CalNex and Cal-Mex campaigns. *Atmos. Environ.* **2014**, *88*, 330–340. [CrossRef]
10. Popovicheva, O.B.; Kireeva, E.D.; Shonija, N.K.; Vojtisek-Lom, M.; Schwarz, J. FTIR analysis of surface functionalities on particulate matter produced by off-road diesel engines operating on diesel and biofuel. *Environ. Sci. Pollut. Res.* **2015**, *22*, 4534–4544. [CrossRef]
11. Popovicheva, O.B.; Irimiea, C.; Carpentier, Y.; Ortega, I.K.; Kireeva, E.D.; Shonija, N.K.; Schwarz, J.; Vojtíšek-Lom, M.; Focsa, C. Chemical composition of diesel/biodiesel particulate exhaust by ftir spectroscopy and mass spectrometry: Impact of fuel and driving cycle. *Aerosol Air Qual. Res.* **2017**, *17*, 1717–1734. [CrossRef]
12. Weingartner, E.; Keller, C.; Stahel, W.; Burtscher, H.; Baltensperger, U. Aerosol emission in a road tunnel. *Atmos. Environ.* **1997**, *31*, 451–462. [CrossRef]
13. Platt, S.M.; El Haddad, I.; Pieber, S.; Zardini, A.; Suarez-Bertoa, R.; Clairotte, M.; Daellenbach, K.; Huang, R.-J.; Slowik, J.; Hellebust, S. Gasoline cars produce more carbonaceous particulate matter than modern filter-equipped diesel cars. *Sci. Rep.* **2017**, *7*, 4926. [CrossRef] [PubMed]
14. Lammers, K.; Arbuckle-Keil, G.; Dighton, J. FT-IR study of the changes in carbohydrate chemistry of three New Jersey pine barrens leaf litters during simulated control burning. *Soil Biol. Biochem.* **2009**, *41*, 340–347. [CrossRef]
15. Fitzpatrick, E.; Ross, A.; Bates, J.; Andrews, G.; Jones, J.; Phylaktou, H.; Pourkashanian, M.; Williams, A. Emission of oxygenated species from the combustion of pine wood and its relation to soot formation. *Process Saf. Environ. Prot.* **2007**, *85*, 430–440. [CrossRef]
16. Popovicheva, O.B.; Kozlov, V.S.; Rakhimov, R.F.; Shmargunov, V.P.; Kireeva, E.D.; Persiantseva, N.M.; Timofeev, M.A.; Engling, G.; Elephteriadis, K.; Diapouli, L.; et al. Optical-microphysical and physical-chemical characteristics of Siberian biomass burning: Small-scale fires in an aerosol chamber. *Atmos. Ocean. Opt.* **2016**, *29*, 492–500. [CrossRef]
17. Maria, S.F.; Russell, L.M.; Turpin, B.J.; Porcja, R.J. FTIR measurements of functional groups and organic mass in aerosol samples over the Caribbean. *Atmos. Environ.* **2002**, *36*, 5185–5196. [CrossRef]
18. Coury, C.; Dillner, A.M. A method to quantify organic functional groups and inorganic compounds in ambient aerosols using attenuated total reflectance FTIR spectroscopy and multivariate chemometric techniques. *Atmos. Environ.* **2008**, *42*, 5923–5932. [CrossRef]
19. Kirchstetter, T.W.; Novakov, T.; Hobbs, P.V. Evidence that the spectral dependence of light absorption by aerosols is affected by organic carbon. *J. Geophys. Res. Atmos.* **2004**, *109*. [CrossRef]
20. Healy, R.; Sofowote, U.; Su, Y.; Debosz, J.; Noble, M.; Jeong, C.-H.; Wang, J.; Hilker, N.; Evans, G.; Doerksen, G. Ambient measurements and source apportionment of fossil fuel and biomass burning black carbon in Ontario. *Atmos. Environ.* **2017**, *161*, 34–47. [CrossRef]

21. Diapouli, E.; Kalogridis, A.-C.; Markantonaki, C.; Vratolis, S.; Fetfatzis, P.; Colombi, C.; Eleftheriadis, K. Annual variability of black carbon concentrations originating from biomass and fossil fuel combustion for the suburban aerosol in Athens, Greece. *Atmosphere* **2017**, *8*, 234. [CrossRef]
22. Popovicheva, O.B.; Shonija, N.K.; Persiantseva, N.; Timofeev, M.; Diapouli, E.; Eleftheriadis, K.; Borgese, L.; Nguyen, X.A. Aerosol Pollutants during Agricultural Biomass Burning: A Case Study in Ba Vi Region in Hanoi, Vietnam. *Aerosol Air Qual. Res.* **2017**, *17*, 2762–2779. [CrossRef]
23. Elliott, G.N.; Worgan, H.; Broadhurst, D.; Draper, J.; Scullion, J. Soil differentiation using fingerprint Fourier transform infrared spectroscopy, chemometrics and genetic algorithm-based feature selection. *Soil Biol. Biochem.* **2007**, *39*, 2888–2896. [CrossRef]
24. Cadet, F.; Robert, C.; Offmann, B. Simultaneous determination of sugars by multivariate analysis applied to mid-infrared spectra of biological samples. *Appl. Spectrosc.* **1997**, *51*, 369–375. [CrossRef]
25. Hori, R.; Sugiyama, J. A combined FT-IR microscopy and principal component analysis on softwood cell walls. *Carbohydr. Polym.* **2003**, *52*, 449–453. [CrossRef]
26. Kulbachevsky, A.O. About environment state. In *Moscow City in 2017*; Department of Environmental Management and Protection: Moscow, Russia, 2018; 358 pages.
27. Elansky, N.F.; Ponomarev, N.A.; Verevkin, Y.M. Air quality and pollutant emissions in the Moscow megacity in 2005–2014. *Atmos. Environ.* **2018**, *175*, 54–64. [CrossRef]
28. Cheng, Z.; Luo, L.; Wang, S.; Wang, Y.; Sharma, S.; Shimadera, H.; Wang, X.; Bressi, M.; de Miranda, R.M.; Jiang, J.; et al. Status and characteristics of ambient PM2.5 pollution in global megacities. *Environ. Int.* **2016**, *89–90*, 212–221. [CrossRef]
29. Popovicheva, O.B.; Kistler, M.; Kireeva, E.; Persiantseva, N.; Timofeev, M.; Kopeikin, V.; Kasper-Giebl, A. Physicochemical characterization of smoke aerosol during large-scale wildfires: Extreme event of August 2010 in Moscow. *Atmos. Environ.* **2014**, *96*, 405–414. [CrossRef]
30. Chubarova, N.E.; Androsova, E.E.; Kirsanov, A.A.; Vogel, B.; Vogel, H.; Popovicheva, O.B.; Rivin, G.S. Aerosol and its radiative effects during the aeroradcity 2018 moscow experiment. *Geogr. Environ. Sustain.* **2019**, *12*, 114–131. [CrossRef]
31. Stein, A.; Draxler, R.; Rolph, G.; Stunder, B.; Cohen, M.; Ngan, F. NOAA's HYSPLIT atmospheric transport and dispersion modeling system. *Bull. Am. Meteorol. Soc.* **2015**, *96*, 2059–2077. [CrossRef]
32. Vojtisek-Lom, M.; Pechout, M.; Dittrich, L.; Beránek, V.; Kotek, M.; Schwarz, J.; Vodička, P.; Milcová, A.; Rossnerová, A.; Ambrož, A. Polycyclic aromatic hydrocarbons (PAH) and their genotoxicity in exhaust emissions from a diesel engine during extended low-load operation on diesel and biodiesel fuels. *Atmos. Environ.* **2015**, *109*, 9–18. [CrossRef]
33. Coates, J. Interpretation of infrared spectra, a practical approach. *Encycl. Anal. Chem. Appl. Theory Instrum.* **2006**. [CrossRef]
34. Maria, S.; Russell, L.; Turpin, B.; Porcja, R.; Campos, T.; Weber, R.; Huebert, B. Source signatures of carbon monoxide and organic functional groups in Asian Pacific Regional Aerosol Characterization Experiment (ACE-Asia) submicron aerosol types. *J. Geophys. Res. Atmos.* **2003**, *108*. [CrossRef]
35. Cain, J.P.; Gassman, P.L.; Wang, H.; Laskin, A. Micro-FTIR study of soot chemical composition—Evidence of aliphatic hydrocarbons on nascent soot surfaces. *Phys. Chem. Chem. Phys.* **2010**, *12*, 5206–5218. [CrossRef]
36. Bladt, H.; Schmid, J.; Kireeva, E.D.; Popovicheva, O.B.; Perseantseva, N.M.; Timofeev, M.A.; Heister, K.; Uihlein, J.; Ivleva, N.P.; Niessner, R. Impact of Fe content in laboratory-produced soot aerosol on its composition, structure, and thermo-chemical properties. *Aerosol Sci. Technol.* **2012**, *46*, 1337–1348. [CrossRef]
37. Rosen, H.; Novakov, T. Optical transmission through aerosol deposits on diffusely reflective filters: A method for measuring the absorbing component of aerosol particles. *Appl. Opt.* **1983**, *22*, 1265–1267. [CrossRef]
38. Bond, T.C.; Bergstrom, R.W. Light absorption by carbonaceous particles: An investigative review. *Aerosol Sci. Technol.* **2006**, *40*, 27–67. [CrossRef]
39. Popovicheva, O.B.; Engling, G.; Ku, I.-T.; Timofeev, M.A.; Shonija, N.K. Aerosol emissions from long-lasting smoldering of boreal peatlands: Chemical composition, markers, and microstructure. *Aerosol Air Qual. Res.* **2019**, *19*, 484–503. [CrossRef]
40. Popovicheva, O.; Timofeev, M.; Persiantseva, N.; Jefferson, M.A.; Johnson, M.; Rogak, S.N.; Baldelli, A. Microstructure and chemical composition of particles from small-scale gas flaring. *Aerosol Air Qual. Res.* **2019**, *19*, 2205–2221. [CrossRef]

41. Popovicheva, O.B.; Kireeva, E.D.; Steiner, S.; Rothen-Rutishauser, B.; Persiantseva, N.M.; Timofeev, M.A.; Shonija, N.K.; Comte, P.; Czerwinski, J. Microstructure and chemical composition of diesel and biodiesel particle exhaust. *Aerosol Air Qual. Res.* **2014**, *14*, 1392–1401. [CrossRef]
42. Popovicheva, O.B.; Engling, G.; Lin, K.-T.; Persiantseva, N.; Timofeev, M.; Kireeva, E.; Voelk, P.; Hubert, A.; Wachtmeister, G. Diesel/biofuel exhaust particles from modern internal combustion engines: Microstructure, composition, and hygroscopicity. *Fuel* **2015**, *157*, 232–239. [CrossRef]
43. Gentner, D.R.; Isaacman, G.; Worton, D.R.; Chan, A.W.; Dallmann, T.R.; Davis, L.; Liu, S.; Day, D.A.; Russell, L.M.; Wilson, K.R. Elucidating secondary organic aerosol from diesel and gasoline vehicles through detailed characterization of organic carbon emissions. *Proc. Natl. Acad. Sci. USA* **2012**, *109*, 18318–18323. [CrossRef] [PubMed]
44. Giakoumis, E.G.; Rakopoulos, C.D.; Dimaratos, A.M.; Rakopoulos, D.C. Exhaust emissions of diesel engines operating under transient conditions with biodiesel fuel blends. *Prog. Energy Combust. Sci.* **2012**, *38*, 691–715. [CrossRef]
45. Kalogridis, A.C.; Popovicheva, O.B.; Engling, G.; Diapouli, E.; Kawamura, K.; Tachibana, E.; Ono, K.; Kozlov, V.S.; Eleftheriadis, K. Smoke aerosol chemistry and aging of Siberian biomass burning emissions in a large aerosol chamber. *Atmos. Environ.* **2018**, *185*, 15–28. [CrossRef]
46. Popovicheva, O.B.; Engling, G.; Diapouli, E.; Saraga, D.; Persiantseva, N.M.; Timofeev, M.A.; Kireeva, E.D.; Shonija, N.K.; Chen, S.-H.; Nguyen, D.-L.; et al. Impact of smoke intensity on size-resolved aerosol composition and microstructure during the biomass burning season in northwest Vietnam. *Aerosol Air Qual. Res.* **2016**, *16*, 2635–3654. [CrossRef]
47. Arıl, I.; Golcuk, K.; Karaca, F. ATR-FTIR spectroscopic study of functional groups in aerosols: The contribution of a Saharan dust transport to urban atmosphere in Istanbul, Turkey. *Water Air Soil Pollut.* **2014**, *225*, 1898. [CrossRef]
48. Ravisankar, R.; Kiruba, S.; Eswaran, P.; Senthilkumar, G.; Chandrasekaran, A. Mineralogical characterization studies of ancient potteries of Tamilnadu, India by FT-IR spectroscopic technique. *J. Chem.* **2010**, *7*, S185–S190. [CrossRef]
49. Martınez, J.; Ruiz, F.; Vorobiev, Y.V.; Pérez-Robles, F.; González-Hernández, J. Infrared spectroscopy analysis of the local atomic structure in silica prepared by sol-gel. *J. Chem. Phys.* **1998**, *109*, 7511–7514. [CrossRef]
50. Wold, S.; Esbensen, K.; Geladi, P. Principal component analysis. *Chemom. Intell. Lab. Syst.* **1987**, *2*, 37–52. [CrossRef]
51. Olson, M.R.; Victoria Garcia, M.; Robinson, M.A.; Van Rooy, P.; Dietenberger, M.A.; Bergin, M.; Schauer, J.J. Investigation of black and brown carbon multiple-wavelength-dependent light absorption from biomass and fossil fuel combustion source emissions. *J. Geophys. Res. Atmos.* **2015**, *120*, 6682–6697. [CrossRef]
52. Fanning, P.E.; Vannice, M.A. A DRIFTS study of the formation of surface groups on carbon by oxidation. *Carbon* **1993**, *31*, 721–730. [CrossRef]
53. Akhter, M.; Chughtai, A.; Smith, D. The structure of hexane soot I: Spectroscopic studies. *Appl. Spectrosc.* **1985**, *39*, 143–153. [CrossRef]
54. Sauvain, J.-J.; Rossi, M.J. Quantitative aspects of the interfacial catalytic oxidation of dithiothreitol by dissolved oxygen in the presence of carbon nanoparticles. *Environ. Sci. Technol.* **2016**, *50*, 996–1004. [CrossRef] [PubMed]
55. Kouyoumdjian, H.; Saliba, N. Ion concentrations of PM10 and PM2.5 aerosols over the eastern Mediterranean region: Seasonal variation and source identification. *Atmos. Chem. Phys. Discuss. Eur. Geosci. Union* **2005**, *5*, 13053–13073. [CrossRef]
56. Bauer, H.; Claeys, M.; Vermeylen, R.; Schueller, E.; Weinke, G.; Berger, A.; Puxbaum, H. Arabitol and mannitol as tracers for the quantification of airborne fungal spores. *Atmos. Environ.* **2008**, *42*, 588–593. [CrossRef]
57. Pey, J.; Querol, X.; De la Rosa, J.; González-Castanedo, Y.; Alastuey, A.; Gangoiti, G.; de la Campa, A.S.; Alados-Arboledas, L.; Sorribas, M.; Pio, C. Characterization of a long range transport pollution episode affecting PM in SW Spain. *J. Environ. Monit.* **2008**, *10*, 1158–1171. [CrossRef]
58. Agarwal, S.; Aggarwal, S.G.; Okuzawa, K.; Kawamura, K. Size distributions of dicarboxylic acids, ketoacids, α-dicarbonyls, sugars, WSOC, OC, EC and inorganic ions in atmospheric particles over Northern Japan: Implication for long-range transport of Siberian biomass burning and East Asian polluted aerosols. *Atmos. Chem. Phys.* **2010**, *10*, 5839–5858. [CrossRef]

59. Cuccia, E.; Piazzalunga, A.; Bernardoni, V.; Brambilla, L.; Fermo, P.; Massabò, D.; Molteni, U.; Prati, P.; Valli, G.; Vecchi, R. Carbonate measurements in PM10 near the marble quarries of Carrara (Italy) by infrared spectroscopy (FT-IR) and source apportionment by positive matrix factorization (PMF). *Atmos. Environ.* **2011**, *45*, 6481–6487. [CrossRef]

 © 2020 by the authors. Licensee MDPI, Basel, Switzerland. This article is an open access article distributed under the terms and conditions of the Creative Commons Attribution (CC BY) license (http://creativecommons.org/licenses/by/4.0/).

Article

Source Apportionment of Fine Organic and Inorganic Atmospheric Aerosol in an Urban Background Area in Greece

Manousos Ioannis Manousakas [1,*], **Kalliopi Florou** [1] **and Spyros N. Pandis** [1,2,3]

[1] Institute of Chemical Engineering Sciences, (FORTH/ICE-HT), 26504 Patras, Greece; kalli@chemeng.upatras.gr (K.F.); spyros@chemeng.upatras.gr (S.N.P.)
[2] Department of Chemical Engineering, University of Patras, 26504 Patras, Greece
[3] Department of Chemical Engineering, Carnegie Mellon University, Pittsburgh, PA 15213, USA
* Correspondence: manosman@ipta.demokritos.gr

Received: 7 March 2020; Accepted: 26 March 2020; Published: 29 March 2020

Abstract: Fine particulate matter (PM) originates from various emission sources and physicochemical processes. Quantification of the sources of PM is an important step during the planning of efficient mitigation strategies and the investigation of the potential risks to human health. Usually, source apportionment studies focus either on the organic or on the inorganic fraction of PM. In this study that took place in Patras, Greece, we address both PM fractions by combining measurements from a range of on- and off-line techniques, including elemental composition, organic and elemental carbon (OC and EC) measurements, and high-resolution Aerosol Mass Spectrometry (AMS) from different techniques. Six fine $PM_{2.5}$ sources were identified based on the off-line measurements: secondary sulfate (34%), biomass burning (15%), exhaust traffic emissions (13%), nonexhaust traffic emissions (12%), mineral dust (10%), and sea salt (5%). The analysis of the AMS spectra quantified five factors: two oxygenated organic aerosols (OOA) factors (an OOA and a marine-related OOA, 52% of the total organic aerosols (OA)), cooking OA (COA, 11%) and two biomass burning OA (BBOA-I and BBOA-II, 37% in total) factors. The results of the two methods were synthesized, showcasing the complementarity of the two methodologies for fine PM source identification. The synthesis suggests that the contribution of biomass burning is quite robust, but that the exhaust traffic emissions are not due to local sources and may also include secondary OA from other sources.

Keywords: source apportionment; Positive Matrix Factorization (PMF); AMS; chemical composition

1. Introduction

Atmospheric particulate matter (PM) has serious adverse effects on human health and the environment [1–4]. PM originates from various emission sources but is also the product of gas-to-particle processes in the atmosphere (secondary PM). The identification of the different types of PM sources and the quantification of their contributions, including the role of natural and anthropogenic sources, is a critical step in our efforts to reduce the PM effects on human health [5]. The process of identifying PM sources is called source apportionment and there is a wide range of methods that have been proposed based on the available measurements [6]. A widely used source apportionment approach is the use of various types of source-receptor models [7–11]. Most receptor models are based on the assumption that at the receptor site, the time dependence of the concentration of PM components originating from the same source will be the same. The concentrations of the relevant PM chemical components are measured in a large number of samples gathered at the receptor site over time. Components with similar temporal variability are grouped together in a minimum number of factors that explain the variability of the complete dataset. It is assumed that each factor is associated with a source or source type. Those factors can be considered as the chemical fingerprint of the PM sources and lead to their identification. Positive Matrix Factorization (PMF) is one of the most useful source-receptor analysis approaches [12]. PMF introduces a weighting scheme taking into account errors of the individual measurements. Adjustment of the corresponding error estimates also allows it to handle missing and below detection limit data. One of its major advantages compared to other approaches is that it obtains only factors with positive elements using appropriate constraints.

Usually, source-apportionment studies focus either on the organic [13,14] or on the inorganic components of PM [15,16]. To fully assess the particulate related pollution in an area, it is important to examine the sources of both the organic and the inorganic fraction of PM. In the current study that took place in Patras, Greece, we focus on both the sources of organic and inorganic fractions using appropriate measurement techniques.

Greece is located in the Eastern Mediterranean, which is characterized as an air pollution hotspot located at the crossroad of air masses coming from Asia, Europe, and Africa [17]. Forest fires and agricultural burning emissions affect the area during the dry season, while the high radiation intensity all year long and the relatively high temperatures enhance the formation of secondary PM. Patras is a medium-sized city located in the northwest Peloponnese. Previous studies have shown that PM_1 in the area is composed mainly of OA (organic aerosols) 45% and sulfate 38% [18,19]. Transportation, biomass burning (residential and agricultural), and shipping emissions have been found to be major local sources in the past [8,20], but long-range transport is an important source, especially during the summer, spring, and fall. Almost all of the sulfates and 40–90%, depending on the season, of the organic aerosol are believed to be transported to the city from other areas [21]. However, the situation in Patras and the rest of Greece has changed significantly during the last few years due to the financial crisis. This has led to an apparent reduction of the emissions of some sources (transportation, industry), but also to the increase of others (residential biomass burning). Organic aerosol (OA) and especially residential wood burning, have dramatically increased due to the Greek financial crisis [22]. A 30% increase in the $PM_{2.5}$ mass concentration as well as a 2–5 fold increase in the concentration of wood smoke tracers, including potassium, levoglucosan, mannosan, and galactosan has been reported for the city of Thessaloniki [23], while the concentrations of fuel oil tracers (e.g., Ni and V), have declined by 20–30%. One of the goals of the present study is to try to quantify these changes using recent measurements and improved source-apportionment techniques.

2. Experiments

2.1. Experimental Design

The sampling site was located in a suburban area, around 7 km northeast of the city center (38°18′ N, 21°47′ E) in the Institute of Chemical Engineering Sciences (FORTH/ICE-HT) and at an elevation of 85 m above sea level. The sampling station is surrounded by olive trees, while the coast is approximately 3 km away. The closest major road is found at a distance of 1 km. Sampling took place from late October until the start of November 2018 and late February to June 2019.

The size-resolved chemical composition of the submicrometer aerosol (organics, nitrate, sulfate, chloride, and ammonium) was monitored using a High-Resolution Time-of-Flight Aerosol Mass Spectrometer (HR–ToF–AMS) from Aerodyne [24]. The sampled aerosol during this study was not dried, while the vaporizer surface temperature was set at 600 °C. The HR–ToF–AMS measurements were collected with a 3 min temporal resolution. The software toolkit SQUIRREL v1.57I (Sueper, D, ToF-AMS Data Analysis Software) in Igor Pro (Wavemetrics) was used for all the HR–ToF–AMS data analysis combined with the Peak Integration by Key Analysis (PIKA v1.16I package) (Sueper, D, ToF-AMS Data Analysis Software) software. For the calculation of the elemental ratios measured by the HR–AMS, the improved ambient approach was used [25].

A Super Speciation Air Sampling System Sampler (Met One SuperSASS) operating at a flow rate of 6.7 L min^{-1} was used to collect $PM_{2.5}$ samples every 24 h, both on PTFE (Whatman, PTFE, White

Plain, 0.2 µm × 47 mm) and on quartz (Pall 2500 QAO-UP 47 mm) filters The Super SASS sampler is based on designs that have been field tested for several years in the US; eight years with three years of testing in the EPA program. These filters were used for the measurement of the daily average $PM_{2.5}$ mass concentration and its chemical composition. The $PM_{2.5}$ mass concentration of each filter collected was determined by gravimetric analysis using a six-digit microbalance at 35% relative humidity (RH). Both sampling and gravimetric analysis were compliant with US EPA requirements. The $PM_{2.5}$ mass concentration measured at 35% relative humidity includes less water than measurements at higher RH values.

The concentrations of organic carbon (OC) and elemental carbon (EC) were measured by thermo-optical transmittance (TOT) analysis (Lab OC-EC Aerosol Analyzer, Sunset Laboratory, Inc., Tigard, Washington, DC, USA) [26]. Quartz filter samples were heated first up to 650 °C in He and then up to 850 °C in a mixture of 2% O_2 in He, using the controlled heating ramps of the EUSAAR2 thermal protocol [27]. OC evolved in the inert atmosphere, while EC was oxidized in the He–O_2 atmosphere. A charring correction was applied by monitoring the sample transmittance throughout the heating process. The limit of detection for the TOT analysis is 0.2 µg cm^{-2}. Before sampling, the quartz fiber filters were pretreated at 500 °C for 6 h to remove possible organic contaminants. Laboratory and field blanks were collected and analyzed following the same procedures as those adopted for the samples. The QA/QC procedures described in EN 16,909:2017 were followed during the OC/EC analysis.

A high energy, polarization geometry energy-dispersive XRF spectrometer (Epsilon 5 by PANalytical, Almelo, the Netherlands) was used for the elemental analysis of $PM_{2.5}$ samples. Epsilon 5 is constructed with optimized Cartesian-triaxial geometry composed of several secondary targets for attaining lower spectral background and with extended K line excitation allowing it to operate the X-ray tube at a high voltage of 100 kV. The secondary target-XRF spectrometer includes a side-window low power X-ray tube with a W/Sc anode (spot size 1.8–2.1 cm, 100 kV maximum voltage, 6 mA current, 600 W maximum power consumption). A Ge X-ray detector detects the characteristic X-rays emitted from the sample with 140 eV FWHM at MnKα, 30 mm^2, and 5-mm thick Ge crystal with an 8 µm Be window. The spectrometer used is equipped with eight secondary targets (Al, CaF_2, Fe, Ge, Zr, Mo, Al_2O_3, KBr, and LaB_6) that can polarize the X-ray tube-generated incident radiation through Barkla scattering. The methodology that was used for the elemental composition analysis is described in detail elsewhere [8]. The 20 elements determined by the ED–XRF method were Na, Mg, Al, Si, S, Cl, K, Ca, Ti, V, Cr, Mn, Fe, Co, Ni, Cu, Zn, Br, Sr, and Pb.

In total, 55 $PM_{2.5}$ samples were collected for each filter type (Teflon and quartz). The sampling campaign took place during five periods, October–November 2018 (12 samples), February–March 2019 (13 samples), April 2019 (10 samples), May 2019 (10 samples), and June 2019 (10 samples).

The AMS measurements cover the PM_1 range while the filter samples and the corresponding analysis were in the $PM_{2.5}$ range. This small discrepancy in size ranges was unavoidable because of the characteristics of the AMS used in this study. The AMS measurements are mainly used in this study for the organic aerosol characterization and source apportionment. Given that most of the organic aerosol in the $PM_{2.5}$ is in submicrometer particles, this size discrepancy does not affect our conclusions and the synthesis of the results from the two types of measurements.

2.2. PMF Analysis

The basic equation for the solution of the mass balance problem is common for all the utilized multivariate receptor models, including PMF:

$$X_{ij} = \sum_{k=1}^{p} g_{ik} f_{kj} + e_{ij} \quad (1)$$

where X_{ij} is the concentration of species j measured on sample i, p is the number of factors contributing to the measured PM, f_{kj} is the concentration of species j in factor profile k, g_{ik} is the relative contribution of factor k to sample i, and e_{ij} is error of the PMF model for the jth species measured in sample i. The values of g_{ik} and f_{kj} are adjusted until a minimum value of Q for a given p is found where Q is defined as:

$$Q = \sum_{j=1}^{m}\sum_{i=1}^{n} \frac{e_{ij}^2}{s_{ij}^2} \qquad (2)$$

where s_{ij} is the uncertainty of the jth species concentration in sample i, n is the number of samples, and m is the number of species. Additional terms are added to the Q term in Equation (2) to make sure that only non-negative solutions are obtained.

2.3. Trajectory Analysis and Long-Range Transport

To identify the origin of the various aerosol components, air mass back-trajectories arriving at the site were analyzed by statistical methods. For the analysis of the back trajectories, the OPENAIR package was used [28]. The trajectories were calculated using the National Oceanic and Atmospheric Administration (NOAA) HYSPLIT 4.0 model [29,30].

In this study, 120 h (5 days) backward air mass trajectories arriving at the sampling site were computed. Global data assimilation system (GDAS) meteorological files were used as inputs for the backward trajectory computation. The backward trajectories were computed every three hours (ending at 00:00, 03:00, 06:00, 09:00, 12:00, 15:00, 18:00, and 00:00 UTC) for each day of the period of study at 500 m above ground level. The measured concentration of the PM component during the sampling period was assigned to all trajectories during that period.

To identify the possible position of sources that are affecting the concentration of PM chemical components measured, the Potential Source Contribution Function (PSCF) [31] was calculated. The PSCF is calculated as:

$$\text{PSCF} = \frac{m_{ij}}{n_{ij}} \qquad (3)$$

where n_{ij} is the number of times that the trajectories passed through the cell (i, j) and m_{ij} is the number of times that a source concentration was high when the trajectories passed through the cell (i, j). The criterion for determining m_{ij} is based on the distribution of the measured values (e.g., upper quartile).

3. Results

3.1. PM$_{2.5}$ Chemical Composition

The average PM$_{2.5}$ concentration measured was 12.4 ± 6.4 µg m^{-3} (Table 1). The highest average concentration was observed during October–November 2018 (14.8 µg m^{-3}) followed by March 2019 (14.3 µg m^{-3}). The lowest PM$_{2.5}$ concentrations were recorded during April and May 2019, with average concentrations of 8.3 and 9.5 µg m^{-3}, respectively.

Table 1. Average PM$_{2.5}$ mass concentration and composition during the study (all in µg m^{-3}).

	PM$^{2.5}$	EC	OC	Na	Mg	Al	Si	S	Cl	K	Ca
Average	12.40	0.38	2.85	0.35	<0.10	0.29	0.71	0.92	0.12	0.17	0.10
Standard Deviation	6.46	0.20	1.63	0.27	-	0.29	0.64	0.47	0.31	0.14	0.13
	Ti	V	Cr	Mn	Fe	Ni	Cu	Zn	Br	Sr	Pb
Average	0.02	<0.01	0.01	0.02	0.10	0.02	0.06	0.05	<0.05	<0.05	0.04
Standard Deviation	0.02	-	0.00	0.00	0.11	0.01	0.03	0.05	-	-	0.01

The average OC and EC concentrations were 2.85 and 0.38 µg m^{-3}. To calculate OA from OC concentration, the PM$_1$ OA/OC was determined based on the AMS measurements and found to be equal to 2.25. This ratio was then used for the conversion of the PM$_{2.5}$ OC to OA. The OA represented on average 52% of the PM$_{2.5}$ concentration with the EC contributing another 3%. When compared to results from other Greek cities, these concentrations are comparable to the ones reported for the Athens suburbs [26] and lower than those reported for urban and traffic sites in Athens and Thessaloniki [26,32,33]. Recent studies conducted in another European city, Lisbon, Portugal, have also reported higher concentrations of OC and EC (4.02 and 6.61 µg m^{-3} respectively) [34]. The highest average OC concentrations were recorded during November 2018 (3.77 µg m^{-3}) and March 2019 (4.19 µg m^{-3}). These two months also had the highest EC levels with averages of 0.6 and 0.59 µg m^{-3}, respectively. The average OC/EC ratio was 7.5, suggesting relatively low contribution from traffic emissions [26]. In general, the OC/EC ratio values reported for other European cities show OC/EC ratios of less than 2 at curbside sites, 2–9 at urban background sites, and increase with decreasing fresh anthropogenic emissions, reaching values above 10 at rural sites [33,35].

Fifteen elements had concentrations consistently higher than the detection limit (Table 1). Sulfur had the highest average concentration at 0.92 µg m^{-3}. Assuming that all was in the form of sulfate, it represented 22.2% of the PM$_{2.5}$. Silicon, sodium, aluminum, potassium, chloride, calcium, and iron also had relatively high average concentrations.

3.2. PM$_{2.5}$ Source Apportionment

Six factors (Figure 1) were identified by the EPA PMF model using 55 samples and the 17 aerosol components shown in Table 1. No species were down-weighted. Because the number of samples used was close to the lower acceptable limit for PMF analysis, the solution was thoroughly evaluated to make sure that the overall and rotational ambiguity uncertainties were low. The low number of samples may have limited the number of identified sources and also increased the uncertainty of the analysis [36]. The number of factors was decided after the investigation of a range of possible solutions starting from four factors up to nine. Six factors were found to be the highest number of factors with a physical meaning, and resulted to the lower unaccounted mass estimation. The differences between Q_{true} and Q_{rob} and also between Q_{true} and Q_{the} were less than 5%. One hundred simulations were performed and the differences in scaled residuals between the different simulations were very low. The results of the diagnostic tools offered by EPA PMF revealed that the solution was robust with low rotational ambiguity. Bootstrap results indicated that the factors were reproduced at a minimum level of 85% of the produced resamples, while displacement and bootstrap-displacement showed no factor swaps for the minimum dQ level. The PM mass reconstruction was satisfactory. The R^2 between modeled and measured PM$_{2.5}$ mass was 0.75 and the slope equal to 0.97. The composition of the six factors and the contribution of each factor to the concentration of each species are shown in Figure 1.

Factor 1 is characterized by the relatively high levels of Cu, Zn, Pb, Cr, Mn, Fe, and EC. All these elements are known to originate from the mechanical abrasion of different moving parts of the vehicles, such as tire and brake-wear, as well as the use of lubricant oils [37,38]. This factor was therefore assigned to non-exhaust transportation emissions. The average contribution of this source to PM$_{2.5}$ was 12% (1.4 µg m^{-3}).

Factor 2 is attributed to biomass burning because it is characterized by the relatively high levels of EC, OC, and K in the factor. The presence of tracers like Cr, Zn, and Pb in this factor may indicate some mixing with other sources. Since the contributions of these tracers in the factor profile are small, these mixing uncertainties should have only a minor effect on the estimated factor concentrations. The average contribution of the source to PM$_{2.5}$ is 15% (1.9 µg m^{-3}). Contribution of biomass burning was found to be higher than the contribution reported for the same site in 2011 (11%), which might be an effect of the economic crisis in Greece.

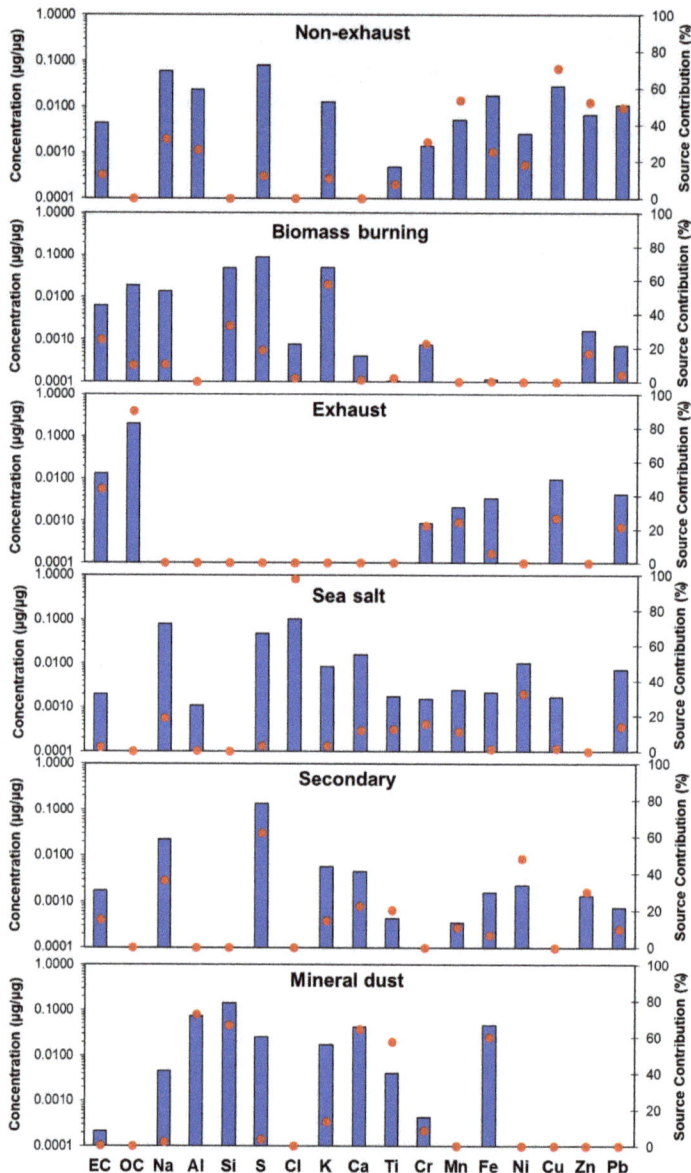

Figure 1. Factor profiles obtained by Positive Matrix Factorization (PMF) using the EC, OC, and elemental concentrations. The bars represent the normalized concentration of the species in the profile and the circles show the contribution of the source to the average species concentration.

The contribution of this factor was quite high during the high $PM_{2.5}$ periods of November 2018 and March 2019 (Figure 2). This was probably due to the burning of agricultural waste (mainly olive tree branches), but also to some domestic heating (fireplaces and woodstoves). PMF cannot separate agricultural burning from residential burning since their chemical profiles are very similar to each other. Some additional insights about the contributions of these two sources (agricultural burning versus residential heating) may be obtained from the analysis of the temporal profiles of the OA concentrations based on the AMS measurements. These will be discussed in a subsequent section.

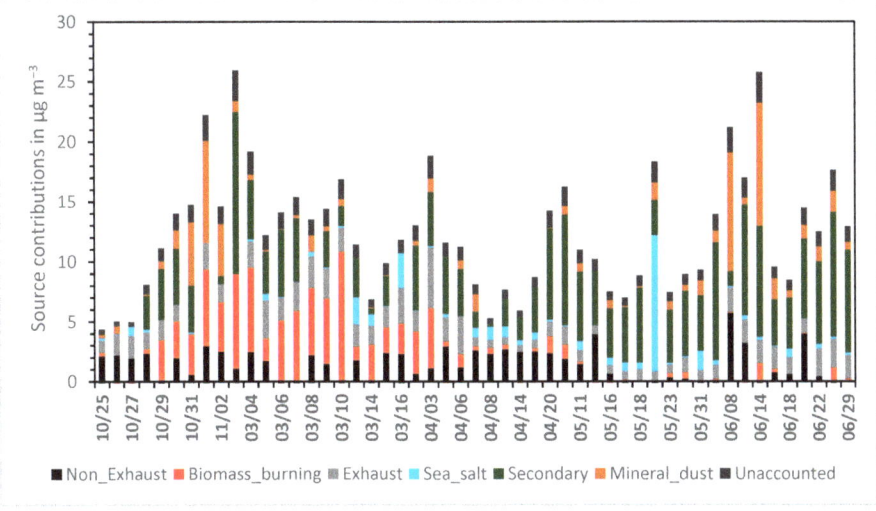

Figure 2. Daily average source contributions to PM$_{2.5}$ (µg m^{-3}) determined by PMF of the elemental composition and OC/EC levels.

Factor 3 is characterized by high levels of OC and EC and assigned to transportation exhaust emissions. Some elements originating from non-exhaust emissions, such as Cr, Mn, Cu, and Pb, are also present in the factor, but their contribution to the overall mass concentration is minor. The average contribution of this factor to PM$_{2.5}$ is 13% (1.6 µg m^{-3}) during the study period. The contribution is relatively stable, something expected given the lack of a seasonal pattern for these emissions.

Factor 4 is identified as sea salt and characterized by the presence of Na and Cl. Ni is also present in the factor, which might be an indication that this factor also includes some shipping emissions. Previous studies conducted in Patras have estimated that shipping emissions contributed approximately 4% to PM$_{10}$ [8]. The concentration of vanadium, a good tracer of shipping emissions, was below the detection limit for a large number of samples, and could not be used reliably as a tracer in the PMF analysis. The average contribution of this marine source to PM$_{2.5}$ was 5% (0.6 µg m^{-3}). The high sea salt contribution observed on 19 May might be associated with high sea breezes or favorable meteorological conditions for sea salt generation and transport to the site.

The next factor (Factor 5) is characterized by the high contribution of sulfur and is identified as secondary sulfate. EC, Ni, and Na are also present in this factor, indicating that it may also include some contribution from shipping emissions. The combustion of residual oil used for shipping will produce particles containing V and Ni. Vanadium reacts with the oxygen from the combustion air surplus creating V_2O_5 that form layers on the heat exchanger and other boiler and stack surfaces. The V_2O_5 acts as a catalyst accelerating the SO_3 formation, and therefore sulfate formation [39]. This factor has the highest contribution of 34% (4.2 µg m^{-3}) to the average PM$_{2.5}$ in the area. Its contribution can be compared with the concentration of 3.8 µg m^{-3} for sulfate, based on the measured sulfur.

Factor 6 is identified as mineral dust, and it is characterized by Al, Si, Ca, Ti, and Fe. All of these elements are related to mineral dust. The average contribution of the factor is 10%, and it contributes 1.3 µg m^{-3} to the average PM$_{2.5}$. This is lower than the contribution of the sum of oxides above (19%), something reasonable given that some of the elements used for the calculation of the sum are attributed by PMF to sources other than mineral dust. This sum of oxides can be thus viewed only as an upper limit of the mineral dust contribution. The contribution of mineral dust was high on 11/01, 06/08, and

06/14. For those days the HYSPLIT backward trajectories indicated dust transport events from Africa. These Sahara dust events are common to Southern Europe in general and in Greece in particular [40].

The uncertainty of the solution was thoroughly evaluated using standard methods of the PMF analysis, since the number of samples used was relatively low. The results of the uncertainty analysis can be found in the Supplementary Material (Text S1 and Table S1). The unaccounted mass concentration was 1.4 µg m^{-3}, which corresponds to 11% of the total PM$_{2.5}$ mass. This is indicative of the order of magnitude of the uncertainty of our analysis.

3.3. The Role of Long-Range Transport

To identify if the high-intensity events for the mineral dust, secondary sulfate, and sea salt factors were related to long-range transport, the PSCFs of their source contributions for the highest 15% concentrations were computed (Figure 3). The corresponding maps show the probability that a trajectory passing through the site was associated with one of the highest (top 15%) concentrations of the factor in the receptor. Sea salt during the high concentration period appears to have been transported to the sampling site from the Mediterranean Sea to the west. This is reasonable given that this factor should have a marine origin. The high concentrations of sulfates, on the other hand, were associated with transport from northwestern and northern Turkey, as well as the Black Sea region. There are a number of lignite-fired power plants with high emissions of SO$_2$ in that area [41]. Emissions from the same areas were found to affect the concentrations of sulfates in different locations of Greece in the past [18].

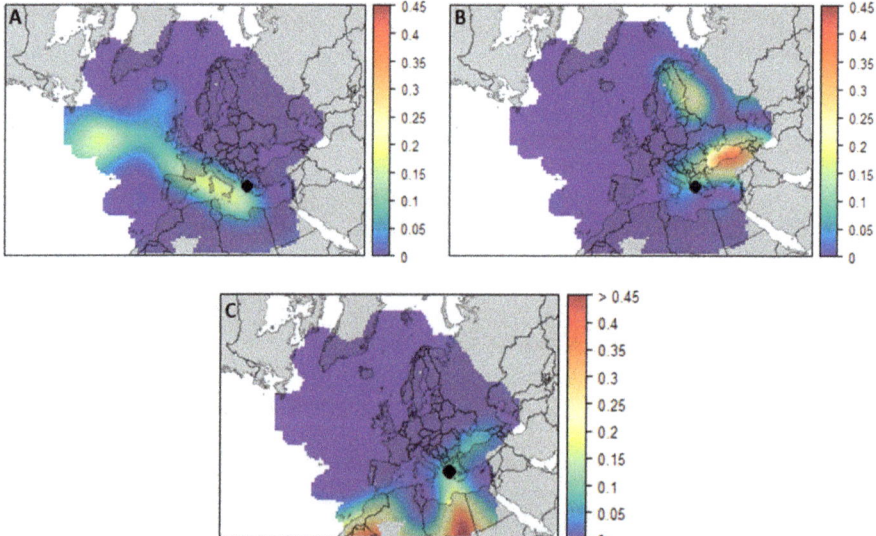

Figure 3. Potential Source Contribution Functions (PSCFs) for the 85th percentile of (**A**) sea salt, (**B**) secondary sulfate, and (**C**) mineral dust factors.

Finally, the highest concentrations of mineral dust appear to have been related to transport from northern Africa. These Saharan dust events are known to affect Greece [15,40,42], as well as other areas in Southern Europe [43,44]. It should be noted that our PSCF results help identify source areas that contribute to the highest concentrations. Of course, there are also local contributions to the corresponding concentrations.

3.4. Nonrefractory PM$_1$ Chemical Composition

The average concentrations of the major nonrefractory PM$_1$ (NR-PM$_1$) components during the measurement periods are presented in Table 2. The average total NR-PM$_1$ concentration during the sampling campaign was 7.7 µg m^{-3} ranging from 0.3 to 26 µg m^{-3} on an hourly average basis. The lowest NR-PM$_1$ mass concentration values were recorded during May 2019 with an average concentration of 3.8 µg m^{-3}. The highest concentrations were measured during the October–November 2018 period with an average NR-PM$_1$ concentration of 9.9 µg m^{-3}.

Table 2. Average concentrations of nonrefractory PM$_1$ (NR-PM$_1$) components for the sampling periods.

	Ammonium (µg m^{-3})	Sulfate (µg m^{-3})	Organics (µg m^{-3})	Nitrate (µg m^{-3})	Chloride (µg m^{-3})	NR-PM$_1$ (µg m^{-3})
October–November 2018	1.2	3.7	4.7	0.24	0.05	9.9
February–March 2018	0.95	2.5	4.7	0.45	0.09	8.7
April 2019	1.04	2.8	4.2	0.22	0.03	8.3
May 2019	0.6	1.6	1.6	0.09	0.02	3.8
June 2019	0.96	2.7	3.1	0.1	0.01	6.9
Average	0.96	2.7	3.7	0.23	0.04	7.7

Organics (49%) and sulfate (35%) were the two major components of NR-PM$_1$, followed by ammonium (13%) and nitrate (3%). These results are in general consistent with those from the analysis of the PM$_{2.5}$ filters considering the differences in the size ranges (PM$_{2.5}$ and PM$_1$) and also the fact that the AMS does not measure the refractory PM components. OA was highest during November and March 2019, reaching hourly values as high as 20 µg m^{-3}. Ammonium was present in sufficient concentrations to neutralize sulfate and had an average concentration of 0.96 µg m^{-3}.

The nitrate concentration was relatively low with an average of 0.23 µg m^{-3} with the highest levels during the February–March period (0.45 µg m^{-3}). Some of the differences between the PM$_{2.5}$ and the NR-PM$_1$ are due to the fact that elements like chloride often exist mainly in the 1–2.5 µm size range (in reality in the coarse mode) and are therefore part of the former and not the latter [45].

The average diurnal profiles of the OA during the sampling periods are shown in Figure 4. During May 2019, the OA concentrations are the lowest with a rather flat diurnal profile.

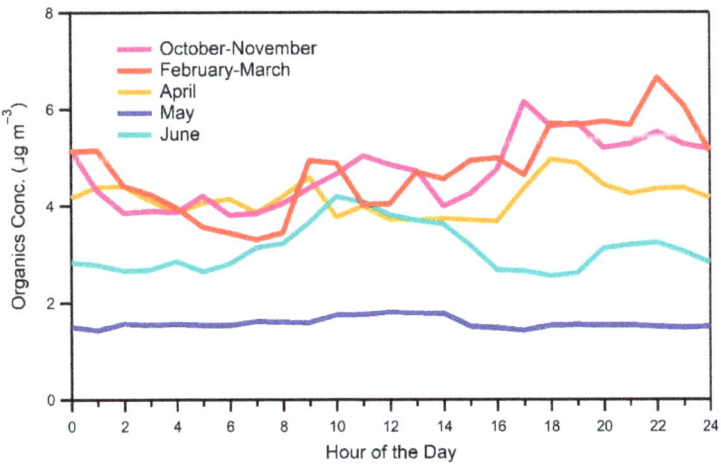

Figure 4. Average diurnal profiles of the OA during different periods of the study based on the Aerosol Mass Spectrometer (AMS) measurements.

This suggests that they are mainly the result of long-range transport with small contributions of local sources or local production. The high concentrations of OA during October–November had a peak of around 6 µg m^{-3} at 17:00 and a minimum of around 4 µg m^{-3} in the early morning hours. During February–March, the high OA concentrations were during the night (from 18:00 to midnight), reaching a maximum of around 7 µg m^{-3} at 22:00 LT. The OA during June peaked at 10:00 in the morning while in April there was a morning peak at 9:00 and an afternoon peak at 18:00. These different patterns indicate the different sources and processes affecting the OA during different periods of the year and will be discussed further in the next section using the AMS spectra.

Additional insights about the contribution of biomass burning to the OA can be gained by examining the behavior of the *m/z* 60 (levoglucosan-related marker) in the AMS spectrum. The *m/z* 60 is a tracer of biomass burning OA and its diurnal average profile for the various periods of the study is shown in Figure 5. For May and June, the *m/z* 60 values are low and the diurnal profile is flat indicating that local biomass burning was not a major source of PM$_1$. However, there is some regional biomass burning influence. In contrast, for the other periods (fall, winter, and spring) the *m/z* 60 has a notable diurnal cycle, with two peaks, one during the morning (9:00–11:00 LT) probably connected to local agricultural burning, and a second one during the evening.

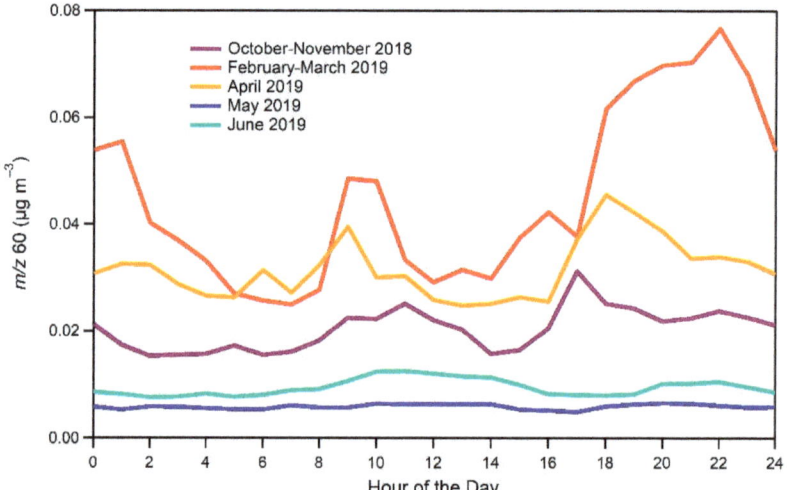

Figure 5. Diurnal variation of *m/z* 60 (related to levoglucosan), a tracer of biomass burning, in µg/m^3.

The evening peak moves from the late afternoon (16:00–18:00 LT) during the spring and fall to the evening (23:00 LT) during the winter. This is consistent with agricultural burning being a significant source during the spring and fall, while residential heating increases in importance during the winter [22]. The diurnal AMS profiles add valuable information about the location of the biomass burning activities and some information about its sources that is not available from the analysis of the daily filters.

The O:C atomic ratio of the OA provides information both about its sources but also the processes influencing its concentration and chemical composition. The OA was in general quite oxidized (O:C equal to 0.82 on average) (Figure 6), a sign of important contributions of both long-range transport (aged OA) and secondary OA formation.

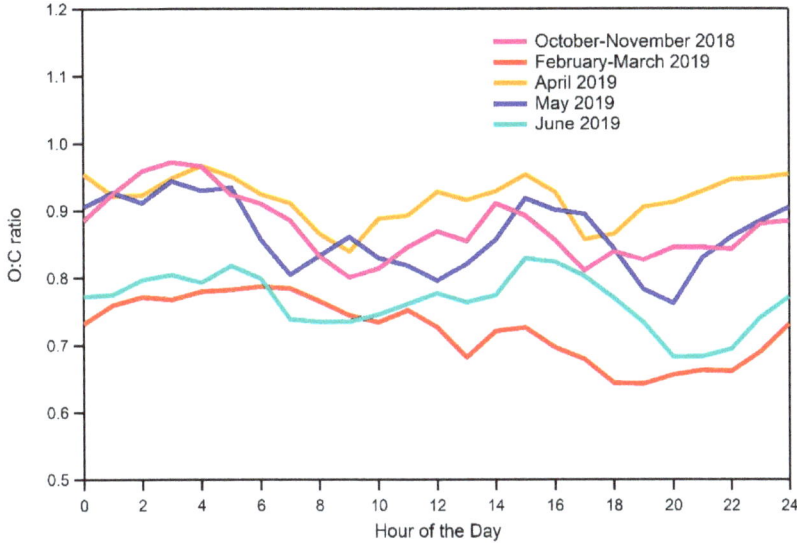

Figure 6. Diurnal variation of O:C of the organic aerosol based on the AMS measurements for the different periods of the study.

The lowest O:C values were observed for the February–March period with an average value of 0.73. The O:C peaked during the early morning (7:00 LT) at 0.79 and reached its minimum of 0.65 in the early evening (18:00). The highest average O:C of 0.91 was recorded during April. During this period, the O:C peaked during the afternoon (15:00 LT), a period associated with local photochemistry and at 4:00 LT. This second peak is more difficult to explain as it may be due to a reduction of local emissions of fresh, less oxygenated OA but also due to transport of more oxidized OA. The behavior of O:C during May and October–November was quite similar to that during April but with a little lower O:C value. Finally, the O:C during June also exhibited the two peaks (at 5:00 and 15:00 LT) but with even lower O:C values (the average was 0.77).

The density of OA was estimated using the Kuwata et al. equation [46] and the AMS O:C and H:C measurements, and was found to average 1.47 ± 0.09 g cm^{-3}.

3.5. Sources of OA

PMF was applied to the high-resolution AMS organic mass spectra (m/z up to 200) for all the sampling periods. The determination of the PMF solution was based on the comparison of the spectra of the calculated factors to those of the literature, their physical meaning, their diurnal cycles, the correlation of the factor's resulting time series with each other, and many other variables. The PMF resulted in a five-factor solution, including two oxygenated OA factors (an OOA, 30% and a marine-related OOA, 22%), a cooking OA (COA, 11%), and two biomass burning OA factors (BBOA-I, 19% and BBOA-II, 18%). The corresponding factor profiles are presented in Figure 7.

The two biomass burning OA factors contributed on average 37% to the total OA. Given that OA was approximately half of the NR-PM$_1$, biomass burning was responsible for 19% of the NR-PM$_1$ based on the analysis of the AMS data. This is quite consistent with the 17% contribution of the biomass burning to the PM$_{2.5}$ estimated from the analysis of the filter samples. This contribution was quite variable, ranging from 57% during the February–March period to 16% during June (Figure 8). The different chemical composition of the two BBOA factors was probably due to a combination of different sources and different atmospheric processing (chemical aging). The BBOA-I factor was responsible for more than 75% of the BBOA during the spring and summer, approximately half of the BBOA during February–March, and just 13% during the fall.

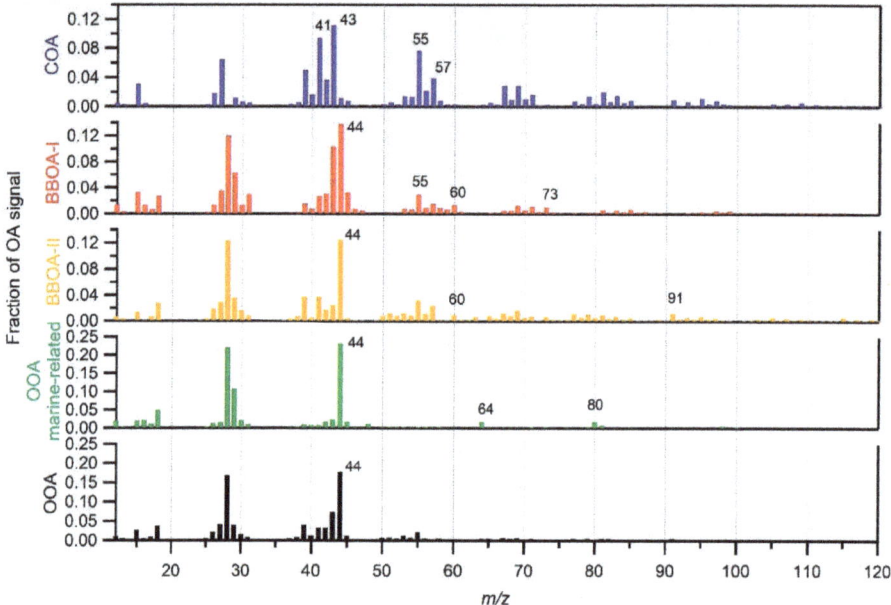

Figure 7. Source profiles of OA factors determined by PMF of the AMS organic spectra.

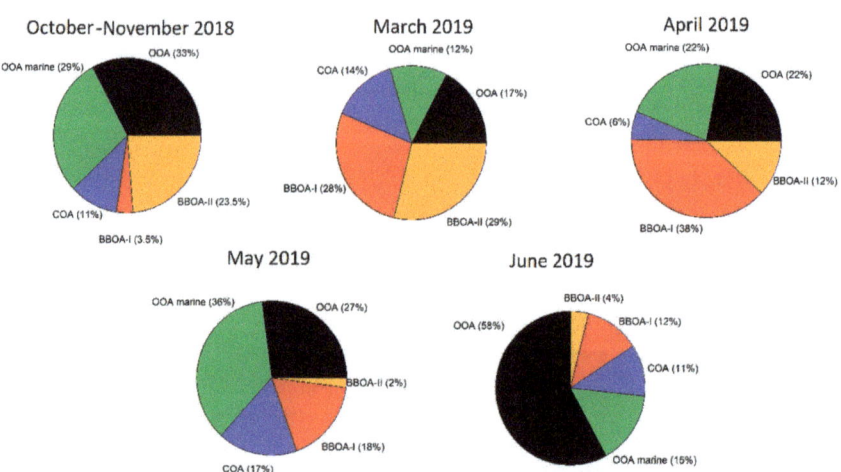

Figure 8. Average contributions of the various OA factors to total OA for the different sampling periods.

The cooking OA factor contributed approximately 10% to the average OA from 6% to 17% of the OA depending on the sampling period. The maximum fractional contribution was recorded during May (17%) and minimum (6%) during April.

Most of the OA, according to the analysis of the AMS results, was oxygenated OA representing on average approximately half (52%) of the OA in the site. This oxygenated OA was either fresh secondary OA or aged primary and secondary OA transported to the site from other areas.

The OOA dominated the OA composition during the warmer months: 73% of the OA during June, 63% during May, and 62% during October–November. It represented 44% of the OA during April and 29% in February–March. This is consistent with its photochemical source. One of the two OOA factors resolved by PMF had contributions at m/z 64 and 80 (Figure 7), associated with marine OA and more specifically methanesulfonic acid (MSA) [47]. MSA is a product of the atmospheric oxidation of dimethyl sulfide (DMS) emitted by marine phytoplankton. This marine OOA was 40–60% of the total OOA during all periods with the exception of June, when it dropped to 20% of the OOA.

The average diurnal variations of all factors during the study are depicted in Figure 9. The OOA factor is the major contributor to PM_1 with OA peaking at 12:00–14:00 LT. The second OOA factor peaks a little later at 16:00–17:00 LT. The behavior of both factors is consistent with the photochemical origin. Both BBOA-I and BBOA-II have a daytime peak (9:00–10:00 LT) and a second one at night (18:00–19:00 LT).

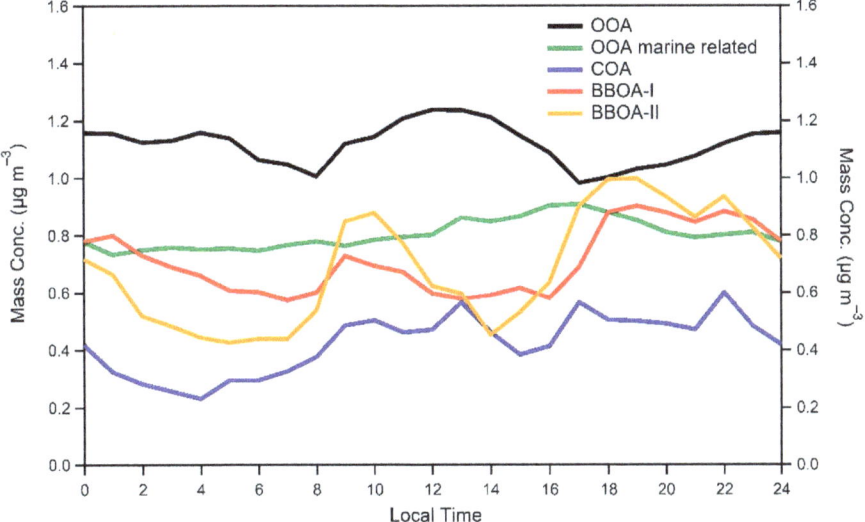

Figure 9. Diurnal variation of OA factors determined by the AMS analysis for the whole study period.

The COA factor has several small peaks throughout the day (10:00, 13:00, 17:00, and 22:00 LT). Some of them are related to Greek cooking hours, but some are not. The COA factor may also include the OA from local transportation, which was not identified by the PMF analysis of the AMS measurements. A hydrocarbon-like OA (HOA) factor was not identified even when we tried a six-factor solution. The COA factor AMS spectrum was quite similar to the others in the literature. As a consistency check, this factor reached high concentrations levels (around 8 µg m^{-3}) during noon and the early afternoon of Fat Thursday, (28 February 2019) a day during which grilling meat is practiced throughout the city.

4. Conclusions

In the current study that took place in a suburban area of Patras, Greece, we quantified the sources of fine PM and its components based on both the elemental composition and EC/OC of the $PM_{2.5}$, but also the sources of OA based on the high-resolution AMS spectra. The study covered periods during the whole year (from October 2018 to June 2019). The $PM_{2.5}$ levels were modest with an average concentration of 12.4 µg m^{-3}. The highest average $PM_{2.5}$ was recorded during October–November (14.8 µg m^{-3}) and the lowest during April (8.3 µg m^{-3}).

Organics were the dominant PM component representing approximately 50% of both the $PM_{2.5}$ and the nonrefractory PM_1. Sulfates were the second most important component representing 35% of the NR-PM$_1$ (based on the AMS) and approximately 25% of the PM$_{2.5}$ (based on the XRF analysis of the filter samples). Ammonium was 12.6% of the NR-PM$_1$ and nitrate levels were quite low, just 3% of the NR-PM$_1$. The average EC was 0.38 µg m^{-3} contributing 3% to the PM$_{2.5}$. The concentrations of 14 additional elements were quantified in the PM$_{2.5}$ samples. Several of these components were expected to be mostly in the coarse PM mode and to contribute a lot more to PM$_{2.5}$ (where the lower tail of the coarse mode is found) than to PM$_1$.

The positive matrix factorization of the PM$_{2.5}$ elemental composition together with the OC/EC data suggested six sources during the study period: secondary sulfate (34% of the PM$_{2.5}$), biomass burning (15%), nonexhaust traffic emissions (12%), exhaust traffic emissions (13%), mineral dust (10%), and sea salt (5%). The PSCF analysis suggested that the high concentration periods of secondary sulfates were related to long-range transport of from northern and northwestern Turkey and the Black Sea region, of mineral dust transported from the Sahara, and of seasalt transported from the Mediterranean to the west of the site The relatively low number of filter samples increased the uncertainty of our PMF analysis. The unaccounted PM$_{2.5}$ mass concentration was 1.4 µg m^{-3}, which corresponds to 11% of the total.

The PMF analysis of the OA AMS mass spectra supported the importance of the biomass burning source. Two biomass burning OA factors were identified, contributing 37% on average to the PM$_1$ OA. Given that OA was approximately half of the PM$_{2.5}$, this estimate is consistent with that of the PMF, based on the elemental analysis. Both approaches suggested that the contribution of biomass burning was high during many of the high PM$_{2.5}$ mass concentrations periods. Burning of agricultural waste and residential burning for heating contributed to the corresponding PM. Based on the diurnal profiles of the factors, agricultural waste burning was more prominent during the fall and spring, while residential burning increased in importance during the winter. The two factors identified may be due to different BBOA sources or a different degree of atmospheric processing (chemical aging). This should be investigated in future work.

The analysis of the AMS spectra suggested that approximately half of the OA was oxidized. Part of this OA was secondary and part was due to the oxidation of semivolatile primary OA emissions coming to the site from other areas. A part of the oxidized OA also included MSA, suggesting that it was also influenced by marine sources. This marine OOA should also be investigated in the future. Approximately 11% of the PM$_1$ OA was due to cooking. The OOA and COA factors represented on average 62% of the OA, based on the AMS and they were not directly identified by the PMF of the elemental analysis, which included most of the OA in a factor traditionally attributed to nonexhaust transportation emissions. Interestingly, the analysis of the AMS data could not identify a fresh transportation OA factor (usually called hydrocarbon-like OA or HOA). The AMS analysis has difficulties in identifying sources that contribute less than 10% of the OA, so the HOA was probably low and was included in the COA but also in the BBOA or OOA factors. One possible explanation for the apparent difference in the predictions of the two methods is that a lot of the OOA was due to transportation sources, but not from the area of the study. If it was due to transport from other areas, there should have been enough time for the production of secondary OA from volatile and intermediate volatility organic compounds, but also for the oxidation of the semivolatile components of primary OA. A second explanation is that this factor characterized as nonexhaust transportation in many studies is not mainly due to transportation in this case, but rather is mostly secondary OA with some contributions from transportation. This is an important finding of this work that clearly deserves additional attention in future studies.

This study showcases the importance of using data for both the organic and inorganic components of PM to fully assess their sources. The next step should be the combination of that information into one single dataset, which might lead to the identification of more source-specific chemical profiles that assist in identifying and quantifying PM sources with higher accuracy. Inclusion of additional information such as PM size distribution could also be helpful.

Supplementary Materials: The following are available online at http://www.mdpi.com/2073-4433/11/4/330/s1, Text S1: Uncertainty Estimation of PM$_{2.5}$ Source Apportionment, Table S1: Results of EPA PMF 5.0 uncertainty tools for source contributions to PM$_{2.5}$ mass concentration (μg m^{-3}).

Author Contributions: Conceptualization, S.N.P. and M.I.M.; methodology, M.I.M. and K.F.; validation, S.N.P., M.I.M., and K.F.; writing—original draft preparation, M.I.M. and K.F.; writing—review and editing, S.N.P.; supervision, S.N.P. All authors have read and agreed to the published version of the manuscript.

Funding: This research was financed by the European Union and Greek national funds through the program "Support for Researchers with Emphasis on Young Researchers" (call code: EDBM34), operational program "Development of Human Resources, Education and Lifelong Learning", NSRF 2014–2020, and under the research title: "Source apportionment of fine atmospheric aerosols and their organic component in urban and regional background areas in West Greece, for assisting air pollution and climate change long term management," MIS 5005098.

Conflicts of Interest: The authors declare no conflict of interest.

References

1. Samoli, E.; Stafoggia, M.; Rodopoulou, S.; Ostro, B.; Declercq, C.; Alessandrini, E.; Díaz, J.; Karanasiou, A.; Kelessis, A.G.; Le Tertre, A.; et al. Associations between fine and coarse particles and mortality in Mediterranean cities: Results from the MED-PARTICLES Project. *Environ. Health Perspect.* **2013**, *121*, 932–938. [CrossRef]
2. Samoli, E.; Stafoggia, M.; Rodopoulou, S.; Ostro, B.; Alessandrini, E.; Basagaña, X.; Díaz, J.; Faustini, A.; Gandini, M.; Karanasiou, A.; et al. Which specific causes of death are associated with short term exposure to fine and coarse particles in Southern Europe? Results from the MED-PARTICLES project. *Environ. Int.* **2014**, *67*, 54–61. [CrossRef]
3. Katsouyanni, K.; Touloumi, G.; Samoli, E.; Gryparis, A.; Le Tertre, A.; Monopolis, Y.; Rossi, G.; Zmirou, D.; Ballester, F.; Boumghar, A.; et al. Confounding and effect modification in the short-term effects of ambient particles on total mortality: Results from 29 European cities within the APHEA2 Project. *Epidemiology* **2001**, *12*, 521–531. [CrossRef]
4. Taghvaee, S.; Sowlat, M.H.; Diapouli, E.; Manousakas, M.I.; Vasilatou, V.; Eleftheriadis, K.; Sioutas, C. Source apportionment of the oxidative potential of fine ambient particulate matter (PM2.5) in Athens, Greece. *Sci. Total Environ.* **2019**, *653*, 1407–1416. [CrossRef] [PubMed]
5. Cesari, D.; Donateo, A.; Conte, M.; Contini, D. Inter-comparison of source apportionment of PM10 using PMF and CMB in three sites nearby an industrial area in central Italy. *Atmos. Res.* **2016**, *182*, 282–293. [CrossRef]
6. Belis, C.A.; Karagulian, F.; Larsen, B.R.; Hopke, P.K. Critical review and meta-analysis of ambient particulate matter source apportionment using receptor models in Europe. *Atmos. Environ.* **2013**, *69*, 94–108. [CrossRef]
7. Amato, F.; Alastuey, A.; Karanasiou, A.; Lucarelli, F.; Nava, S.; Calzolai, G.; Severi, M.; Becagli, S.; Gianelle, V.L.; Colombi, C.; et al. AIRUSE-LIFE +: A harmonized PM speciation and source apportionment in five southern European cities. *Atmos. Chem. Phys.* **2016**, *16*, 3289–3309. [CrossRef]
8. Manousakas, M.; Diapouli, E.; Papaefthymiou, H.; Kantarelou, V.; Zarkadas, C.; Kalogridis, A.-C.; Karydas, A.G.; Eleftheriadis, K. XRF characterization and source apportionment of PM10 samples collected in a coastal city. *X-ray Spectrom.* **2018**, *47*, 190–200. [CrossRef]
9. Belis, C.A.; Pernigotti, D.; Pirovano, G.; Favez, O.; Jaffrezo, J.L.; Kuenen, J.; Denier van Der Gon, H.; Reizer, M.; Riffault, V.; Alleman, L.Y.; et al. Evaluation of receptor and chemical transport models for PM10 source apportionment. *Atmos. Environ. X* **2020**, *5*, 100053. [CrossRef]
10. Gunchin, G.; Manousakas, M.; Osan, J.; Karydas, A.G.; Eleftheriadis, K.; Lodoysamba, S.; Shagjjamba, D.; Migliori, A.; Padilla-Alvarez, R.; Streli, C.; et al. Three-year long source apportionment study of airborne particles in Ulaanbaatar using X-ray fluorescence and Positive Matrix Factorization. *Aerosol Air Qual. Res.* **2019**, *5*, 1056–1067. [CrossRef]
11. Favez, O.; El Haddad, I.; Piot, C.; Borave, A.; Abidi, E.; Marchand, N.; Jaffrezo, J.L.; Besombes, J.L.; Personnaz, M.B.; Sciare, J.; et al. Inter-comparison of source apportionment models for the estimation of wood burning aerosols during wintertime in an Alpine city (Grenoble, France). *Atmos. Chem. Phys.* **2010**, *10*, 5295–5314. [CrossRef]
12. Paatero, P.; Tappert, U. Positive Matrix Factorization: A non-negative factor model with optimal utilization of error stimates of data values. *Environmetrics* **1994**, *5*, 111–126. [CrossRef]

13. Crippa, M.; Canonaco, F.; Lanz, V.A.; Äijälä, M.; Allan, J.D.; Carbone, S.; Capes, G.; Ceburnis, D.; Dall'Osto, M.; Day, D.A.; et al. Organic aerosol components derived from 25 AMS data sets across Europe using a consistent ME-2 based source apportionment approach. *Atmos. Chem. Phys.* **2014**, *14*, 6159–6176. [CrossRef]
14. Petit, J.E.; Favez, O.; Sciare, J.; Canonaco, F.; Croteau, P.; Močnik, G.; Jayne, J.; Worsnop, D.; Leoz-Garziandia, E. Submicron aerosol source apportionment of wintertime pollution in Paris, France by double positive matrix factorization (PMF2) using an aerosol chemical speciation monitor (ACSM) and a multi-wavelength Aethalometer. *Atmos. Chem. Phys.* **2014**, *14*, 13773–13787. [CrossRef]
15. Manousakas, M.; Diapouli, E.; Papaefthymiou, H.; Migliori, A.; Karydas, A.G.; Padilla-Alvarez, R.; Bogovac, M.; Kaiser, R.B.; Jaksic, M.; Bogdanovic-Radovic, I.; et al. Source apportionment by PMF on elemental concentrations obtained by PIXE analysis of PM10 samples collected at the vicinity of lignite power plants and mines in Megalopolis, Greece. *Nucl. Instrum. Methods Phys. Res. Sect. B Beam Interact. Mater. Atoms* **2015**, *349*, 114–124. [CrossRef]
16. Karanasiou, A.; Querol, X.; Alastuey, A.; Perez, N.; Pey, J.; Perrino, C.; Berti, G.; Gandini, M.; Poluzzi, V.; Ferrari, S.; et al. Particulate matter and gaseous pollutants in the Mediterranean Basin: Results from the MED-PARTICLES project. *Sci. Total Environ.* **2014**, *488*, 297–315. [CrossRef]
17. Karanasiou, A.; Mihalopoulos, N. Air quality in urban Environments in the Eastern Mediterranean. In *The Handbook of Environmental Chemistry 26*; Viana, M., Ed.; Springer: Berlin/Heidelberg, Germany, 2013; pp. 219–238.
18. Tsiflikiotou, M.A.; Kostenidou, E.; Papanastasiou, D.K.; Patoulias, D.; Zarmpas, P.; Paraskevopoulou, D.; Diapouli, E.; Kaltsonoudis, C.; Florou, K.; Bougiatioti, A.; et al. Summertime particulate matter and its composition in Greece. *Atmos. Environ.* **2019**, *213*, 597–607. [CrossRef]
19. Kostenidou, E.; Florou, K.; Kaltsonoudis, C.; Tsiflikiotou, M.; Vratolis, S.; Eleftheriadis, K.; Pandis, S.N. Sources and chemical characterization of organic aerosol during the summer in the eastern Mediterranean. *Atmos. Chem. Phys.* **2015**, *15*, 11355–11371. [CrossRef]
20. Kostenidou, E.; Kaltsonoudis, C.; Tsiflikiotou, M.; Louvaris, E.; Russell, L.M.; Pandis, S.N. Burning of olive tree branches: A major organic aerosol source in the Mediterranean. *Atmos. Chem. Phys. Discuss.* **2013**, *13*, 7223–7266. [CrossRef]
21. Pikridas, M.; Tasoglou, A.; Florou, K.; Pandis, S.N. Characterization of the origin of fine particulate matter in a medium size urban area in the Mediterranean. *Atmos. Environ.* **2013**, *80*, 264–274. [CrossRef]
22. Florou, K.; Papanastasiou, D.K.; Pikridas, M.; Kaltsonoudis, C.; Louvaris, E.; Gkatzelis, G.I.; Patoulias, D.; Mihalopoulos, N.; Pandis, S.N. The contribution of wood burning and other pollution sources to wintertime organic aerosol levels in two Greek cities. *Atmos. Chem. Phys.* **2017**, *17*, 3145–3163. [CrossRef]
23. Saffari, A.; Daher, N.; Samara, C.; Voutsa, D.; Kouras, A.; Manoli, E.; Karagkiozidou, O.; Vlachokostas, C.; Moussiopoulos, N.; Shafer, M.M.; et al. Increased biomass burning due to the economic crisis in Greece and its adverse impact on wintertime air quality in Thessaloniki. *Environ. Sci. Technol.* **2013**, *47*, 13313–13320. [CrossRef] [PubMed]
24. Canagaratna, M.R.; Jayne, J.T.; Jimenez, J.L.; Allan, J.D.; Alfarra, M.R.; Zhang, Q.; Onasch, T.B.; Drewnick, F.; Coe, H.; Middlebrook, A.; et al. Chemical and microphysical characterization of ambient aerosols with the aerodyne aerosol mass spectrometer. *Mass Spectrom. Rev.* **2007**, *26*, 185–222. [CrossRef]
25. Canagaratna, M.R.; Jimenez, J.L.; Kroll, J.H.; Chen, Q.; Kessler, S.H.; Massoli, P.; Hildebrandt Ruiz, L.; Fortner, E.; Williams, L.R.; Wilson, K.R.; et al. Elemental ratio measurements of organic compounds using aerosol mass spectrometry: Characterization, improved calibration, and implications. *Atmos. Chem. Phys.* **2015**, *15*, 253–272. [CrossRef]
26. Diapouli, E.; Manousakas, M.; Vratolis, S.; Vasilatou, V.; Maggos, T.; Saraga, D.; Grigoratos, T.; Argyropoulos, G.; Voutsa, D.; Samara, C.; et al. Evolution of air pollution source contributions over one decade, derived by PM 10 and PM 2.5 source apportionment in two metropolitan urban areas in Greece. *Atmos. Environ.* **2017**, *164*, 416–430. [CrossRef]
27. Cavalli, F.; Viana, M.; Yttri, K.E.; Genberg, J.; Putaud, J.-P. Toward a standardised thermal-optical protocol for measuring atmospheric organic and elemental carbon: The EUSAAR protocol. *Atmos. Meas. Tech.* **2010**, *3*, 79–89. [CrossRef]
28. Carslaw, D.C.; Ropkins, K. Openair—An R package for air quality data analysis. *Environ. Model. Softw.* **2012**, *27*, 52–61. [CrossRef]

29. Stein, A.F.; Draxler, R.R.; Rolph, G.D.; Stunder, B.J.B.; Cohen, M.D.; Ngan, F. Noaa's hysplit atmospheric transport and dispersion modeling system. *Bull. Am. Meteorol. Soc.* **2015**, *96*, 2059–2077. [CrossRef]
30. Rolph, G.; Stein, A.; Stunder, B. Real-time environmental applications and display system: READY. *Environ. Model. Softw.* **2017**, *95*, 210–228. [CrossRef]
31. Stohl, A. Trajectory statistics—A new method to establish source-receptor relationships of air pollutants and its application to the transport of particulate sulfate in Europe. *Atmos. Environ.* **1996**, *30*, 579–587. [CrossRef]
32. Remoundaki, E.; Kassomenos, P.; Mantas, E.; Mihalopoulos, N.; Tsezos, M. Composition and mass closure of PM2.5 in urban environment (Athens, Greece). *Aerosol Air Qual. Res.* **2013**, *13*, 72–82. [CrossRef]
33. Samara, C.; Voutsa, D.; Kouras, A.; Eleftheriadis, K.; Maggos, T.; Saraga, D.; Petrakakis, M. Organic and elemental carbon associated to PM10 and PM 2.5 at urban sites of northern Greece. *Environ. Sci. Pollut. Res. Int.* **2014**, *21*, 1769–1785. [CrossRef] [PubMed]
34. Martins, V.; Faria, T.; Diapouli, E.; Manousakas, M.I.; Eleftheriadis, K.; Viana, M.; Almeida, S.M. Relationship between indoor and outdoor size-fractionated particulate matter in urban microenvironments: Levels, chemical composition and sources. *Environ. Res.* **2020**, *183*, 109203. [CrossRef] [PubMed]
35. Viana, M.; Chi, X.; Maenhaut, W.; Querol, X.; Alastuey, A.; Mikuska, P.; Vecera, Z. Organic and elemental carbon concentrations in carbonaceous aerosols during summer and winter sampling campaigns in Barcelona, Spain. *Atmos. Environ.* **2006**, *40*, 2180–2193. [CrossRef]
36. Belis, C.A.; Larsen, B.R.; Amato, F.; Haddad, I.E.; Favez, O.; Harrison, R.M.; Hopke, P.K.; Nava, S.; Paatero, P.; Prévôt, A.; et al. *Air Pollution Source Apportionment with Receptor Models*; Europena Union: Ispra, Italy, 2014.
37. Manousakas, M.; Papaefthymiou, H.; Diapouli, E.; Migliori, A.; Karydas, A.G.; Bogdanovic-Radovic, I.; Eleftheriadis, K. Assessment of PM2.5 sources and their corresponding level of uncertainty in a coastal urban area using EPA PMF 5.0 enhanced diagnostics. *Sci. Total Environ.* **2017**, *574*, 155–164. [CrossRef] [PubMed]
38. Pateraki, S.; Manousakas, M.; Bairachtari, K.; Kantarelou, V.; Eleftheriadis, K.; Vasilakos, C.; Assimakopoulos, V.D.; Maggos, T. The traffic signature on the vertical PM profile: Environmental and health risks within an urban roadside environment. *Sci. Total Environ.* **2019**, *646*, 448–459. [CrossRef]
39. Kim, E.; Hopke, P.K. Source characterization of ambient fine particles at multiple sites in the Seattle area. *Atmos. Environ.* **2008**, *42*, 6047–6056. [CrossRef]
40. Vasilatou, V.; Manousakas, M.; Gini, M.; Diapouli, E.; Scoullos, M.; Eleftheriadis, K. Long term flux of Saharan dust to the Aegean sea around the Attica region, Greece. *Front. Mar. Sci.* **2017**, *4*, 42. [CrossRef]
41. Say, N.P. Lignite-fired thermal power plants and SO2 pollution in Turkey. *Energy Policy* **2006**, *34*, 2690–2701. [CrossRef]
42. Mitsakou, C.; Kallos, G.; Papantoniou, N.; Spyrou, C.; Solomos, S.; Astitha, M.; Housiadas, C. Saharan dust levels in Greece and received inhalation doses. *Atmos. Chem. Phys. Discuss.* **2008**, *8*, 11967–11996. [CrossRef]
43. Nava, S.; Becagli, S.; Calzolai, G.; Chiari, M.; Lucarelli, F.; Prati, P.; Traversi, R.; Udisti, R.; Valli, G.; Vecchi, R. Saharan dust impact in central Italy: An overview on three years elemental data records. *Atmos. Environ.* **2012**, *60*, 444–452. [CrossRef]
44. Rodriguez, S.; Querol, X.; Alastuey, A.; Kallos, G.; Kakaliagou, O. Saharan dust contributions to PM_{10} and TSP levels in Southern and Eastern Spain. *Atmos. Environ.* **2001**, *35*, 2433–2447. [CrossRef]
45. Eleftheriadis, K.; Ochsenkuhn, K.M.; Lymperopoulou, T.; Karanasiou, A.; Razos, P.; Ochsenkuhn-Petropoulou, M. Influence of local and regional sources on the observed spatial and temporal variability of size resolved atmospheric aerosol mass concentrations and water-soluble species in the Athens metropolitan area. *Atmos. Environ.* **2014**, *97*, 252–261. [CrossRef]
46. Kuwata, M.; Zorn, S.R.; Martin, S.T. Using elemental ratios to predict the density of organic material composed of carbon, hydrogen, and oxygen. *Environ. Sci. Technol.* **2012**, *46*, 787–794. [CrossRef]
47. Huang, S.; Wu, Z.; Poulain, L.; Van Pinxteren, M.; Merkel, M.; Assmann, D.; Herrmann, H.; Wiedensohler, A. Source apportionment of the organic aerosol over the Atlantic Ocean from 53° N to 53° S: Significant contributions from marine emissions and long-range transport. *Atmos. Chem. Phys.* **2018**, *18*, 18043–18062. [CrossRef]

© 2020 by the authors. Licensee MDPI, Basel, Switzerland. This article is an open access article distributed under the terms and conditions of the Creative Commons Attribution (CC BY) license (http://creativecommons.org/licenses/by/4.0/).

Article

Local and Remote Sources of Airborne Suspended Particulate Matter in the Antarctic Region

César Marina-Montes [1], Luis Vicente Pérez-Arribas [2], Jesús Anzano [1] and Jorge O. Cáceres [2],*

[1] Laser Lab, Chemistry & Environment Group, Department of Analytical Chemistry, Faculty of Sciences, University of Zaragoza, Pedro Cerbuna 12, 50009 Zaragoza, Spain; cesarmarinamontes@unizar.es (C.M.-M.); janzano@unizar.es (J.A.)
[2] Laser Chemistry Research Group, Department of Analytical Chemistry, Faculty of Chemistry, Complutense University of Madrid, Plaza de Ciencias 1, 28040 Madrid, Spain; lvperez@ucm.es
* Correspondence: jcaceres@ucm.es

Received: 29 March 2020; Accepted: 8 April 2020; Published: 10 April 2020

Abstract: Quantification of suspended particulate matter (SPM) measurements—together with statistical tools, polar contour maps and backward air mass trajectory analyses—were implemented to better understand the main local and remote sources of contamination in this pristine region. Field campaigns were carried out during the austral summer of 2016–2017 at the "Gabriel de Castilla" Spanish Antarctic Research Station, located on Deception Island (South Shetland Islands, Antarctic). Aerosols were deposited in an air filter through a low-volume sampler and chemically analysed using Inductively Coupled Plasma-Mass Spectrometry (ICP-MS) and Inductively Coupled Plasma-Atomic Emission Spectroscopy (ICP-AES). Elements such as Al, Ca, Fe, K, Mg, Na, P, S, Cu, Pb, Sr, Ti, Zn, Hf, Zr, V, As, Ti, Mn, Sn and Cr were identified. The statistical tools together with their correlations (Sr/Na, Al/Ti, Al/Mn, Al/Sr, Al/Pb, K/P) suggest a potentially significant role of terrestrial inputs for Al, Ti, Mn, Sr and Pb; marine environments for Sr and Na; and biological inputs for K and P. Polar contour graphical maps allowed reproducing wind maps, revealing the biological local distribution of K and P (penguin colony). Additionally, backward trajectory analysis confirmed previous affirmations and atmospheric air masses following the Antarctic circumpolar pattern.

Keywords: Antarctic region; Deception Island; atmospheric aerosols; particulate matter; statistical tools; backward trajectories; polar contour maps

---

1. Introduction

The Antarctic region is considered to be one of the most virgin, isolated and remote areas globally. The Antarctic is a polar desert and acts as a global thermostat, controlling the global climate system. Nevertheless, although this area is well distanced from continental regions, anthropogenic aerosol pollution from remote and local sources negatively affects the Antarctic environment [1]. Besides, a growing number of tourists visit Antarctica every year. Port Foster, situated in the caldera of Deception Island, is the second most-visited site in this continent, with more than 20,000 visitors in the 2016–2017 austral summer season [2].

Atmospheric aerosols act as climate drivers, since they are involved in the radiation balance of the Earth [3] and the formation of clouds [4]. However, aerosol studies regarding their origin and interactions are still limited and unclear [5]. Aerosols are produced by natural sources (sea, earth erosion, biogenic emissions, volcanoes, etc.) and anthropogenic sources (fossil fuel combustion, mining, agriculture, etc.), negatively affecting the Antarctic's air quality and ecosystems. Thus, it is important to study them and their impacts on the environment [6].

Atmospheric aerosols have a typical lifetime of one day to two weeks in the troposphere [5]. They transform their size and distribution as they evolve [7], affecting remote places such as the Antarctic [8]. Particulate Matter (PM) transport depends on the wind and meteorological conditions [9,10]. This is important, since the Antarctic region is known for being a trap for aerosols from other places [11].

Air mass back trajectories are precise tools used for investigating the source of the pollutants [12] by estimating the flow patterns of air arriving at the sample collection point. Back trajectory analyses allow for the tracking of the source of aerosols emissions. Long-distance back trajectories are influenced by physical and chemical processes (e.g., deposition and advection), while short-distance trajectories are influenced by emission source zones [13]. However, position inaccuracies of up to 20% of the travel distance can exist [14]. As much as 30% of chemical variability in the troposphere can be associated with transport [15].

Through multivariate analysis, Pérez-Arribas [16] studied a large set of air quality data, aiming to identify elemental correlations and their shared origin. Modelling relationships are important tools to study the relationships between different analytical data, allowing the reproduction of temporal and spatial variations [17,18].

Chemical PM composition has been previously studied in the Antarctic [19]. Nevertheless, despite the importance of PM origin, limited publications using air mass back trajectories exist regarding the Antarctic region [20–22].

The aim of this study is to gain better insight into the potential natural or anthropogenic sources of the measured particulate matter composition and concentration, through air mass back trajectories and polar contour maps. Furthermore, a combination of statistical and graphical approaches is proposed to establish the relationship between the elements and their concentration, in order to reveal the influence of marine environments, human-made activities and penguin colonies on the Antarctic region.

2. Materials and Methods

2.1. Site Description and Sampling

The site description, sampling procedure, and methodology used in the present work, along with the most significant experimental conditions, have been previously described elsewhere [6]. Thus, only the more relevant information to this study is presented here. Atmospheric aerosol particles were collected during the austral summer (from December 2016 to February 2017) at Deception Island (Figure 1), at the Spanish Antarctic Research base "Gabriel de Castilla" (62°58′09″ S, 60°42′33″ W). A total of 37 samples were collected in circular quartz filter paper of 47 mm diameter (Munktell) by a Derenda LVS 3.1 low volume sampler (2.3 m^3/h), and after 24 h, were placed by hand into sterile Petri dishes using sterile tweezers and nitrile globes. Mass concentration was obtained by gravimetry following the European Norm [23]. Additionally, soil samples were taken from the area of interest. After the gravimetric analysis, Hf, Zr, V, As, Ti, Mn, Cu, Sn, Zn and Pb were determined using Inductively Coupled Plasma-Mass Spectrometry (ICP-MS) and Inductively Coupled Plasma-Atomic Emission Spectroscopy (ICP-AES), as previously described in another publication [6].

2.2. Data Processing

Correlation analysis is a statistical method used to provide information about the relationship between elements that form the air PM by means of a mathematical model. Zhu [24] studied the sources of chemical constituents using multidimensional analysis. The study established correlations between the elements in aerosols. Additionally, a principal component analysis (PCA) was used as an element position simplification tool. Pérez-Arribas [16] adopted PCA to allow a group of correlated data to be transformed into a reduced coordinate system. In this study, Statgraphics Centurion 18, version 18.1.06 of 64-bits (Statpoint Technologies, Warrenton, VA, USA) was used for executing the correlation analysis and the principal component analysis (PCA).

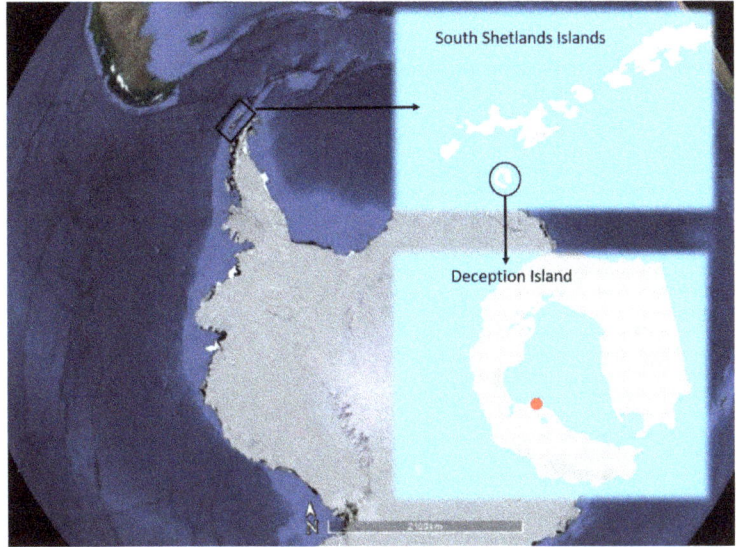

Figure 1. Map of Deception Island with the location of Gabriel de Castilla Spanish Antarctic Research Station. The red circle point indicates the position of the research base.

2.3. Polar Contour Maps

OriginPro 2017, 64-bit (OriginLab Corporation, Northampton, MA, EEUU) was used for studying data relationships through Polar contour maps. These were used as an effective tool for local data visualization. The use of wind direction and speed provided an easy and useful plot of air-pollutants sources. The wind speed is displayed on the x-axis, wind direction on the y-axis of the angle (the radius in degrees), and concentration on the z-axis.

2.4. Air-Mass Back Trajectories

Aiming to distinguish potential remote natural and anthropogenic sources at the sampling location, ten days of air-mass backward trajectories were implemented for an endpoint 600 m above ground level (AGL). This height was selected since the mountainous island has a circular form, and the highest altitude (Mount Pond) is 542 m. In this study, diverse models were used to calculate backward trajectories. Nevertheless, Cabello and Harris [25] pointed out that there are discrepancies between trajectories obtained with different models and different vertical transport methods. The trajectories were calculated with the NOAA HYSPLIT4 (Hybrid Single Particle Lagrangian Integrated Trajectory) model [12,26]. GDAS 1 degree (Global Data Assimilation System) meteorological data from the National Weather Service's National Centre for Environmental Prediction (NCEP) were chosen as the best appropriate trajectory model over those that did not cover the investigated region and had lower-resolution. GDAS has been widely used to perform backward air trajectories in locations other than the Antarctic region [27–30].

3. Results and Discussion

3.1. Statistical Description of the Data Normality Tests

Particulate matter results show an average of 10 ± 4 µg/m^3. This value is similar to those found in low population density areas and places such as the Southern Ocean (13.4 µg/m^3; [31]). However, this result differs from other values found on the Antarctic coast with an average of 1.5 µg/m^3 or 3.4 µg/m^3 [32,33]). The statistical analysis result of total particulate matter (PM10) is shown in Table 1.

The results show a great variability between the sampling days, ranging from 2.9 to 28.2 µg/m^3 with a variation coefficient of 76.4%. The variability is explained by the singular shape of Deception Island (mountainous horseshoe). This peculiar shape keeps aerosols in the air for an extended period of time.

Table 1. Particulate matter (PM) statistical summary together with central tendency, variability, and distribution form.

Total Particulate Matter (µg/m^3)	
Date	PM10 (µg/m^3)
28 December 2016	6.1
30 December 2016	28.2
01 January 2017	21.7
07 January 2017	8.1
21 January 2017	12.2
22 January 2017	2.9
23 January 2017	9.4
01 February 2017	4.3
08 February 2017	3.8
14 February 2017	2.9
17 February 2017	4.7
23 February 2017	12.1
24 February 2017	19.5
25 February 2017	7.4
26 February 2017	6.0
PM10 Statistical Overview	
Number of samples	15
Average	9.95
Standard deviation	7.60
Coefficient of variation	76.4%
Minimum	2.9
Maximum	28.2
Range	25.3
Standard Skewness	2.1299
Standard Kurtosis	0.8392

Both standard skewness and standard kurtosis can be used to determine if the sample has a normal (Gaussian) or different statistical distribution. In general, statistical values outside the range of −2 to + 2 indicate significant deviations from normality, which would tend to invalidate many of the statistical procedures that are usually applied to these data.

It can be seen in Table 1 and Figure 2 that the distribution of PM10 data indicates a certain asymmetry, although not well above the critical value of two. Positive asymmetry is usually an indication of a logarithmic-normal or lognormal type of distribution of data. To verify this possibility, the Kolmogorov–Smirnov test was carried out, with the P-value estimated for a possible lognormal distribution of 0.9983 and 0.6100 estimated for the normal one. Seeing as the two probability values calculated are above the level of usual significance ($\alpha = 0.05$), both statistical distributions are statistically valid. However, as can be seen in Figure 2, the lognormal distribution is better suited to the PM10 data obtained. Figure 2 shows a representation of the distribution of number of days (frequency) and PM concentration (µg/m³).

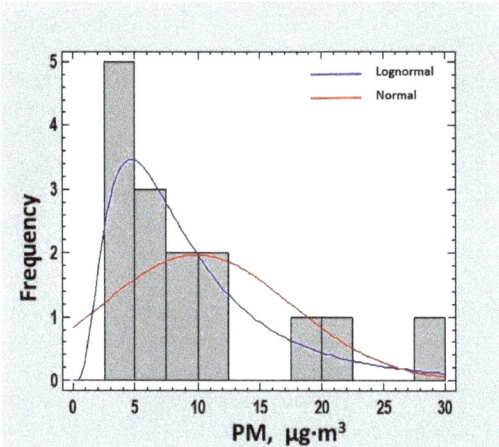

Figure 2. Histogram of normal and lognormal statistical distribution for the total particulate material obtained during the 2016–2017 field campaign. Frequency indicates the number of days.

Once the general PM statistical behaviour was studied, the same procedure was applied to the different pollutants found in PM10. Specific elements found in the aerosol also showed great variability between different sampling days. However, this variability was not homogeneous for all of them, as shown in Table S1 (Supplementary Materials).

Most of the elements have variation coefficients between 100% and 50%, similar to the PM10 variation coefficient of 76.4%. However, some elements show variation coefficients which are very different and higher than this value. Airborne PM is often affected by different climatic conditions (wind direction, speed, humidity, precipitation, etc.). This implies that, generally, the behaviour of PM composition follows the same trends, but this does not occur for some elements such as; P, K and Sn, with coefficient of variation (C.V.) of 249%, 169.9% and 124.5%, respectively; and for Cu and Fe, with C.V. of 112% and 102%, respectively. This means that these elements may be related to emission sources relatively close to the collection place and possibly very localized, probably due to very intense emission sources. Generally, this situation occurs when the distribution changes from normal or Gaussian statistical behavior, to strong positive asymmetry (lognormal distribution). In the case of K and P, its Kolmogorov–Smirnov test is shown in Table 2 compared to their Gaussian distribution.

Table 2. P values obtained for P, K and Cu in PM10 after Kolmogorov–Smirnov test.

	Lognormal	Normal
Phosphorus	0.806278	0.0123709
Potassium	0.735451	0.0267816
Copper	0.288687	0.315736

In both cases, the P-value for a normal distribution is <0.05, indicating that the contents of K and P in the particulate material are not distributed over the sampling days according to a Gaussian model, and that the lognormal distribution model is much more suitable to describe the behaviour of these major elements. Regarding the trace elements found in the Antarctic aerosol, Cu shows marked positive asymmetry, although in this case, its distribution over the days of sampling can be explained both by a normal model and by a lognormal model, since the fit tests show P-value > 0.05 on both models.

3.2. Correlation Analysis

Aiming to explore the chemical/environmental associations among the elements found in the PM10, correlation analysis has been carried out. A statistically significant correlation between elements indicates a common origin of both, so if the source of one of them is identified, it can be presumed that the other comes from the same source of emission. Statistically significant correlations were found between the major constituents in the PM10 [6] and in the case of the elements present at trace levels, 6 significant correlations were found between them [21]. However, it is unknown if some trace elements are correlated to any of the major components. Table S2 (Supplementary Materials) shows the analysis of correlations corresponding to the elements found in the aerosol during the campaign. Those elements that show significant correlation (P-value < 0.05) are shown in shaded cells.

As Table S2 (Supplementary Materials) shows, there is a very significant correlation between Sr and Na (Pearson = 0.9873 and P-value < 0.0001). Since the main source of Na in the particulate material is marine aerosol, this implies that Sr also has a marine origin. The remaining trace elements do not show a significant correlation with Na, so the marine origin of these elements is discarded. Moreover, Al is a characteristic element of the earth's crust. Al and Ti have a very significant, high correlation (Pearson 0.9430; P-value < 0.0001). Therefore, Ti has a crustal origin as well. Along with its marine origin, Sr has a correlation with Al (Pearson 0.5979; P-value = 0.0240). This indicates that some Sr has a terrestrial input. Other trace elements present in the aerosol with possible terrestrial origin are Mn (Pearson 0.8635; P-value = 0.001) and Pb (Pearson 0.7557; P-value 0.0028).

3.3. Multivariate Analysis

Previous correlation analysis has described that there are significant correlations (P-value < 0.05) among some of the elements found in the PM10. Subsequently, PCA has been carried out, aiming to check if these relationships occur globally. Figure 3 shows the scatter diagrams corresponding to the sampling days. Figure 3a presents the scatter plot when all elements found have been computed as variables; in Figure 3b, the dispersion diagram corresponds to the majority of the elements and Figure 3c corresponds to the components that are at trace levels. Given that the elements were found in different concentration levels, the PCA has been carried out after scaling the data by standard deviation (autoscaling). The green circle indicates that there are two values (samples) close to each other, and therefore there is some similarity between them. However, at the same time, these values are different from the rest. The green and red arrows indicate the same, values that are out of the guideline or general behavior.

As can be seen in Figure 3, the distribution of the scores are quite similar. Most of the points are located at the bottom of the diagram and close to the origin of coordinates. This means that the different pollutants maintain a similar pattern. However, there are exceptions. Figure 3a shows that the sample taken on 1 February 2017 is separated from the rest of the points. This indicates that there is a different pattern compared to the rest of the samples. Similarly, this occurs in Figure 3b,c.

The different patterns show the occurrence of K and P (major elements); and V and as (trace elements). These elements are significantly correlated with each other (see correlation Table S2 in Supplementary Materials). The maximum values of these elements were measured on 1 February 2017, while the rest of the elements gave values close to or below the average. On the other hand, these elements do not show a significant correlation with either Na or Al, so their marine or crustal origin is discarded.

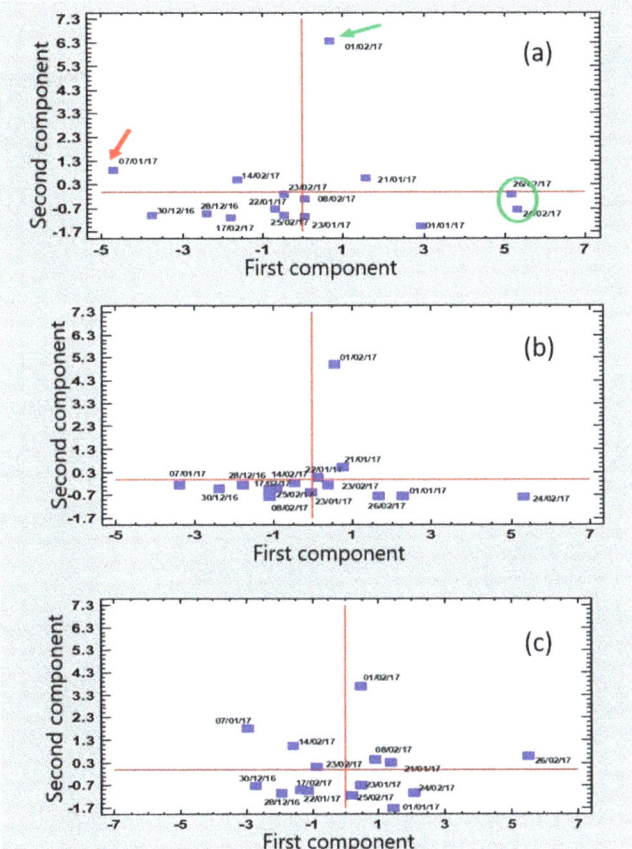

Figure 3. Scatterplot with scores from statistical analysis of the different sampling days by principal component analysis (PCA). (**a**) represents the whole data set computed as variables. (**b**) represents the data of the majority elements. (**c**) represents the data of the trace elements. The green circle, and red and green arrows indicate values that are out of the general behavior.

It can be observed in Figure 3a that for two days (24 February 2017 and 26 February 2017) the elemental composition differs, in relative terms, from the rest of the days. Even though these days appeared close on the distribution diagram, they have different characteristics. One of them corresponds to the presence of majority elements and the other to trace elements. The highest concentration levels were recorded on 24 February 2017 (Figure 3b). The majority elements concentration was recorded to be above the average, except for K and P, on 26 February 2017 (Figure 3c). On this day, a similar situation happened with minority elements, with the exception of V and as (Table S1, Supplementary Materials).

Finally, the point associated to PM collected on 07 January 2017 also deviates from the general pattern. However, this point should be considered anomalous or unrepresentative, since on that day, Fe, Mn, Zn, Sn and Hf were not detected. Furthermore, the rest of the elemental concentration was minimum or largely below the average. Some of these facts could be explained based on the air mass backward trajectories associated with the days prior to the sampling or the sampling meteorological conditions (Figure 4). Generally, air mass backward trajectory analysis showed that PM10 moved following the Antarctic Circumpolar Pattern through the Southern and Pacific Oceans, while for some of the cases air masses originated close to the Antarctic Peninsula at the Weddell and Bellingshausen Seas.

- Air mass backward trajectories associated to 01 February 2017 (Figure 4) show previous winds to this day with origin on the Weddell Sea, passing through the Antarctic continent and the Bellingshausen Sea. These areas are mostly covered by ice and snow. This explains the low concentration levels of crustal or marine elements.
- Backward trajectories associated to 24 February 2017 and 26 February 2017 (Figure 4) show previous days were similar, although on the 26th wind travelled through areas farther north than on the 24th. In both cases, the wind route passed mainly through ice-free areas, so these days prevail the presence of Na and the elements correlated with it, such as Mg. In addition, on 26 February 2017 there was moderate rainfall and high relative humidity, which favoured the deposition of the aerosol.
- On 07 January 2017 (Figure 4), very low levels of all the elements were detected. This cannot be explained either by the air mass backward trajectories or by the sampling weather conditions. Therefore, it should be considered to be anomalous.

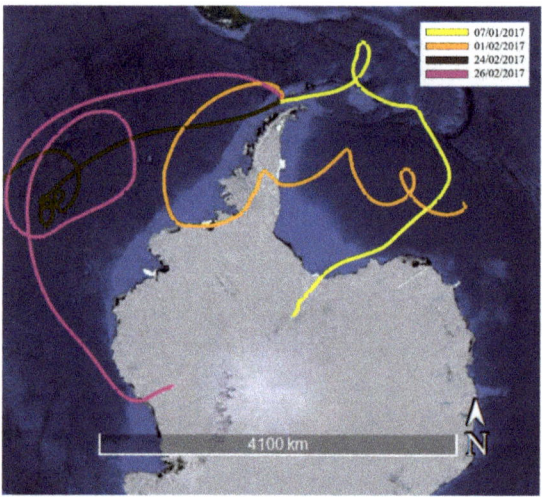

Figure 4. Total 10-day air mass backward trajectories at Gabriel de Castilla Station using GDAS1 (Global data assimilation system) dataset.

3.4. Polar Contour Maps

In the statistical study of the pollutant's distribution, it was found that K and P did not show Gaussian distribution. However, their distribution was a positive asymmetric lognormal type. This type of distribution is relatively frequent in the environmental analysis, mainly when there are high intensity emission sources, relatively close to the sampling site. These sources are usually located somewhere very localized.

Since the transport of pollutants in the air is carried out by the wind, polar contour maps were used to relate the concentration of these pollutants in the air with wind direction and speed. On the other hand, according to the correlation study mentioned above, K and P have an important and significant correlation between them (Pearson 0.9821; P-value < 0.0001). Alternatively, they do not present a significant correlation with either Na or Al. This indicates that both elements have the same non-crustal, marine origin.

If we look at the polar contour maps associated to these two elements (Figure 5), the emission focus of both maps is in the 210° direction (south-west of the sampling site). This implies a natural and biological origin of these elements, since the Punta de la Descubierta penguin colony is located in that direction. This colony is one of the largest penguin colonies in Deception Island, so the presence of these elements in the PM is related to the excrement (guano) in the area.

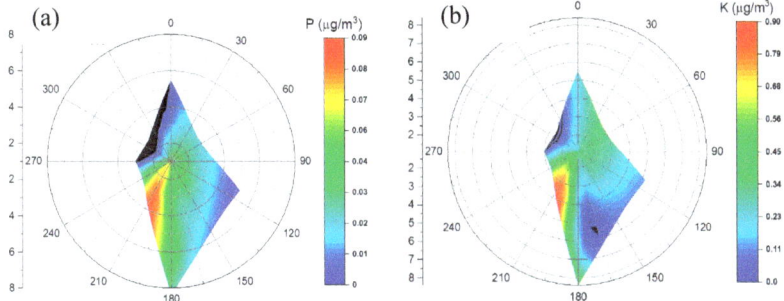

Figure 5. (a) Polar contour maps in relation to wind direction and speed for phosphorus (P) and (b) potassium (K) at the Gabriel de Castilla Spanish Research Station (Deception Island).

4. Conclusions

By using statistical methods combined with polar contour Maps and air mass back trajectories, the different sources of PM collected during the austral summer campaign in 2016–2017 (2.5-month term) at the Spanish Antarctic Research Base "Gabriel de Castilla" located on Deception Island have been determined.

Correlation analysis between major and minor elements showed the potential source of the airborne PM. Sr/Na correlation indicated the influence of marine aerosols, whereas Al/Ti correlation revealed the influence of local crustal sources. Al/Mn, Al/Sr and Al/Pb correlations indicated terrestrial sources. Furthermore, K/P correlation was revealed to have non-crustal/marine origins, since no correlation with Na or Al was found. Multivariate analysis proved the importance of establishing guidelines in the behaviour of PM10 and its composition throughout the sampling period. Score distribution showed a similar pattern for most of the elements, except for K and P. The origin of the K/P correlation was validated by PCA and polar contour maps to be biological (penguin scats from the penguin colony).

Air mass back trajectories were used to confirm the elemental source. These trajectories revealed that both crustal and marine inputs occurred following different pathways and were influenced by the Antarctic Circumpolar pattern.

This work revealed the importance of an improved understanding of the potential origin and behaviour of PM in the Antarctic, through the use of statistical tools, air mass back trajectories and polar contour maps. Consistent aerosol tracking in the Antarctic region is crucial, since atmospheric aerosols play a significant role in the Earth's climate and ecosystems. Further work will be undertaken on Deception Island as well as at other Antarctic stations.

Supplementary Materials: The following are available online at http://www.mdpi.com/2073-4433/11/4/373/s1. Table S1: Statistical summary of major (a; µg/m^3) and minor elements (b; ng/m^3). Those elements that show significant deviation from the Gaussian model (statistical values outside the range −2 to +2) are in red., Table S2: Statistics of correlations between different elements. For each element the first, second and third row correspond to Pearson correlation coefficient, sample size and P-Value, respectively. Those elements that show significant correlation (P-value <0.05) are shown in shaded cells.

Author Contributions: C.M.-M.: Writing-original draft. L.V.P.-A.: Conceptualization, Methodology, Formal analysis. J.A.: Project administration, Funding acquisition, Supervision. J.O.C.: Writing - original draft, Project administration, Funding acquisition, Supervision. C.M.-M., L.V.P.-A., J.A. and J.O.C. All authors have read and agreed to the published version of the manuscript.

Funding: This project forms part of the Ministry of Science, Innovation and Universities (Spain) proposal #CTM2017-82929R in collaboration with the Government of Aragon proposal E49_20R. Financial support from European Social Fund and University of Zaragoza is acknowledged.

Acknowledgments: The authors gratefully acknowledge the Complutense University of Madrid for facilities and material resources and the Air Resources Laboratory (ARL) for provision of the HYSPLIT trajectory model on the READY website. The authors thank the military staff at the Gabriel de Castilla Spanish research base for help with the installation of equipment and sample collection. The ICP-MS and ICP-AES data were obtained from the Institute of Environmental Assessment and Water Research, IDÆA. Trajectory figures have been taken from Google Earth Pro.

Conflicts of Interest: The authors declare no conflict of interest.

References

1. Turner, J.; Barrand, N.E.; Bracegirdle, T.J.; Convey, P.; Hodgson, D.A.; Jarvis, M.; Jenkins, A.; Marshall, G.; Meredith, M.P.; Roscoe, H.; et al. Antarctic climate change and the environment: An update. *Polar Rec.* **2014**, *50*, 237–259. [CrossRef]
2. IAATO. *Report on Iaato Operator Use of Antarctic Peninsula Landing Sites and Atcm Visitor Site Guidelines, 2016-17 Season. Ip 164*; International Association of Antarctica Tour Operators (IAATO): South Kingston, RI, USA, 2017; Available online: https://iaato.org/es/past-iaato-information-papers (accessed on 28 April 2017).
3. Nielsen, I.E.; Skov, H.; Massling, A.; Eriksson, A.C.; Dall'Osto, M.; Junninen, H.; Sarnela, N.; Lange, R.; Collier, S.; Zhang, Q.; et al. Biogenic and anthropogenic sources of aerosols at the High Arctic site Villum Research Station. *Atmos. Chem. Phys.* **2019**, *19*, 10239–10256. [CrossRef]
4. Tomasi, C.; Lanconelli, C.; Mazzola, M.; Lupi, A. Aerosol and Climate Change: Direct and Indirect Aerosol Effects on Climate. In *Atmospheric Aerosols*; Wiley-VCH Verlag GmbH & Co. KGaA: Weinheim, Germany, 2016; pp. 437–551.
5. IPCC. *Climate Change 2013: The Physical Science Basis. Contribution of Working Group i to the Fifth Assessment Report of the Intergovernmental Panel on Climate Change*; Cambridge University Press: Cambridge, UK; New York, NY, USA, 2013.
6. Cáceres, J.O.; Sanz-Mangas, D.; Manzoor, S.; Pérez-Arribas, L.V.; Anzano, J. Quantification of particulate matter, tracking the origin and relationship between elements for the environmental monitoring of the Antarctic region. *Sci. Total Environ.* **2019**, *665*, 125–132. [CrossRef] [PubMed]
7. Mahowald, N.; Albani, S.; Kok, J.F.; Engelstaeder, S.; Scanza, R.; Ward, D.S.; Flanner, M.G. The size distribution of desert dust aerosols and its impact on the Earth system. *Aeolian Res.* **2014**, *15*, 53–71. [CrossRef]
8. Hu, Q.-H.; Xie, Z.-Q.; Wang, X.-M.; Kang, H.; Zhang, P. Levoglucosan indicates high levels of biomass burning aerosols over oceans from the Arctic to Antarctic. *Sci. Rep.* **2013**, *3*, 3119. [CrossRef]
9. Radojevic, M.; Bashkin, V.N. *Practical Environmental Analysis*; RSC: Cambridge, UK, 2006.
10. Reeve, R.N. *Introduction to Environmental Analysis*; John Wiley & Sons: Chichester, UK, 2002.
11. Bargagli, R. Atmospheric chemistry of mercury in Antarctica and the role of cryptogams to assess deposition patterns in coastal ice-free areas. *Chemosphere* **2016**, *163*, 202–208. [CrossRef]
12. Stein, A.F.; Draxler, R.R.; Rolph, G.D.; Stunder, B.J.B.; Cohen, M.D.; Ngan, F. NOAA's HYSPLIT Atmospheric Transport and Dispersion Modeling System. *Bullet. Am. Meteorol. Soc.* **2015**, *96*, 2059–2077. [CrossRef]
13. Fleming, Z.L.; Monks, P.S.; Manning, A.J. Review: Untangling the influence of air-mass history in interpreting observed atmospheric composition. *Atmos. Res.* **2012**, *104*, 1–39. [CrossRef]
14. Hondula, D.M.; Sitka, L.; Davis, R.E.; Knight, D.B.; Gawtry, S.D.; Deaton, M.L.; Lee, T.R.; Normile, C.P.; Stenger, P.J. A back-trajectory and air mass climatology for the Northern Shenandoah Valley, USA. *Int. J. Climatol.* **2010**, *30*, 569–581. [CrossRef]
15. Moody, J.; Galusky, J.; Galloway, J. The use of atmospheric transport pattern recognition techniques in understanding variation in precipitation chemistry. In *Atmospheric Deposition (Proceedings of the Baltimore Symposium, May 1989)*; IAHS Publ. No. 179; IAHS: Wallingford, UK, 1989.
16. Pérez-Arribas, L.V.; León-González, M.E.; Rosales-Conrado, N. Learning Principal Component Analysis by Using Data from Air Quality Networks. *J. Chem. Educ.* **2017**, *94*, 458–464. [CrossRef]

17. Cristofanelli, P.; Putero, D.; Bonasoni, P.; Busetto, M.; Calzolari, F.; Camporeale, G.; Grigioni, P.; Lupi, A.; Petkov, B.; Traversi, R.; et al. Analysis of multi-year near-surface ozone observations at the WMO/GAW "Concordia" station (75°06′S, 123°20′E, 3280 m a.s.l. – Antarctica). *Atmos. Environ.* **2018**, *177*, 54–63. [CrossRef]
18. Mihalikova, M.; Kirkwood, S. Tropopause fold occurrence rates over the Antarctic station Troll (72° S, 2.5° E). *Ann. Geophys.* **2013**, *31*, 591–598. [CrossRef]
19. Osipov, E.Y.; Osipova, O.P.; Khodzher, T.V. Recent variability of atmospheric circulation patterns inferred from East Antarctica glaciochemical records. *Geochemistry* **2019**, 125554. [CrossRef]
20. Mishra, V.K.; Kim, K.-H.; Hong, S.; Lee, K. Aerosol composition and its sources at the King Sejong Station, Antarctic peninsula. *Atmos. Environ.* **2004**, *38*, 4069–4084. [CrossRef]
21. Marina-Montes, C.; Pérez-Arribas, L.V.; Escudero, M.; Anzano, J.; Cáceres, J.O. Heavy metal transport and evolution of atmospheric aerosols in the Antarctic region. *Sci. Total Environ.* **2020**, *721*, 137702. [CrossRef]
22. Danuta, S.; Sebastian, C.; Małgorzata, S.; Żaneta, P. Analysis of air mass back trajectories with present and historical volcanic activity and anthropogenic compounds to infer pollution sources in the South Shetland Islands (Antarctica). *Bull. Geogr. Phys. Geogr. Ser.* **2018**, *15*, 111–137.
23. European Committee for Standardization (CEN). *UNE EN 12341:2015. AmbientAir. Standard Gravimetric Measurement Method for the Determination of the PM10 and PM2.5 MassConcentration of Suspended Particulate Matter*; European Committee for Standardization: Brussel, Belgium, 2015.
24. Zhu, G.; Guo, Q.; Xiao, H.; Chen, T.; Yang, J. Multivariate statistical and lead isotopic analyses approach to identify heavy metal sources in topsoil from the industrial zone of Beijing Capital Iron and Steel Factory. *Environ. Sci. Pollut. Res.* **2017**, *24*, 14877–14888. [CrossRef]
25. Cabello, M.; Orza, J.A.G.; Galiano, V.; Ruiz, G. Influence of meteorological input data on backtrajectory cluster analysis – a seven-year study for southeastern Spain. *Adv. Sci. Res.* **2008**, *2*, 65–70. [CrossRef]
26. Rolph, G.; Stein, A.; Stunder, B. Real-time Environmental Applications and Display sYstem: READY. *Environ. Model. Softw.* **2017**, *95*, 210–228. [CrossRef]
27. Jorba, O.; Pérez, C.; Rocadenbosch, F.; Baldasano, J. Cluster Analysis of 4-Day Back Trajectories Arriving in the Barcelona Area, Spain, from 1997 to 2002. *J. Appl. Meteorol.* **2004**, *43*, 887–901. [CrossRef]
28. Moore, B.J.; Neiman, P.J.; Ralph, F.M.; Barthold, F.E. Physical Processes Associated with Heavy Flooding Rainfall in Nashville, Tennessee, and Vicinity during 1–2 May 2010: The Role of an Atmospheric River and Mesoscale Convective Systems. *Mon. Weather Rev.* **2012**, *140*, 358–378. [CrossRef]
29. Su, L.; Yuan, Z.; Fung, J.C.H.; Lau, A.K.H. A comparison of HYSPLIT backward trajectories generated from two GDAS datasets. *Sci. Total Environ.* **2015**, *506–507*, 527–537. [CrossRef] [PubMed]
30. Hong, S.-b.; Yoon, Y.J.; Becagli, S.; Gim, Y.; Chambers, S.D.; Park, K.-T.; Park, S.-J.; Traversi, R.; Severi, M.; Vitale, V.; et al. Seasonality of aerosol chemical composition at King Sejong Station (Antarctic Peninsula) in 2013. *Atmos. Environ.* **2019**, 117185. [CrossRef]
31. Budhavant, K.; Safai, P.D.; Rao, P.S.P. Sources and elemental composition of summer aerosols in the Larsemann Hills (Antarctica). *Environ. Sci. Pollut. Res.* **2015**, *22*, 2041–2050. [CrossRef]
32. Budhavant, K.B.; Rao, P.S.P.; Safai, P.D. Size distribution and chemical composition of summer aerosols over Southern Ocean and the Antarctic region. *J. Atmos. Chem.* **2017**, *74*, 491–503. [CrossRef]
33. Mazzera, D.; Lowenthal, D.; Chow, J.; Watson, J.; Grubišić, V. PM10 measurements at McMurdo Station, Antarctica. *Atmos. Environ.* **2001**, *35*, 1891–1902. [CrossRef]

© 2020 by the authors. Licensee MDPI, Basel, Switzerland. This article is an open access article distributed under the terms and conditions of the Creative Commons Attribution (CC BY) license (http://creativecommons.org/licenses/by/4.0/).

Article

Fine and Coarse Carbonaceous Aerosol in Houston, TX, during DISCOVER-AQ

Subin Yoon [1,2], Sascha Usenko [2,3] and Rebecca J. Sheesley [2,*]

[1] Department of Earth and Atmospheric Sciences, University of Houston, Houston, TX 77004, USA; syoon9@central.uh.edu
[2] Department of Environmental Science, Baylor University, Waco, TX 76706, USA; sascha_usenko@baylor.edu
[3] Department of Chemistry and Biochemistry, Baylor University, Waco, TX 76706, USA
* Correspondence: rebecca_sheesley@baylor.edu

Received: 14 April 2020; Accepted: 7 May 2020; Published: 9 May 2020

Abstract: To investigate major sources and trends of particulate pollution in Houston, total suspended particulate (TSP) and fine particulate matter ($PM_{2.5}$) samples were collected and analyzed. Characterization of organic (OC) and elemental (EC) carbon combined with realtime black carbon (BC) concentration provided insight into the temporal trends of $PM_{2.5}$ and coarse PM (subtraction of $PM_{2.5}$ from TSP) during the Deriving Information on Surface Conditions from Column and VERtically Resolved Observations Relevant to Air Quality (DISCOVER-AQ) Campaign in Houston in 2013. Ambient OC, EC, and BC concentrations were highest in the morning, likely due to motor vehicle exhaust emissions associated with the morning rush hour. The morning periods also had the lowest OC to EC ratios, indicative of primary combustion sources. Houston also had significant coarse EC at the downtown site, with an average (±standard deviation) $PM_{2.5}$ to TSP ratio of 0.52 ± 0.18 and an average coarse EC concentration of 0.44 ± 0.24 µg·C·m^{-3}. The coarse EC concentrations were likely associated with less efficient industrial combustion processes from industry near downtown Houston. During the last week (20–28 September, 2013), increases in OC and EC concentrations were predominantly in the fine fraction. Both $PM_{2.5}$ and TSP samples from the last week were further analyzed using radiocarbon analysis. Houston's carbonaceous aerosol was determined to be largely from contemporary sources for both size fractions; however, $PM_{2.5}$ had less impact from fossil sources. There was an increasing trend in fossil carbon during a period with the highest carbonaceous aerosol concentrations (September 24 night and 25 day) that was observed in both the $PM_{2.5}$ and TSP. Overall, this study provided insight into the sources and trends of both fine and coarse PM in a large urban U.S. city impacted by a combination of urban, industrial, and biogenic emissions sources.

Keywords: radiocarbon; carbonaceous aerosol; urban air quality; black carbon; aethalometer

1. Introduction

It is important to improve characterization of carbonaceous aerosols because they impact both human health and global climate. Carbonaceous aerosols can impact climate change directly via absorption and scattering of radiation [1], as well as indirectly due to the aerosols' ability to act as cloud condensation nuclei [2,3]. The complex interactions between carbonaceous aerosols and climate change are still being studied [4,5]. Similarly, the human health impacts of atmospheric aerosols are being investigated in urban settings. Prolonged exposure to respirable PM (PM_{10}; particulates with an aerodynamic diameter of 10 µm and smaller), and more significantly fine PM ($PM_{2.5}$; particulates with an aerodynamic diameter of 2.5 µm and smaller), has been found to cause respiratory and cardiopulmonary diseases and overall increased mortality [6–8]. In the short term, epidemiological studies have observed an increase in nonaccidental, respiratory-related hospital emergency visits on days with enhanced PM [9,10]. Thus, improving understanding of the sources and trends in urban aerosol will enable better mitigation strategies for both climate and human health.

Houston, TX, is the fourth most populous city in the U.S., with 2.3 million residents [11], and has an abundance of anthropogenic and natural emissions. The city's air quality is impacted by urban emissions (e.g., motor vehicle exhaust (MVE), cooking, residential wood burning, etc.) and industrial emissions associated with activities in and around the Houston Ship Channel (HSC) (e.g., heavy-duty diesel exhaust, ship emissions, petrochemical and refinery processes, etc.) [12–15]. Houston, like many southeastern U.S. cities, is also heavily impacted by biogenic emissions from vegetation within the city and forested regions surrounding the city (e.g., Piney Woods) (Figure 1) [16–18]. The objective of this study was to chemically characterize Houston's carbonaceous aerosol in the fine and coarse PM ($PM_{TSP-2.5}$; subtraction of carbon concentration of $PM_{2.5}$ from TSP) fractions to better understand the major sources and trends of PM in the city. For this study, filter samples and measurements were taken during the NASA-sponsored Deriving Information on Surface Conditions from Column and VERtically Resolved Observations Relevant to Air Quality (DISCOVER-AQ) campaign in September 2013 [19]. The purpose of the DISCOVER-AQ campaign was to use ground and airborne measurements to improve the efficiency of satellites in diagnosing ground-level air quality. This study focused on ground-based measurements and sampling at two sites: a primary urban site located near the downtown area, representative of the urban Houston region, while the auxiliary site located southeast of the city was near Trinity Bay and the HSC (Figure 1) [13,15]. Organic and elemental carbon (OC and EC, respectively) concentrations and radiocarbon (^{14}C) analysis of the total carbon (TC: OC + EC) were used to characterize the carbonaceous aerosols. From these measurements, week 4 of the campaign (21–28 September), was designated as the week of interest due to the increased TC concentration in both the $PM_{2.5}$ and TSP. Previous studies also observed increased $PM_{2.5}$ mass and a peak ozone event across the Houston metropolitan area during week 4 [20–22]. In addition to these off-line analyses, real-time measurement of black carbon (BC) was made during this sampling period to evaluate hourly BC trends and compare to filter-based EC measurements. Though BC and EC are both used to describe the refractory fraction of carbonaceous aerosols, BC is defined by the aerosol's optical attenuation [23], while EC is measured based on a thermal-optical approach [24]. More recent aerosol studies in Houston have focused on fine and submicron PM size fractions; however, this study provides an in-depth characterization of both fine and coarse carbonaceous aerosols in Houston in order to identify sources, trends, and relationships between these PM size fractions.

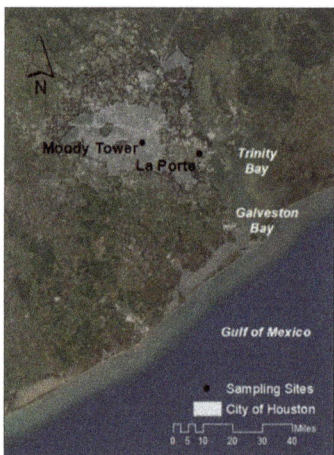

Figure 1. Map of the Moody Tower (MT) and La Porte (LP) sampling sites. Map includes outline of the city of Houston.

2. Methods and Materials

2.1. Sampling

PM filter-sample collection and BC measurement took place from 4 to 28 September and 1 to 30 September 2013, respectively. This sampling campaign was part of the larger DISCOVER-AQ campaign in Houston. The primary sampling site was on top of Moody Tower (MT; 29.7197, −95.3432), a high-rise residence hall (~70 m) located on the University of Houston's campus. This site is approximately 4.4 km southeast of downtown Houston and northwest of the HSC (Figure 1). The auxiliary sampling site, La Porte (LP; 29.6721, −95.0647), was located within a small municipal airport approximately 5.6 km west of Trinity Bay and close to the HSC (Figure 1). The LP site provides spatial comparison to the primary urban site (MT), providing insight into whether trends observed at the urban core were present in other parts of the metropolitan area. The LP site is close to Trinity Bay and closer to the Gulf of Mexico, serving as potential upwind site for marine onshore winds to MT (Figure 1). This LP site is also in close proximity to the highly industrialized region of Houston (i.e., HSC).

During the campaign, $PM_{2.5}$ samples were collected at MT, while TSP samples were collected at MT and LP. A Tisch sampler with a $PM_{2.5}$ inlet (200 L·min^{-1}, Tisch Environmental Inc. Cleves, OH, USA) and a URG medium-volume sampler with a $PM_{2.5}$ cyclone (82 L·m^{-1}, URG Corporation, Chapel Hill, NC, USA) were used for $PM_{2.5}$ sample collection at MT. Two Tisch high-volume samplers (1170 and 1130 L·m^{-1}) were alternated for TSP sample collection at MT. A Tisch high-volume sampler (1000 L·m^{-1}) was also used for TSP sample collection at LP. The $PM_{2.5}$ filter-sample collections at MT included morning (06:30 to 10:00), afternoon (10:00 to 20:00), day (06:30 to 20:00), and 24-h (06:30 to 06:00) samples (Table 1). The TSP filter-sample collections at MT included day and night (20:00 to 06:30) samples. The TSP filter-sample collection at LP included 24-h samples (Table 1). Both filter samples and blanks were collected on 90 mm and 102 mm quartz fiber filters for $PM_{2.5}$ samples, respectively, while 20 × 25 cm quartz fiber filters were used for the TSP samples. All quartz fiber filters were pre-cleaned (i.e., baked at 500 °C for 12 h), wrapped in pre-cleaned aluminum foil packages, sealed in Ziploc bags, and stored in on-sight freezers until they were brought back to Baylor University for permanent storage. A more detailed sampling description of the MT samples has been published previously [25].

The BC measurement was made at MT using a seven-channel AE42 Aethalometer (Magee Scientific, Berkeley, CA, USA). The instrument had a $PM_{2.5}$ impactor at the inlet with a flowrate of 4 L·m^{-1}. The time interval for data logging of the aethalometer was set to 5 m and was averaged for hourly BC data. For comparison to filter-based bulk carbon measurements, the hourly BC concentrations were averaged in agreement to the duration of each filter sample.

Table 1. Description of filter samples for reported filter-based analysis. MT = Moody Tower; LP = La Porte; PM = particulate matter; TSP = total suspended particulate; MV = medium volume; HV = high volume; OC = organic carbon; EC = elemental carbon.

Site	Size Fraction	Sampler Type	Samples	Sample Duration	Analysis
MT	$PM_{2.5}$	Tisch	4–28 September	morning, afternoon, day	OC EC
			21–28 September	day	^{14}C
		MV URG	6–28 September	24-h	OC EC
			23–25 September	24-h	^{14}C
LP	TSP	HV Tisch	4–28 September	morning, afternoon, day, night, 24-h	OC EC
			23–25 September	day and night	^{14}C
		HV Tisch	21–26 September	24-h	OC EC, ^{14}C

2.2. Sample and Measurement Analysis

2.2.1. Bulk Carbon Analysis

All filter-based samples were analyzed for bulk OC and EC concentrations (Table S1) with a thermal-optical-transmittance carbon analyzer (Sunset Laboratory Inc., Tigard, OR, USA) utilizing the National Institute for Occupational Safety and Health 5040 protocol [26–28]. For quality assurance purposes, sucrose spikes were run daily, and triplicate analysis was completed for every tenth sample run with a relative standard deviation of 4.6 and 1.9% for OC and EC, respectively. Each sample was blank-corrected using an average of several field blanks. $PM_{2.5}$ blank filters ($n = 7$) averaged (±standard deviation; SD) 0.83 ± 0.65 µg·C·cm^{-2}, which was an average of 14 ± 9% of the sample OC. TSP blank filters ($n = 4$) averaged 0.24 ± 0.06 µg·C·cm^{-2}, which was an average of 2 ± 1% of the sample OC. There were no EC contributions in either $PM_{2.5}$ and TSP filter blanks.

TSP samples were also analyzed for calcium carbonate (CC) contribution utilizing removal by acid fumigation [29,30] and then reanalysis of the OC and EC. In brief, a 1.5 cm^2 punch of each TSP filter sample was placed in a pre-cleaned glass petri dish and exposed to 1 N hydrochloric acid (Fisher Chemical, Hampton, NH, USA) in a desiccator for 12 h, where CC would be released. These samples were then dried at 60 °C for 1 h and analyzed for CC-free OC and EC on the carbon analyzer. Average percent contributions (±SD) of CC to OC and EC concentrations were −8 ± 7% and 13 ± 22%, respectively. This low CC contribution in the TSP was not considered significant as it was within the OC and EC uncertainty (79 and 88% of CC concentration were within the uncertainty for OC and EC, respectively). Therefore, there was no indication of a positive bias in EC due to CC contribution for these Houston samples. This is in contrast to samples collected in Beijing, where CC contributed from 22–88% of coarse EC [30]. The CC contribution in Houston was more comparable to contributions at urban-industrial sites in the southeastern U.S. cities (less than 10%) [31]. Since there was minimal CC contribution and its concentration was largely within the measure of uncertainty for the OC and EC concentration, no further discussion of CC will be included.

2.2.2. BC Corrections

The AE42 measures the light attenuation of light-absorbing aerosols deposited onto a filter at seven different wavelengths: 370, 470, 520, 590, 660, 880, and 950 nm. The 880 nm wavelength was used as the BC equivalent [32]. However, the aethalometer is biased to multiple light scattering (C) and shadowing effects (R(ATN)). Corrections were made for these biases for the absorption coefficient at 880 nm based on Schmid et al.'s [33] calculation:

$$\sigma_{aeth} = \frac{\sigma_{ATN}}{C\,R(ATN)}. \qquad (1)$$

The calculation for these corrections is further detailed in the Supplementary Materials (S1).

2.2.3. Radiocarbon Analysis of TC

Filter samples, including day and 24-h $PM_{2.5}$ samples from MT, day and night TSP samples from MT, and 24-h TSP samples from LP (Table 1, Table S2), were analyzed for ^{14}C by accelerator mass spectrometry (AMS) at the National Ocean Sciences AMS facility (Woods Hole, MA, USA). All filter preparation was completed at Baylor University. For the ^{14}C measurement, a filter area equivalent to ~60 µg of TC per sample was allocated for the analysis. The collected filter aliquots were stored in pre-cleaned glass Petri dishes. These samples and blank filters were acid fumigated using the same 1 N hydrochloric acid method for the CC protocol. The samples were then shipped to the National Ocean Sciences AMS facility where samples are compressed to graphite and analyzed for ^{14}C abundance by AMS.

The AMS measures the ratio of ^{14}C to ^{12}C for the samples, field blanks, and a modern reference standard, which is 0.95 times the specific activity of oxalic acid, the standard reference material [34]. The National Ocean Sciences AMS report their data as fraction modern (F_m), which is described in Equation (2).

$$F_m = \frac{(^{14}C/^{12}C)_{sample} - (^{14}C/^{12}C)_{blank}}{(^{14}C/^{12}C)_{AD1950} - (^{14}C/^{12}C)_{blank}}. \qquad (2)$$

As in Equation (3), the $\Delta^{14}C$ value can be calculated from the F_m value, where the λ is the inverse of the ^{14}C half-life (i.e., 5730 years). The $\Delta^{14}C$ is corrected for Y_c, and the year the sample was collected.

$$\Delta^{14}C = \left[F_m * e^{\lambda(1950 - Y_c)} - 1\right] * 1000 \qquad (3)$$

2.3. ^{14}C Source Apportionment of TC

The Δ^{14}C value of each sample can be used to apportion the TC to fossil (f_{fossil}) and contemporary carbon (f_{cont}) using Equation (4). For this calculation, a contemporary end member ($\Delta^{14}C_{cont}$) of +67.5‰ (average of +107.5‰ and +28‰ representing wood burning and annual growth, respectively [35]) and a fossil end member ($\Delta^{14}C_{fossil}$) of −1000‰ [36] was used.

$$\Delta^{14}C_{sample} = \left(\Delta^{14}C_{sample}\right)(f_{cont}) + \left(\Delta^{14}C_{fossil}\right)(1 - f_{cont}) \quad (4)$$

Uncertainty for each measurement was calculated based on the instrumental standard error, the relative difference of the F_m blank correction, and the SD between results using contemporary end member separately (+107.5‰ and +28‰) for the Δ^{14}C calculation.

3. Results and Discussion

3.1. Bulk Carbon Measurements

3.1.1. Carbonaceous Aerosols Trends of PM$_{2.5}$ at MT

Based on the bulk carbon measurement (Table S1), two distinct periods of increased TC concentrations were observed: 8–15 September (week 2; W2) and 21–28 September (week 4; W4), respectively (Figure 2a,b). Both weeks began with several days of increasing TC until reaching a peak concentration followed by a few days of declining TC concentration (Figure 2a,b). The ambient OC concentration for both weeks was statistically larger compared to the rest of the sampling period (t-test; $p > 0.05$). The temporal trends observed in the MT TC were also observed in PM$_{2.5}$ mass concentrations measured across the Houston metropolitan area [17,18,20]. The description of these two weeks will be further detailed in a future study [20].

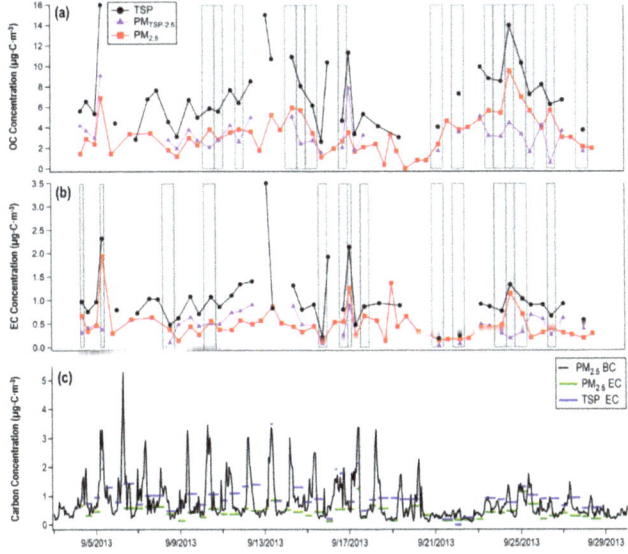

Figure 2. Carbonaceous aerosol concentrations at MT during the full sampling campaign. Figure includes filter-based (**a**) OC and (**b**) EC concentration of TSP, PM$_{2.5}$, and the calculated coarse PM. For (a), (b) The lines between each marker indicate continuous filter collection, while the gray boxes highlight samples where carbon concentrations were higher in the fine than the coarse PM. (**c**) Hourly-averaged black carbon (BC) and filter-based PM$_{2.5}$ and TSP EC concentration. The filter-based samples include morning, afternoon, day, and night samples.

The average $PM_{2.5}$ OC and EC concentrations during the campaign (morning, afternoon, day, and night) are reported in Tables 2 and 3. The highest average OC and EC concentrations (±SD) were from the morning (n = 4) with an average of 3.8 ± 2.3 and 1.3 ± 0.5 µg C·m^{-3}, respectively, while afternoon (n = 3) periods had the lowest average concentrations of OC and EC at 2.06 ± 0.72 and 0.34 ± 0.10 µg·C·m^{-3}, respectively. There was no significant difference between morning and daytime OC concentrations (Mann-Whitney test; $p > 0.05$), while the EC concentration during the morning was significantly higher than the daytime (Mann-Whitney test; $p < 0.05$). The morning sampling periods (06:00–10:00) were shorter and intended to capture the morning rush hour where emission of carbonaceous aerosol was likely increased due to the influx of MVE emission combined with the lower boundary layer; this will be discussed further in Section 3.1.2 with the hourly BC results. The OC and EC concentrations during the day and nighttime were not significantly different (t-test; $p > 0.05$).

Table 2. The average and maximum concentration of OC for the different (morning, afternoon, day, and night) $PM_{2.5}$ and TSP samples at Moody Tower. The number of samples for each sample type and the sample day for maximum OC concentrations are also included.

PM Type	Sample Type	Sample No.	Average OC (±SD) (µg·C·m^{-3})	Max. OC Sample	Max. OC Conc (µg·C·m^{-3})
$PM_{2.5}$	morning	4	3.79 ± 2.27	5 September	6.90 ± 0.71
	afternoon	3	2.06 ± 0.72	4 September	2.89 ± 0.27
	day	19	3.55 ± 1.56	25 September	6.94 ± 0.44
	night [1]	22	3.21 ± 2.25	24 September	9.51 ± 0.89
TSP	morning	5	11.64 ± 4.13	5 September	15.98 ± 1.14
	afternoon	3	6.87 ± 3.67	13 September	10.69 ± 0.78
	day	16	7.24 ± 2.12	14 September	10.92 ± 0.60
	night	16	6.07 ± 2.78	24 September	13.95 ± 0.83

[1] includes calculated night carbon concentration from PM_{TSP} (24-h) and $PM_{2.5}$ (morning and afternoon or day) samples.

Table 3. The average and maximum concentration of EC for the different (morning, afternoon, day, and night) $PM_{2.5}$ and TSP samples at Moody Tower. The number of samples for each sample type and the sample day for maximum EC concentrations are also included.

PM Type	Sample Type	Sample No.	Average EC (±SD) (µg·C·m^{-3})	Max. EC Sample	Max. EC Conc (µg·C·m^{-3})
$PM_{2.5}$	morning	4	1.30 ± 0.52	September 5	1.94 ± 0.40
	afternoon	3	0.34 ± 0.10	September 19	0.43 ± 0.14
	day	19	0.43 ± 0.18	September 13	0.86 ± 0.12
	night [1]	22	0.40 ± 0.24	September 24	1.14 ± 0.21
TSP	morning	5	2.17 ± 0.90	September 13	3.49 ± 0.46
	afternoon	3	0.69 ± 0.18	September 13	0.83 ± 0.28
	day	16	0.87 ± 0.34	September 12	1.40 ± 0.16
	night	16	0.84 ± 0.26	September 11	1.34 ± 0.20

[1] includes calculated night carbon concentration from PM_{TSP} (24-h) and $PM_{2.5}$ (morning and afternoon or day) samples.

The $PM_{2.5}$ OC to EC ratio (OC/EC) was utilized as a qualitative indicator of potential secondary organic aerosol (SOA) contribution [37–39]. As EC is a tracer for primary emissions, a higher OC/EC is indicative of enhanced secondary processes. Biomass burning (BB) sources can also contribute to increased OC/EC [38,40]; however, the emission ratio of OC and EC from BB can be highly variable [41,42]. Urban studies have utilized OC/EC for identifying BB contribution generally during the winters as SOA contribution is more prominent during the summer periods [38,43]. For this study, high OC/EC may indicate increased SOA contribution, but BB cannot be ruled out by this method. Average OC/EC during the full campaign period was 8.7 ± 5.8. The average OC/EC (±SD) of PM$_{2.5}$ for the morning, afternoon, day, and night periods were 2.7 ± 0.6, 6.4 ± 2.3, 9.3 ± 4.7, and 9.8 ± 7.6, respectively. The high nighttime OC/EC is in line with Leong et al. [18] and Bean et al.'s [17] studies, where increased aerosols concentration due to nighttime SOA formation was observed. Unlike nighttime, the morning period (06:00–10:00) was impacted by more primary emissions reflected by increased EC concentration, a lower OC/EC, followed by a near factor of two increase in OC/EC in the afternoon (10:00–20:00) when photochemically driven SOA formation would be more prevalent.

The average OC/EC (±SD) during W2 and W4 was 8.1 ± 3.4 and 14.6 ± 6.4, respectively. The OC/EC was significantly higher during W4 compared to W2 (t-test; $p < 0.05$). The W4, which included an ozone event, was influenced by high atmospheric processing. Previously published studies identified an increase in processed aerosols and highly oxygenated organic aerosols across the metropolitan area during W4 [17,18]. Southeastern U.S. cities are well known to be impacted by SOA contributions associated with high biogenic emissions in this region [44]. Even so, overall OC/EC from this study was higher than reported for other south and southeastern U.S. cities, including Dallas, TX (summer average: 5.56) [28], Atlanta, GA (3.05), and Centreville, AL (6.31) [45]. They were also much higher than annual and seasonal averages in Los Angeles, CA (annual average: 2.03) [46], and New York City, NY (summer average: 4.0) [47], respectively.

3.1.2. BC and EC Comparison

The average BC concentration (±SD) during the full sampling period was 0.80 ± 0.69 µg·C·m^{-3}. Based on hourly-averaged BC measurements for the entire campaign (Table S3), the highest BC concentrations were between 04:00 and 10:00, peaking at 07:00 (Figure S1). This captures Houston's morning rush hour [48]. The early start may also be associated with industrial activity near the HSC, including trains and other combustion sources. The morning peak in BC concentration was also present in EC results, where the morning samples had the highest EC concentration (Figure 2b,c). The BC concentration was significantly higher during W2 (average: 0.93 ± 0.21 µg·C·m^{-3}) compared to W4 (0.55 ± 0.22 µg·C·m^{-3}) (t-test; $p < 0.05$). Defined morning peaks of BC can be observed during W2 but were not present in W4 (Figure 2c). Windrose plots of the BC concentrations in W2 and W4 reveal differences in wind direction between the two weeks (Figure 3). The high BC concentrations, which occurred in the mornings during W2, all come from the East, towards the HSC. This pattern was not observed in the fine EC, where there was no significant difference between W2 (average: 0.43 ± 0.18 µg·C·m^{-3}) and W4 (average: 0.36 ± 0.25 µg·C·m^{-3}) (t-test; $p > 0.05$). Despite this, the BC was well correlated to the fine and coarse EC with a linear regression r^2 value of 0.72 and 0.74, respectively, and a slope of 1.17 with the fine EC. It is not clear why the BC deviates from fine EC with respect to higher concentrations in W2; however, the daily filter change occurred at 06:00–06:30 every morning. This small gap in the filter data record during the peak BC hours may have resulted in a low bias for W2 EC.

The observed daily trend of BC was comparable to measurements made in the spring (April to May) of 2009 during the Study of Houston Atmospheric Radical Precursors (SHARP) field campaign. BC concentrations during SHARP were also high in the early morning with a peak concentration at 07:00 [49]. Average BC and EC concentrations (±SD) during SHARP were 0.31 ± 0.22 and 0.38 ± 0.19 µg·C·m^{-3}, respectively, which is lower than the BC and EC concentrations (±SD) measured for this study in 2013 (average: 0.80 ± 0.69 and 0.48 ± 0.34 µg·C·m^{-3}, respectively). SHARP measurements were made during late spring, while this study was during late summer/fall. EC measurement in both Houston and Dallas have reported higher EC concentrations during the fall and winter periods relative to spring and summer [28], so this may just represent a seasonal difference between SHARP and DISCOVER-AQ.

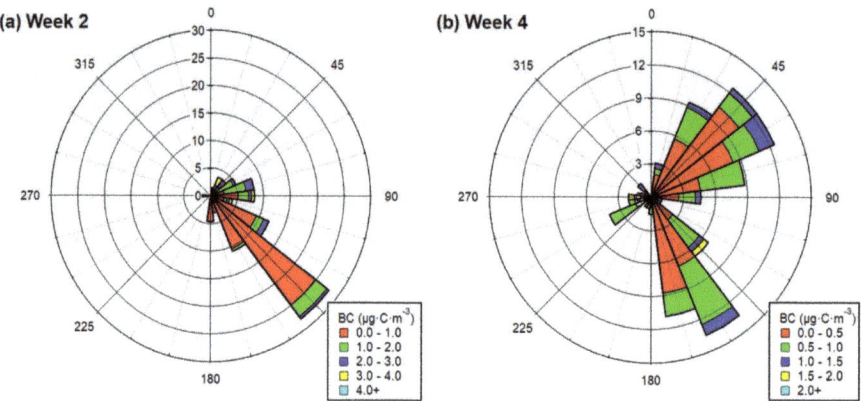

Figure 3. Windrose plots for BC concentration during (**a**) week 2 and (**b**) week 4 at MT. The radius axis indicates the percent contribution of wind direction sector relative to total week. The BC concentration is indicated by color, with different scales for Week 2 and Week 4.

3.1.3. Carbonaceous Aerosol Trends of TSP

The average TSP OC and EC concentrations (±SD) for the morning, afternoon, day, and night samples are reported in Table 2. The TSP TC concentrations during the day and night were not significantly different (t-test; $p > 0.05$); however, OC and EC concentrations during the morning were significantly higher than the night (Mann-Whitney test: $p < 0.05$). This is the same trend observed in the $PM_{2.5}$ EC and BC. The TSP OC and EC concentrations were not statistically different between the two weeks of interest (W2 and W4) for either daytime (t-test; $p > 0.05$) or nighttime (Mann-Whitney test; $p > 0.05$). During the campaign period, the largest TSP OC concentration (±SD) was on the morning of September 5 (16 ± 1.1 µg·C·m^{-3}), while the largest EC concentration was on the morning of September 13 (3.5 ± 0.46 µg·C·m^{-3}, Table 2). The highest non-morning TSP OC concentration (±SD) was during the night of September 24, with a concentration of 13.95 ± 0.83 µg·C·m^{-3} (Table 2).

The average OC/EC for TSP during the full campaign was 8.2 ± 4.6. The average OC/EC (±SD) for the morning, afternoon, day, and night TSP were 5.5 ± 0.94, 9.5 ± 3.1, 10.1 ± 6.4, and 7.3 ± 2.4, respectively. Unlike the $PM_{2.5}$, average OC/EC was higher during the day than at night, indicating a larger contribution of non-combustion aerosols during the day. Like the $PM_{2.5}$, the day-night difference was not significant (t-test; $p > 0.05$). Sources of coarse OC include resuspension of soil, as well as primary biological aerosol particles (e.g., lignan polymers), which were found to be a relatively important source of coarse aerosol in Houston [50]. The average OC/EC (± SD) for W2 and W4 were 7.5 ± 2.6 and 12 ± 6.6, respectively. Like the $PM_{2.5}$, the average OC/EC during W4 was significantly higher than the average OC/EC measured during the W2 (t-test; $p < 0.05$). The carbonaceous aerosol in the W4 was likely impacted by increased secondary processing in the $PM_{2.5}$, while the TSP also was influenced by a change in atmospheric processing and/or sources in the later week.

3.1.4. Comparison of Carbonaceous Aerosols Between $PM_{2.5}$ and TSP

Direct comparison between $PM_{2.5}$ and TSP was possible for 33 samples (three morning, two afternoon, 15 day, and 13 night samples; Figure 2a,b). Although there were few morning and afternoon samples, these were included in the comparison. The OC and EC concentrations for $PM_{2.5}$ and TSP were strongly correlated (r^2 = 0.70 and 0.72, respectively). The $PM_{2.5}$ to TSP ratio ($PM_{2.5}$/TSP) for TC, OC, and EC concentrations were each 0.52 with SDs of 0.12, 0.14, and 0.18, respectively. Average $PM_{2.5}$/TSP (±SD) for OC was slightly higher at night, 0.57 ± 0.16, compared to the day, 0.52 ± 0.09. Within the daytime, the $PM_{2.5}$/TSP for OC was higher in the afternoon, 0.46 ± 0.03, compared to the mornings, 0.33 ± 0.09. The $PM_{2.5}$/TSP for OC was also significantly higher during W2 and W4

(0.55 ± 0.13) compared to the other sampling days (0.41 ± 0.10) (t-test; $p < 0.05$). During these periods of enhanced TC concentrations (W2 and W4), the enhancement in the OC was driven by an increase in fine PM relative to the coarse PM (Figure 2a). In Figure 2a,b, the gray boxes highlight periods when the carbon concentration of the fine PM was greater than the coarse PM. In general, the W2 and W4 periods both had relatively higher OC/EC ($PM_{2.5}$) and $PM_{2.5}$/TSP. The results of these qualitative tests support that $PM_{2.5}$ OC during these periods was enhanced due to secondary processes. SOA formation is via oxidation of gas-phase precursors (e.g., volatile organic compounds (VOC)) and/or condensation of semi-volatile organic compounds. Previous studies, at varying study sites (i.e., urban, marine, and forests), have generally found these photochemically-produced aerosols in the fine and ultrafine aerosol fractions [51,52].

EC is formed from the incomplete combustion of either fossil fuel or BB sources and is typically distributed in the fine to ultrafine particulate fraction [53], but this was not the case for this study. Average $PM_{2.5}$/TSP (± SD) of EC was largest in the morning (0.70 ± 0.13) relative to afternoon (0.47 ± 0.03), day (0.50 ± 0.17) and night (0.50 ± 0.19) periods. The measured morning periods had higher concentrations of fine than coarse EC (Figure 2b). This was likely due to enhanced contribution of fine EC from MVE or activity associated with the HSC. However, the overall average $PM_{2.5}$/TSP ratio for EC was 0.52 ± 0.18, which indicates a significant contribution of EC from the coarse fraction. Concentrations of the coarse EC ranged from 0.04–0.90 $\mu g \cdot C \cdot m^{-3}$ with an average of 0.44 ± 0.24 $\mu g \cdot C \cdot m^{-3}$. Coarse EC has been measured in high concentrations in Karachi, Pakistan (2.9 $\mu g \cdot C \cdot m^{-3}$); Lahore, India (~6.3 $\mu g \cdot C \cdot m^{-3}$); and Beijing, China (2.0 $\mu g \cdot C \cdot m^{-3}$) [54]. Studies have attributed coarse EC to open field or other uncontrolled BB and/or use of less efficient/older technology for industrial combustion processes, including coke ovens, steelmaking, and transportation [31,54,55]. Average coarse EC concentration from this study (0.44± 0.24 $\mu g \cdot C \cdot m^{-3}$) was larger than coarse EC ($PM_{10-2.5}$) measured in other U.S. cities, including Atlanta, GA (urban site; 0.21 ± 0.13 $\mu g \cdot C \cdot m^{-3}$), and Centerville, AL (rural site; 0.27 ± 0.16 $\mu g \cdot C \cdot m^{-3}$). However, Houston's coarse EC was significantly less than North Birmingham, AL (urban/industrial site) with an average of 2.70 ± 3.52 $\mu g \cdot C \cdot m^{-3}$ [31]. The large concentration of coarse EC in North Birmingham was attributed to local industrial processes, including coke ovens and steel making [31]. The coarse EC in Houston may also be due to the different industrial activities that are close in proximity to the MT site, including a petroleum coke facility along with other industrial operations in the HSC, located approximately 10 km southeast of MT.

3.2. ^{14}C-based Apportionment of $PM_{2.5}$ and TSP

Fossil and Contemporary Carbon for TSP versus $PM_{2.5}$

To better understand the contribution of different sources to the TC, ^{14}C-based source apportionment was performed on $PM_{2.5}$ and TSP TC for day and night samples during W4 (Table S2). Overall, the contribution of contemporary carbon was larger than fossil carbon for both size fractions at MT and LP. Aside from the daytime $PM_{2.5}$ on September 27 at MT where contemporary contribution (±SD) was 48 ± 3%, all contemporary contribution of TC ($PM_{2.5}$ and TSP) was above 50% (Figure 4). For MT $PM_{2.5}$, the average daytime contemporary carbon contribution and concentration (±SD) during W4 was 61 ± 10% and 2.7 ± 1.1 $\mu g \cdot C \cdot m^{-3}$, respectively, while the nighttime (23–25 September) was 65 ± 5% and 4.7 ± 1.2 $\mu g \cdot C \cdot m^{-3}$, respectively. In general, daytime carbonaceous aerosols were more impacted by fossil fuel sources, while nighttime aerosols were more impacted by contemporary sources (Figure 4), which could include biogenic SOA or BB. A more detailed examination of the daytime ^{14}C is included in a future manuscript [20]. For MT TSP (day and night samples from 23–25 September), the average contemporary carbon contribution and concentration were 58 ± 5% and 6.2 ± 1.3 $\mu g \cdot C \cdot m^{-3}$, respectively. Like the diurnal trends observed in the $PM_{2.5}$, average contemporary carbon contribution and concentration was greater during the night (61 ± 4% and 6.7 ± 1.8 $\mu g \cdot C \cdot m^{-3}$, respectively) than the day (55 ± 5% and 5.75 ± 0.25 $\mu g \cdot C \cdot m^{-3}$, respectively). When considering source differences between fine and coarse aerosol, the average ± SD $PM_{2.5}$/TSP for contemporary carbon was 0.69 ± 0.09, while the fossil carbon was 0.56 ± 0.08. Considered as a percent contribution, the coarse PM had larger contribution from fossil carbon, ranging from 46 to 53%, compared to fine PM, ranging from 30 to 45% for 23–25 September (Figure 4). It is not clear if the source of this coarse fossil TC was associated with industrial activities or is linked to soil/crustal PM.

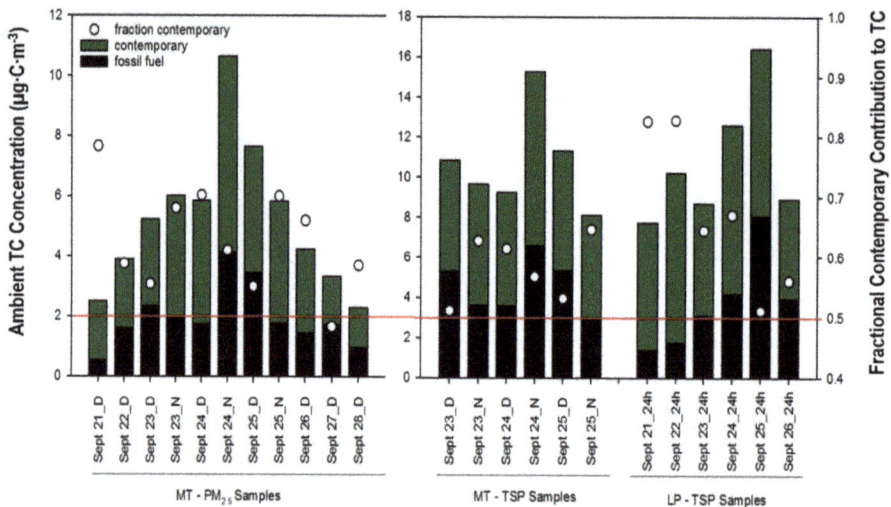

Figure 4. Radiocarbon-based apportionment of ambient total carbon concentration (left-axis) and fractional contemporary contribution (right-axis) of $PM_{2.5}$ day (D) and calculated night (N) samples and TSP D and N samples at MT. Apportioned TSP 24-h filter samples from LP are also included. Red line is a marker for the 0.5 contemporary carbon contribution of total carbon (TC).

Within W4, September 24 day to September 25 day has a different trend. The OC and EC concentrations in the fine were greater than coarse PM (Figure 2a,b), and the contemporary carbon contribution decreased. The carbonaceous aerosol during this period was likely driven by secondary processing of fossil carbon emissions specifically impacting the fine particulate fraction.

Source apportionment results from the auxiliary site, LP, reported a larger contribution of contemporary carbon contribution relative to the urban core site at MT. The average ± SD contemporary contribution and concentration of TSP at LP was 67 ± 13% and 7.1 µg·C·m^{-3}, respectively. The average ± SD fossil contribution and concentration for LP was 33 ± 13% and 3.7 ± 2.4 µg·C·m^{-3}, respectively. Though the LP site is closer to the HSC, the contemporary carbon contribution was higher than in MT (day and night) (Figure 4). However, the general trends observed during the W4 at MT for both $PM_{2.5}$ and TSP were also observed in the LP TSP (Figure 4). The largest contribution and concentration (±SD) of fossil carbon at LP was on September 25 with 49 ± 3% and 8.0 µg·C·m^{-3} (Figure 4). Both fossil carbon contribution and concentration were higher at LP than at MT (day and night) on 25 September. Previous studies have attributed the high pollution event on 25 September to the bay breeze, where a re-circulation of air mass via continental offshore winds transports it out to the Gulf and then back to Houston city (crossing over the HSC), where the resultant wind speeds across the Houston metropolitan area were low, producing stagnant conditions [22,56]. Point source emissions from Houston's HSC was another important factor for the high pollution event on 25 September [56,57]. The larger contribution and concentration of fossil carbon observed at LP than at MT on 25 September could be due to LP proximity to this industrial region of Houston.

4. Conclusions

In this study, a detailed characterization of the carbonaceous aerosols for $PM_{2.5}$ and TSP provided a better understanding of PM sources and trends in Houston. This study (1) identified important diurnal and temporal trends and (2) distinguished major sources and aerosol size fractions contributing to poor air quality days (i.e., increased PM and/or ozone levels) in Houston. Initial bulk carbon analysis identified the highest OC and EC concentrations during mornings for both $PM_{2.5}$ and TSP. Real-time BC measurements confirmed these morning peaks (i.e., 04:00–10:00). The enhanced OC, EC, and BC concentrations during the morning period were likely due to a combination of increased emissions and favorable meteorological conditions (i.e., low boundary layer). For these mornings, as expected, EC was more enhanced in the fine than the coarse PM, likely due to the incomplete combustion from MVE. The large contribution of coarse EC was also observed during this project, likely from less efficient industrial processes near the Houston MT site.

Overall, the OC/EC was relatively high, especially in the $PM_{2.5}$ fraction. The OC/EC was high during both W4 and W2. During both weeks, the OC concentration in the fine was larger than the coarse fraction, particularly during W4. The enhanced concentration of carbonaceous aerosol during W4 was driven by OC and EC in the fine fraction. The high OC/EC during this period supports enhanced contribution from secondary processes and/or biomass burning.

Further analysis of W4 was accomplished using ^{14}C analysis to distinguish the contribution and concentration of contemporary and fossil carbon. Overall, Houston aerosol was largely from contemporary sources in both $PM_{2.5}$ and TSP. However, the coarse TC had more impact from fossil sources than the fine TC, with an average ± SD of 51 ± 3% for coarse and 38 ± 7% for fine aerosol. The LP TSP, relative to MT TSP, had larger variability in contemporary and fossil carbon concentrations. Depending on meteorology and wind patterns, the LP site can be impacted by either strong, clean onshore winds or by industrial emissions. For days with poor air quality, 24 to 25 September, the carbonaceous aerosols were impacted by an increase in fossil carbon contribution at both MT and LP sites but driven by fine PM. This study has identified differences in coarse and fine sources of EC in Houston. Further study is needed to identify the sources of coarse EC and monitor potential seasonal trends.

Supplementary Materials: The following is available online at http://www.mdpi.com/2073-4433/11/5/482/s1, S1: Detailed calculation for BC correction, Figure S1: Hourly averaged ambient BC concentration, including 1 standard deviation error bars; Table S1: Bulk carbon, including OC and EC (and, respectively, uncertainty) ambient concentration for all $PM_{2.5}$ and TSP samples from MT. Each sample is identified by the sample date (YYMMDD) followed by the sampling period which includes morning (M), afternoon (A), day (D), night (N), and 24-h samples; Table S2: Contemporary and fossil carbon (and respective uncertainty) ambient concentration of TC from ^{14}C analysis and its uncertainty for all $PM_{2.5}$ and TSP samples from MT. Samples from Moody Tower (MT) and La Porte (LP) are included. Each sample is identified by the sample date (YYMMDD) followed by the sampling period which includes day (D), night (N) and 24-h samples; Table S3: Hourly-averaged ambient BC concentration during sampling period. Dates are formatted as YYMMDD.

Author Contributions: Conceptualization, S.Y. and R.J.S.; formal analysis, S.Y.; data curation, S.Y.; investigation, S.Y.; writing—original draft, S.Y.; writing—review and editing, S.U. and R.J.S.; supervision, R.J.S.; project administration, R.J.S.; funding acquisition, S.U. and R.J.S. All authors have read and agreed to the published version of the manuscript.

Funding: The preparation of this manuscript was financed through a grant from the Texas Commission on Environmental Quality (TCEQ, grant number 12-032 and 14-029), administered by the University of Texas at Austin, Center for Energy and Environmental Resources (CEER) through the Air Quality Research Program (AQRP). The contents findings, opinions, and conclusions are the work of the authors and do not necessarily represent findings, opinions, or conclusions of the TCEQ. Additional funding was provided by the C. Gus Glasscock, Jr. Endowed Fund for Excellence in Environmental Sciences.

Acknowledgments: The authors would like to acknowledge the U.S. Environmental Protection Agency, especially Rachelle Duvall and Russell Long, for providing a Tisch $PM_{2.5}$ sampler used at MT and for sample collection at LP. The authors would also like to thank James Schauer of the University of Wisconsin–Madison for providing two Tisch high volume TSP samplers used at MT and LP. In addition, the authors would like to thank Raj B. Nadkarni and Jim Thomas at Texas Commission on Environmental Quality for site access and support.

Conflicts of Interest: The authors declare no financial conflict of interest.

References

1. Chung, C.E.; Ramanathan, V.; Decremer, D. Observationally constrained estimates of carbonaceous aerosol radiative forcing. *Proc. Natl. Acad. Sci. USA* **2012**, *109*, 11624–11629. [CrossRef]
2. Spracklen, D.; Carslaw, K.; Pöschl, U.; Rap, A.; Forster, P. Global cloud condensation nuclei influenced by carbonaceous combustion aerosol. *Atmos. Chem. Phys.* **2011**, *11*, 9067–9087. [CrossRef]
3. Novakov, T.; Penner, J. Large contribution of organic aerosols to cloud-condensation-nuclei concentrations. *Nature* **1993**, *365*, 823–826. [CrossRef]
4. Kanakidou, M.; Seinfeld, J.; Pandis, S.; Barnes, I.; Dentener, F.; Facchini, M.; Dingenen, R.V.; Ervens, B.; Nenes, A.; Nielsen, C. Organic aerosol and global climate modelling: A review. *Atmos. Chem. Phys.* **2005**, *5*, 1053–1123. [CrossRef]
5. Tsigaridis, K.; Daskalakis, N.; Kanakidou, M.; Adams, P.; Artaxo, P.; Bahadur, R.; Balkanski, Y.; Bauer, S.; Bellouin, N.; Benedetti, A. The AeroCom evaluation and intercomparison of organic aerosol in global models. *Atmos. Chem. Phys.* **2014**, *14*, 10845–10895. [CrossRef]
6. Dockery, D.W.; Pope, C.A. Acute respiratory effects of particulate air pollution. *Ann. Rev. Public Health* **1994**, *15*, 107–132. [CrossRef]
7. Brook, R.D.; Rajagopalan, S.; Pope, C.A.; Brook, J.R.; Bhatnagar, A.; Diez-Roux, A.V.; Holguin, F.; Hong, Y.; Luepker, R.V.; Mittleman, M.A. Particulate matter air pollution and cardiovascular disease: An update to the scientific statement from the American Heart Association. *Circulation* **2010**, *121*, 2331–2378. [CrossRef]
8. Laden, F.; Neas, L.M.; Dockery, D.W.; Schwartz, J. Association of fine particulate matter from different sources with daily mortality in six US cities. *Environ. Health Perspect.* **2000**, *108*, 941. [CrossRef]
9. Delfino, R.J.; Murphy-Moulton, A.M.; Becklake, M.R. Emergency room visits for respiratory illnesses among the elderly in Montreal: Association with low level ozone exposure. *Environ. Res.* **1998**, *76*, 67–77. [CrossRef]
10. Chen, R.; Zhao, Z.; Kan, H. Heavy smog and hospital visits in Beijing, China. *Am. J. Respir. Crit. Care Med.* **2013**, *188*, 1170–1171. [CrossRef]
11. United States Census Bureau: QuickFacts Houston City, Texas. Available online: https://www.census.gov/quickfacts/fact/table/houstoncitytexas/PST045219 (accessed on 28 January 2020).
12. Sullivan, D.W.; Price, J.H.; Lambeth, B.; Sheedy, K.A.; Savanich, K.; Tropp, R.J. Field study and source attribution for $PM_{2.5}$ and PM_{10} with resulting reduction in concentrations in the neighborhood north of the Houston Ship Channel based on voluntary efforts. *J. Air Waste Manag. Assoc.* **2013**, *63*, 1070–1082. [CrossRef] [PubMed]
13. Wallace, H.W.; Sanchez, N.P.; Flynn, J.H.; Erickson, M.H.; Lefer, B.L.; Griffin, R.J. Source apportionment of particulate matter and trace gases near a major refinery near the Houston Ship Channel. *Atmos. Environ.* **2018**, *173*, 16–29. [CrossRef]
14. Zhang, X.; Craft, E.; Zhang, K. Characterizing spatial variability of air pollution from vehicle traffic around the Houston Ship Channel area. *Atmos. Environ.* **2017**, *161*, 167–175. [CrossRef]
15. Schulze, B.C.; Wallace, H.W.; Bui, A.T.; Flynn, J.H.; Erickson, M.H.; Alvarez, S.; Dai, Q.; Usenko, S.; Sheesley, R.J.; Griffin, R.J. The impacts of regional shipping emissions on the chemical characteristics of coastal submicron aerosols near Houston, TX. *Atmos. Chem. Phys.* **2018**, *18*, 14217–14241. [CrossRef]
16. Park, C.; Schade, G.W.; Boedeker, I. Characteristics of the flux of isoprene and its oxidation products in an urban area. *J. Geophys. Res. Atmos.* **2011**, *116*, D21303. [CrossRef]
17. Bean, J.K.; Faxon, C.B.; Leong, Y.J.; Wallace, H.W.; Cevik, B.K.; Ortiz, S.; Canagaratna, M.R.; Usenko, S.; Sheesley, R.J.; Griffin, R.J. Composition and Sources of Particulate Matter Measured near Houston, TX: Anthropogenic-Biogenic Interactions. *Atmosphere* **2016**, *7*, 73. [CrossRef]
18. Leong, Y.; Sanchez, N.; Wallace, H.; Cevik, B.K.; Hernandez, C.; Han, Y.; Flynn, J.; Massoli, P.; Floerchinger, C.; Fortner, E. Overview of surface measurements and spatial characterization of submicrometer particulate matter during the DISCOVER-AQ 2013 campaign in Houston, TX. *J. Air Waste Manag. Assoc.* **2017**, *67*, 854–872. [CrossRef]
19. DISCOVER-AQ. Available online: https://discover-aq.larc.nasa.gov/ (accessed on 1 August 2015).
20. Yoon, S.; Ortiz, S.; Clark, A.E.; Barrett, T.E.; Usenko, S.; Duvall, R.M.; Ruiz, L.H.; Bean, J.K.; Faxon, C.B.; Flynn, J.; et al. Apportioned primary and secondary organic aerosol during pollution events of DISCOVER-AQ Houston. *Atmos. Environ.* under review.

21. Mazzuca, G.M.; Ren, X.; Loughner, C.P.; Estes, M.; Crawford, J.H.; Pickering, K.E.; Weinheimer, A.J.; Dickerson, R.R. Ozone production and its sensitivity to NOx and VOCs: Results from the DISCOVER-AQ field experiment, Houston 2013. *Atmos. Chem. Phys.* **2016**, *16*, 14463–14474. [CrossRef]
22. Baier, B.C.; Brune, W.H.; Lefer, B.L.; Miller, D.O.; Martins, D.K. Direct ozone production rate measurements and their use in assessing ozone source and receptor regions for Houston in 2013. *Atmos. Environ.* **2015**, *114*, 83–91. [CrossRef]
23. Hansen, A.; Rosen, H.; Novakov, T. *Aethalometer—An Instrument for the Real-Time Measurement of Optical Absorption by Aerosol Particles*; Lawrence Berkeley Lab.: Berkeley, CA, USA, 1983.
24. Chow, J.C.; Watson, J.G.; Crow, D.; Lowenthal, D.H.; Merrifield, T. Comparison of IMPROVE and NIOSH carbon measurements. *Aerosol Sci. Technol.* **2001**, *34*, 23–34. [CrossRef]
25. Clark, A.E.; Yoon, S.; Sheesley, R.J.; Usenko, S. Pressurized liquid extraction technique for the analysis of pesticides, PCBs, PBDEs, OPEs, PAHs, alkanes, hopanes, and steranes in atmospheric particulate matter. *Chemosphere* **2015**, *137*, 33–44. [CrossRef]
26. Birch, M.; Cary, R. Elemental carbon-based method for monitoring occupational exposures to particulate diesel exhaust. *Aerosol Sci. Technol.* **1996**, *25*, 221–241. [CrossRef]
27. Schauer, J.J. Evaluation of elemental carbon as a marker for diesel particulate matter. *J. Expo. Sci. Environ. Epidemiol.* **2003**, *13*, 443–453. [CrossRef] [PubMed]
28. Barrett, T.E.; Sheesley, R.J. Urban impacts on regional carbonaceous aerosols: Case study in central Texas. *J. Air Waste Manag. Assoc.* **2014**, *64*, 917–926. [CrossRef] [PubMed]
29. Cachier, H.; Bremond, M.-P.; Buat-Menard, P. Determination of atmospheric soot carbon with a simple thermal method. *Tellus B* **1989**, *41*, 379–390. [CrossRef]
30. Tian, S.; Pan, Y.; Liu, Z.; Wen, T.; Wang, Y. Reshaping the size distribution of aerosol elemental carbon by removal of coarse mode carbonates. *Atmos. Environ.* **2019**, *214*, 116852. [CrossRef]
31. Edgerton, E.S.; Casuccio, G.S.; Saylor, R.D.; Lersch, T.L.; Hartsell, B.E.; Jansen, J.J.; Hansen, D.A. Measurements of OC and EC in coarse particulate matter in the southeastern United States. *J. Air Waste Manag. Assoc.* **2009**, *59*, 78–90. [CrossRef]
32. Snyder, D.C.; Schauer, J.J. An inter-comparison of two black carbon aerosol instruments and a semi-continuous elemental carbon instrument in the urban environment. *Aerosol Sci. Technol.* **2007**, *41*, 463–474. [CrossRef]
33. Schmid, O.; Artaxo, P.; Arnott, W.; Chand, D.; Gatti, L.V.; Frank, G.; Hoffer, A.; Schnaiter, M.; Andreae, M. Spectral light absorption by ambient aerosols influenced by biomass burning in the Amazon Basin. I: Comparison and field calibration of absorption measurement techniques. *Atmos. Chem. Phys.* **2006**, *6*, 3443–3462. [CrossRef]
34. Stuiver, M.; Polach, H.A. Discussion; reporting of C-14 data. *Radiocarbon* **1977**, *19*, 355–363. [CrossRef]
35. Zotter, P.; El-Haddad, I.; Zhang, Y.; Hayes, P.L.; Zhang, X.; Lin, Y.H.; Wacker, L.; Schnelle-Kreis, J.; Abbaszade, G.; Zimmermann, R. Diurnal cycle of fossil and nonfossil carbon using radiocarbon analyses during CalNex. *J. Geophys. Res. Atmos.* **2014**, *119*, 6818–6835. [CrossRef]
36. Gustafsson, Ö.; Kruså, M.; Zencak, Z.; Sheesley, R.J.; Granat, L.; Engström, E.; Praveen, P.; Rao, P.; Leck, C.; Rodhe, H. Brown clouds over South Asia. Biomass or fossil fuel combustion? *Science* **2009**, *323*, 495–498. [CrossRef] [PubMed]
37. Chow, J.C.; Watson, J.G.; Fujita, E.M.; Lu, Z.; Lawson, D.R.; Ashbaugh, L.L. Temporal and spatial variations of $PM_{2.5}$ and PM_{10} aerosol in the Southern California air quality study. *Atmos. Environ.* **1994**, *28*, 2061–2080. [CrossRef]
38. Benetello, F.; Squizzato, S.; Hofer, A.; Masiol, M.; Khan, M.B.; Piazzalunga, A.; Fermo, P.; Formenton, G.M.; Rampazzo, G.; Pavoni, B. Estimation of local and external contributions of biomass burning to PM 2.5 in an industrial zone included in a large urban settlement. *Environ. Sci. Pollut. Res.* **2017**, *24*, 2100–2115. [CrossRef]
39. Turpin, B.J.; Huntzicker, J.J. Identification of secondary organic aerosol episodes and quantitation of primary and secondary organic aerosol concentrations during SCAQS. *Atmos. Environ.* **1995**, *29*, 3527–3544. [CrossRef]
40. Ram, K.; Sarin, M. Day–night variability of EC, OC, WSOC and inorganic ions in urban environment of Indo-Gangetic Plain: Implications to secondary aerosol formation. *Atmos. Environ.* **2011**, *45*, 460–468. [CrossRef]
41. Zhang, Y.; Zotter, P.; Perron, N.; Prévôt, A.; Wacker, L.; Szidat, S. Fossil and non-fossil sources of different carbonaceous fractions in fine and coarse particles by radiocarbon measurement. *Radiocarbon* **2013**, *55*, 1510–1520. [CrossRef]

42. Hong, L.; Liu, G.; Zhou, L.; Li, J.; Xu, H.; Wu, D. Emission of organic carbon, elemental carbon and water-soluble ions from crop straw burning under flaming and smoldering conditions. *Particuology* **2017**, *31*, 181–190. [CrossRef]
43. Viana, M.; Maenhaut, W.; Brink, H.T.; Chi, X.; Weijers, E.; Querol, X.; Alastuey, A.; Mikuška, P.; Večeřa, Z. Comparative analysis of organic and elemental carbon concentrations in carbonaceous aerosols in three European cities. *Atmos. Environ.* **2007**, *41*, 5972–5983. [CrossRef]
44. Zeng, T.; Wang, Y. Nationwide summer peaks of OC/EC ratios in the contiguous United States. *Atmos. Environ.* **2011**, *45*, 578–586. [CrossRef]
45. Blanchard, C.L.; Hidy, G.M.; Tanenbaum, S.; Edgerton, E.; Hartsell, B.; Jansen, J. Carbon in southeastern US aerosol particles: Empirical estimates of secondary organic aerosol formation. *Atmos. Environ.* **2008**, *42*, 6710–6720. [CrossRef]
46. Kim, B.M.; Teffera, S.; Zeldin, M.D. Characterization of PM_{25} and PM_{10} in the South Coast air basin of Southern California: Part 1—Spatial variations. *J. Air Waste Manag. Assoc.* **2000**, *50*, 2034–2044. [CrossRef] [PubMed]
47. Rattigan, O.V.; Felton, H.D.; Bae, M.-S.; Schwab, J.J.; Demerjian, K.L. Multi-year hourly $PM_{2.5}$ carbon measurements in New York: Diurnal, day of week and seasonal patterns. *Atmos. Environ.* **2010**, *44*, 2043–2053. [CrossRef]
48. Czader, B.H.; Choi, Y.; Li, X.; Alvarez, S.; Lefer, B. Impact of updated traffic emissions on HONO mixing ratios simulated for urban site in Houston, Texas. *Atmos. Chem. Phys.* **2015**, *15*, 1253–1263. [CrossRef]
49. Levy, M.E.; Zhang, R.; Khalizov, A.F.; Zheng, J.; Collins, D.R.; Glen, C.R.; Wang, Y.; Yu, X.Y.; Luke, W.; Jayne, J.T. Measurements of submicron aerosols in Houston, Texas during the 2009 SHARP field campaign. *J. Geophys. Res. Atmos.* **2013**, *118*, 10518–10534. [CrossRef]
50. Shakya, K.M.; Louchouarn, P.; Griffin, R.J. Lignin-derived phenols in Houston aerosols: Implications for natural background sources. *Environ. Sci. Technol.* **2011**, *45*, 8268–8275. [CrossRef]
51. Kavouras, I.G.; Stephanou, E.G. Particle size distribution of organic primary and secondary aerosol constituents in urban, background marine, and forest atmosphere. *J. Geophys. Res. Atmos.* **2002**, *107*, AAC 7-1–AAC 7-12. [CrossRef]
52. Shiraiwa, M.; Yee, L.D.; Schilling, K.A.; Loza, C.L.; Craven, J.S.; Zuend, A.; Ziemann, P.J.; Seinfeld, J.H. Size distribution dynamics reveal particle-phase chemistry in organic aerosol formation. *Proc. Natl. Acad. Sci. USA* **2013**, *110*, 11746–11750. [CrossRef]
53. Offenberg, J.H.; Baker, J.E. Aerosol size distributions of elemental and organic carbon in urban and over-water atmospheres. *Atmos. Environ.* **2000**, *34*, 1509–1517. [CrossRef]
54. Shahid, I.; Kistler, M.; Mukhtar, A.; Ghauri, B.M.; Ramirez-Santa Cruz, C.; Bauer, H.; Puxbaum, H. Chemical characterization and mass closure of PM_{10} and $PM_{2.5}$ at an urban site in Karachi–Pakistan. *Atmos. Environ.* **2016**, *128*, 114–123. [CrossRef]
55. Lee, T.; Yu, X.-Y.; Ayres, B.; Kreidenweis, S.M.; Malm, W.C.; Collett, J.L., Jr. Observations of fine and coarse particle nitrate at several rural locations in the United States. *Atmos. Environ.* **2008**, *42*, 2720–2732. [CrossRef]
56. Caicedo, V.; Rappenglueck, B.; Cuchiara, G.; Flynn, J.; Ferrare, R.; Scarino, A.; Berkoff, T.; Senff, C.; Langford, A.; Lefer, B. Bay Breeze and Sea Breeze Circulation Impacts on the Planetary Boundary Layer and Air Quality From an Observed and Modeled DISCOVER-AQ Texas Case Study. *J. Geophys. Res. Atmos.* **2019**, *124*, 7359–7378. [CrossRef]
57. Dunker, A.M.; Koo, B.; Yarwood, G. Source apportionment of organic aerosol and ozone and the effects of emission reductions. *Atmos. Environ.* **2019**, *198*, 89–101. [CrossRef]

 © 2020 by the authors. Licensee MDPI, Basel, Switzerland. This article is an open access article distributed under the terms and conditions of the Creative Commons Attribution (CC BY) license (http://creativecommons.org/licenses/by/4.0/).

Article

Variation of the Distribution of Atmospheric *n*-Alkanes Emitted by Different Fuels' Combustion

Sofia Caumo [1,*], Roy E. Bruns [2] and Pérola C. Vasconcellos [1]

1. Instituto de Química, Universidade de São Paulo, São Paulo 05508-000, Brazil; perola@iq.usp.br
2. Instituto de Química, Universidade Estadual de Campinas, Campinas 13083-861, Brazil; bruns@iqm.unicamp.br
* Correspondence: sofia.caumo@usp.br

Received: 20 April 2020; Accepted: 28 May 2020; Published: 16 June 2020

Abstract: This study presents the emission profiles of *n*-alkanes for different vehicular sources in two Brazilian cities. Atmospheric particulate matter was collected in São Paulo (Southeast) and in Salvador (Northeast) to determine *n*-alkanes. The sites were impacted by bus emissions and heavy and light-duty vehicles. The objective of the present study is to attempt to differentiate the profile of *n*-alkane emissions for particulate matter (PM) collected at different sites. PM concentrations ranged between 73 and 488 $\mu g\,m^{-3}$, and the highest concentration corresponded to a tunnel for light and heavy duty vehicles. At sites where diesel-fueled vehicles are dominant, the *n*-alkanes show a unimodal distribution, which is different from the bimodal profile observed in the literature. Carbon preference index values corresponded to anthropogenic sources for most of the sites, as expected, but C_{max} varied comparing to literature and a source signature was difficult to observe. The main sources to air pollution were indicated by principal component analysis (PCA). For PCA, a receptor model often used as an exploratory tool to identify the major sources of air pollutant emissions, the principal factors were attributed to mixed sources and to bus emissions. Chromatograms of four specific samples showed distinct profiles of unresolved complex mixtures (UCM), indicating different contributions of contamination from petroleum or fossil fuel residues, which are unable to resolve by gas chromatography. The UCM area seemed higher in samples collected at sites with the abundance of heavy vehicles.

Keywords: atmospheric particulate matter; fossil fuel burning; *n*-alkanes; tunnel measurements; principal component analysis

1. Introduction

The Brazilian fleet corresponds to about 50 million vehicles [1] and, since 2003, a new generation of vehicles, known as flexible fuel vehicles, has been introduced in the country. Their flexible engines are adapted to use gasohol (gasoline with anhydrous ethanol) as well as pure ethanol [2]. In addition, diesel and biodiesel are also being used. The combustion of all these fuels leads to the rise of species in both gaseous and particulate phases in the atmosphere and their composition has not yet been elucidated in detail [3].

Particulate matter emitted by anthropogenic activities might cause many environmental problems, including climate effects, and also impacts human health [4]. Among the compounds emitted by fossil fuel burning, the organic fraction is predominant, and organic pollutants as *n*-alkanes are often present [5,6].

Despite the fact that aliphatic hydrocarbons do not present significant adverse effects on biological systems, as polycyclic aromatic hydrocarbons and their nitro- and oxy-derivatives, their role in atmospheric chemistry is important. Aliphatic hydrocarbons contribute to a significant portion of the total organic fraction which is present in the atmospheric particulate matter, besides containing several important markers, which can provide important information about sources that may be acting at a site [6,7]. Their homologous distribution may indicate different pollution sources [8].

Depending on the ambient temperature, aliphatic hydrocarbons in the particulate phase might volatilize and increase in concentration in the gaseous phase. The initial reaction of the aliphatic hydrocarbon in the atmosphere is generally with hydroxyl radicals, producing water and an alkyl radical, via hydrogen abstraction. The reaction rates at room temperature are in the range between 10^{-15} and 10^{-11} molecule cm^{-3} s^{-1}, increasing with growth in chain size and structural complexity of the aliphatic hydrocarbons [9,10]. Once formed, alkyl radicals are converted through reactions into alkyl peroxy (ROO·) and alkoxy (RO·) radicals which, in sequence, will act as formation precursors of ozone, alkyl nitrates, and carbonyl compounds. These compounds can influence the vapor-to-particle distribution [11]. The toxic effects of these photochemical oxidants are widely recognized in the literature [12,13]. When it is inhaled, they produce reactive oxygen species, which may induce cellular oxidative stress and consequently become the first step in the development of many diseases including respiratory diseases such as asthma and cardiovascular problems [9,12].

Studies performed in smog chambers demonstrated that the organic fraction of aerosol-containing nitro groups, produced when aliphatic and aromatic hydrocarbons are exposed to NOx and solar radiation in the atmosphere, can be potentially mutagenic to organisms [14]. Considering that, the atmospheric reactions occurring with some of the emitted compounds may significantly increase the genotoxic potential of the ambient air particles [15].

Sicre et al. [16] suggested that *n*-alkanes emitted by higher plants are found predominantly in larger particles and, on the other hand, those originating from anthropogenic sources, such as incomplete fuel combustion, lubricating oil volatilization, tire debris and road dust [17–19], are major sources of fine particles. Regarding origin and distribution, natural *n*-alkanes can be derived from biological sources such as plant wax, soil, marine bacteria, phytoplankton, and biomass burning and often present an odd carbon chain prevalence in relation to the odd/even distribution [7]. In urban areas the homologues emitted by anthropogenic sources do not present a characteristic distribution. Therefore, the observed *n*-alkane profiles can contribute to the identification of emission sources [20].

São Paulo and Salvador are big cities with approximately 12 million and 3 million inhabitants, respectively [21], and have ongoing problems concerning air pollution. Local emissions are considered as the major contributors of organic aerosols [22,23]. According to previous studies, São Paulo presented a total concentration of 36 ng m^{-3} for *n*-alkanes [22], while Salvador showed an even higher value, 62 ng m^{-3} [23].

In an attempt to differentiate the emissions from different engines, measurements were undertaken in three tunnels, in a bus station, in an urban area impacted mostly by heavy vehicles, and in a truck parking depot. In this study *n*-alkanes in aerosols were identified and quantified; the samples were collected in two different Brazilian states strongly affected by vehicular emissions in their most populated cities, São Paulo and Salvador.

2. Experiments

2.1. Sampling Site Descriptions

Sampling was conducted in two cities located in two different Brazilian states. The first one was São Paulo city (Southeastern Brazil, Figure 1) where five places were chosen: three tunnels, one road site and one truck shipping corporation site. The first tunnel, Jânio Quadros (JQT), is located in the southwest area/region and is 1900 m long. The traffic is only permitted in one direction and the movement of diesel-fueled vehicles is restricted. Thus, light vehicles are predominant at this site. JQT presented intense traffic during weekdays. The second tunnel, Maria Maluf (MMT) is located in the south area and is 800 m long, connecting with important highways. In MMT, the emission varied between gasohol, ethanol and diesel vehicles [24]. The third tunnel, Rodoanel (ROD) is 1700 m long. It is an important road for heavy-duty vehicles, especially heavy trucks transporting loads. It is a beltway with a radius of approximately 23 km from geographical center of the city downtown. Sampling at the three sites was conducted as follows: at JQT in 2001 and 2011, at MMT in 2001, at ROD in 2011.

The fourth sampling site in São Paulo is located inside the campus of the University of São Paulo (ARN) near city center, at the intersection of two roads with intense traffic of buses and cars during weekdays. The samples at ARN were collected in 2012.

The fifth sampling in São Paulo was conducted at a truck shipping company (TRA) in 2015. This site is used for mechanical repair and fuel supply for trucks. It is located in the North region of the city, close to several important highways. As road transport is the most important way of trading goods in Brazil, companies specialized in the transport of payloads predominantly use heavy trucks, which are fueled with biodiesel (B5% = blend volume of 5% of pure biodiesel and 95% of diesel) and diesel.

The second city was Salvador, capital of the Bahia State (Northeastern Brazil, Figure 1). The sampling was conducted at Lapa Bus Station (LAP), located in the downtown area (Figure 1), a region with heavy commerce and service activities, in close proximity to several office buildings, stores and a big shopping mall. Sampling was conducted at the underground level, where ventilation is very poor, thus impairing air circulation. The site is impacted by emissions from buses of a fleet of about 510 vehicles operating at an average frequency of 325 buses per hour and with a daily circulation of approximately 300,000 people. The samplings were conducted in 2010 (LAP10) and in 2013 (LAP13).

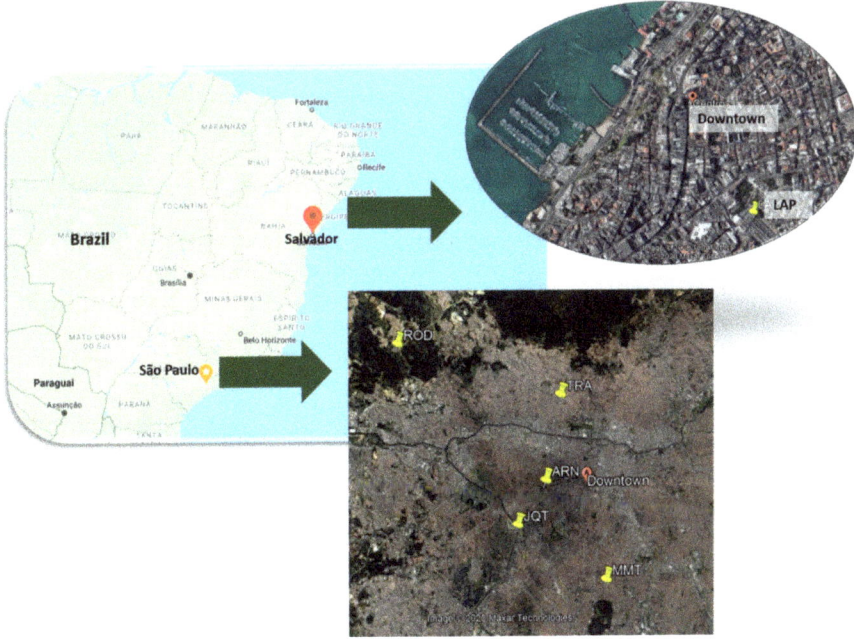

Figure 1. Map of Brazil showing the location of sampling sites in Salvador and São Paulo.

2.2. Aerosol Collection and Chemical Analysis

Before sampling, quartz fiber filters were pre-heated at 600 °C for 6 h. Samples were obtained with a high-volume sampler at a flow rate of 1.13 m^3 min^{-1}, collecting total suspended particles (TSP) (Energética, Brazil and Thermo Scientific, USA), PM$_{10}$ or PM$_{2.5}$ (Thermo Scientific, USA) on quartz fibers.

In the tunnels of São Paulo, 15 samples were collected for a period of 6 h, during the day [2,24] between 2001 and 2011. Owing to the expected high loads of PM inside the tunnels that could overcharge filters, sampling periods were shorter than at the other sites. At TRA and ARN, 11 samples were collected during a 24-h sampling period. Fifty-two samples were collected in Salvador at the Lapa Bus Station, for eight hours during the daytime in 2010 and 2013. After sampling, the filters were kept at 4 °C in a refrigerator. Table 1 summarizes the size of particles, numbers of samples, sampling durations, months and years for each campaign.

Table 1. City, particulate matter (PM) size, number of samples (N), sampling durations (in hours), number of sampling days, and months and years for each sampling campaign.

City	Site	PM Size	N	Sampling Duration (h)	Days of Sampling	Month	Year
São Paulo	JQT01	TSP	3	6	2	August	2001
	MMT	TSP	2	6	2	August	2001
	JQT11	$PM_{2.5}$	6	6	3	May	2011
	ROD	$PM_{2.5}$	4	6	2	July	2011
	ARN	$PM_{2.5}$	7	24	7	September	2013
	TRA	$PM_{2.5}$	4	24	4	May	2015
Salvador	LAP10	PM_{10}	27	8	9	April–May	2010
	LAP13	PM_{10}	25	8	8	October	2013

As the chemical analyzes were conducted in different periods by different research groups, the methodology chosen for the samples from São Paulo and Salvador was not the same. However, analytical parameters related to these methodologies have been published previously [22,23,25]. Furthermore, all samples were analyzed close to the collection period. Samples collected in São Paulo were extracted by an ultrasonic bath (Q335D2, Quimis, Brazil) with 80mL of dichloromethane, three times, and concentrated to 1 mL in a rotatory evaporator with N_2 (Air Liquide) flux. A column packed with silica and alumina was used to separate n-alkanes from the other organic compounds [26]. The eluent used to obtain n-alkanes was n-hexane.

Gas chromatography coupled with a mass spectrometry (GC–MS) analyzer (GC-7820A/MS-5975, Agilent, Palo Alto, CA, USA) was used to identify the compounds, with helium as the carrier gas. The chromatographic conditions were: 60 °C for 1 min, up to 150 °C for 6 °C min^{-1}; next to 280 °C at 5 °C min^{-1}, holding for 15 min. The *splitless* mode was used to make the injections (1 µL) and the injector temperature was set to 300 °C. The standard used for the calibration curve was purchased from Sigma-Aldrich (C_7–C_{40} alkanes standard).

For samples collected in Salvador, the extraction method of the n-alkanes from the filters was based on the work of Pereira et al. [3]. Briefly, samples from atmospheric particulate matter were placed in amber glass vials and 5 mL of a dichloromethane/methanol (1:1 v/v) mixture was added. The vials were then closed and placed into an ultrasonic bath for 10 min. After that, each sample was filtered through a 0.45 µm polytetrafluoroethylene membrane (PTFE); 500 µL aliquot was dried in a gentle nitrogen stream. The dry extract was completed with 50 µL of dichloromethane, for the subsequent GC–MS analysis. The quantification was based on external standard calibration curves using stock solution of 500 mg L^{-1} C_{10}–C_{40} mixture of n-alkanes (AccuStandard, New Haven, CT, USA). The analyses were performed with a GC–MS system (Varian 431/200, Walnut Creek, CA, USA) with an auto sampler (Varian CP-8410, Palo Alto, CA, USA). The column used was BPX5 MS (30 m × 0.25 mm ID × 0.25 µm; 5% phenyl/95% polydimethylsiloxane) with helium as the carrier gas at a 1 mL min^{-1} flow rate. The oven temperature program was: 60 °C for 1 min; increased up to 90 °C at 4 °C min^{-1}; up to 140 °C at 12 °C min^{-1}; up to 180 °C at 9 °C min^{-1}; and up to 330 °C at 6 °C min^{-1}, remaining at this temperature for 6 min and finishing with a total run time of 48 min. Injections (1 µL) were made in the *splitless* mode and the injector temperature was set at 300 °C. Recovery tests were conducted in both experiments, the values ranging between 80% and 98%.

2.3. Data Analysis

Information on the source index used in this study for data analyses are described hereafter. The carbon preferential index (CPI) is a diagnostic tool that represents a relation of proportionality between alkanes with odd and even carbon chains (Equation (1)) [7]. While the contribution of petrogenic sources leads to CPI values close to 1, values greater than 3 are indicative of contributions from biogenic sources [6,27].

$$CPI = \sum C_{odd} \div \sum C_{even} \qquad (1)$$

Two CPI parameters were adopted in the present study to discriminate between petrogenic and biogenic influences, and they include [28]:

Whole range of n-alkanes (Equation (2)):

$$CPI_1 = \sum_{i=17}^{29} C_i \div \sum_{k=16}^{30} C_k \qquad (2)$$

Petrogenic n-alkanes (Equation (3)):

$$CPI_2 = \sum_{i=17}^{25} C_i \div \sum_{k=16}^{24} C_k \qquad (3)$$

Another parameter that can be used to indicate the source is called average chain length (ACL), especially for lipid components produced by plants. ACL is calculated as a mean number of carbon atoms per molecule based on the abundance of odd high homologs ($C_n \geq C_{23}$, Equation (4)) [27]:

$$ACL = (n \times [C_n] + (n+2) \times [C_{n+2}] + \ldots + (n+m) \times ([C_{n+m}])/([C_n] + [C_{n+2}] + \ldots + [C_{n+m}]) \qquad (4)$$

The percentage of petrogenic n-alkanes (%PNA, Equation (5)) was deduced from the percentage of WNA (wax n-alkanes), which is used to estimate the biogenic contribution of wax n-alkanes and was calculated by subtracting the concentration of next higher and lower even carbon-numbered homologs from the average (Equation (6)) [5,28]:

$$\%PNA = 100 - \%WNA \qquad (5)$$

$$\%WNA = \frac{\sum [C_n - (C_{n+1} + C_{n-1})/2]}{\sum C_{total}} \times 100 \qquad (6)$$

The equation derived for computing the ratio of homolog concentrations is called odd-to-even predominance (OEP). It has been adopted as a tool to confirm CPI values, but OEP is calculated for each individual n-alkane, while CPI is calculated for the whole range. Thereby, mathematical improvements have made it possible for OEP to eliminate the limitations found in CPI, as previous studies have reported [29]. This parameter can bring more realistic information about the sources than CPI [30]. The OEP ratios were calculated according to Equation (7). Kavouras et al. [31] suggested adopting the range between C_{14} and C_{34} for this calculation.

$$OEP = \left(\frac{C_{n-2} + 6 \times C_n + C_{n+2}}{4 \times C_{n-1} + 4 \times C_{n+1}}\right)^{(-1)^{(n-1)}} \qquad (7)$$

3. Results and Discussion

3.1. Particulate Matter and n-Alkane Concentrations in the Tunnels and at the Truck Depot

Table 2 shows minima, maxima and mean concentrations, standard deviations and sampling numbers for all sites. The samples collected in 2001 at JQT and MMT presented mean concentrations of TSP of 488 and 253 µg m^{-3} (Table 2), respectively. The truck traffic was allowed in this area and the fleet was constituted by a mixed of gasohol, ethanol and diesel fueled vehicles until 2008 [1]. At this time, flex vehicles were not circulating and biodiesel was not added to the diesel; this addition is often related to better fuel burning and smaller emissions of particulate matter [32,33]. The limits defined in the Brazilian Atmospheric Standards for TSP are related to 24 h of exposure, however, even with sampling time below 24 h, the average concentrations in both tunnels was higher than the limit of 240 µg m^{-3} established in Brazil. A large difference in traffic characteristics between the tunnels MMT and JQT01 was observed. An intense traffic of heavy-duty vehicles (HDV) occurred in MMT during the afternoon and light-duty vehicles (LDV) circulation was constant during the morning and afternoon. In the JQT, the LDV were predominant and the vehicle density increased from morning to afternoon [34].

In JQT11, where LDV were predominant, the PM$_{2.5}$ mean concentration was over 52 µg m^{-3}, and at ROD, where the particulate sources are LDV and HDV, the mean concentration was over 233 µg m^{-3} (Table 2). ROD is approximately 23 km from São Paulo's city downtown and it is an important route for HDV that circulate with diesel and biodiesel, especially for trucks transporting shipments to the entire state. Besides that, these values are comparable to the Standard establish by the São Paulo Environmental Agency, CETESB, for 24 h of exposure (60 µg m^{-3}) [35]. JTQ11 had a mean value below the legislation value. On the other hand, ROD presented a mean concentration almost four times higher than the CETESB standard.

Comparing the results obtained in São Paulo tunnels with other studies, the values obtained in the JQT11 and ROD were lower than the concentration found in a tunnel in Lisbon (Portugal) characterized by the predominance of gasoline vehicles (58%) followed by diesel cars (42%). The PM$_{2.5}$ concentrations ranged from 450 to 1061 µg m^{-3} in October 2008 [36]. For PM$_{10}$ samples collected in Shanghai, China at a road tunnel with intense traffic of gasoline-powered vehicles (91–98%), between October 2011 and May 2012, the mean concentration was 670 µg m^{-3} [37], much higher than the values obtained in the tunnel samples in the present study.

The TSP concentration at JQT for the samples collected in 2001 was 488 µg m^{-3} (JQT01) and for those collected 10 years later at the same site (JQT11), the PM$_{2.5}$ mean concentration was over 52 µg m^{-3}. The restriction of truck circulation, implemented in 2008 by local government, might have influenced this PM reduction [1], since HDV are known to significantly contribute to the PM emissions [38].

The mean concentrations of the homologues found in the tunnel samples in 2001 (JQT01) were over 767 ng m^{-3} and even much higher 1680 ng m^{-3} (Figure 2) at MMT. The C$_{max}$, carbon number of the most abundant n-alkane [39], was C$_{21}$ for JQT01 and C$_{24}$ for MMT. According to the literature, this parameter can be used for a relative source input [28,39]. C$_{max}$ lower than C$_{25}$ is characteristic of anthropogenic emissions, more specifically by fossil fuel burnings [6,40]. Simoneit et al. [40], reports n-alkanes emitted by diesel vehicles and auto engine exhausts (species not burned) maximizing at C$_{22}$–C$_{23}$. Among the sites impacted by diesel burning emissions (ARN, TRA, JQT, LAP, MMT, and ROD), only JQT and MMT presented C$_{max}$ (Table 3) similar to that reported for areas impacted by diesel burning [40].

For the samples collected at tunnels 10 years later, JQT11 (PM$_{2.5}$) presented an n-alkane mean concentration higher (1670 ng m^{-3}) than ROD (1276 ng m^{-3}) (Figure 2). This value was higher than obtained at JQT01 (767 ng m^{-3}) 10 years before. In 2011, JQT11 registered 1806 vehicles/hour, while ROD recorded 1152 vehicles/hour [2].

Comparing tunnel results with strong contributions from heavy duty-vehicles, ROD (1276 ng m^{-3}) presented a smaller n-alkane mean concentration than MMT (1680 ng m^{-3}), probably due to a smaller number of vehicles, because according to Vasconcellos et al. [41], the traffic volume in MMT was 2917 vehicles/hour.

The TRA site is a place where a truck company offers maintenance services and is an open area used for parking, located in an industrial neighborhood of São Paulo. This site presented PM$_{2.5}$ average concentrations of 92 µg m^{-3} almost four times higher than the value recommended by World Health Organization for PM$_{2.5}$ (25 µg m^{-3}) [42] and higher than the standard established for São Paulo State [35]. The total n-alkane mean concentration at TRA was 90 ng m^{-3}. These samples showed a predominant distribution between C$_{26}$ and C$_{30}$. The C$_{max}$ was C$_{28}$, showing the influence from engine exhausts (lubricating oil emissions) in addition to the fossil fuel burning [7].

3.2. Particulate Matter and n-Alkanes Concentration in Bus Station (LAP) and Bus Corridor (ARN)

The PM$_{10}$ concentrations at the LAP13 bus station (Table 2), were lower than those found three years before (LAP10), considering either the weekday or the day period of sampling. The average concentrations in the morning, afternoon and night for 2010 samples were 162, 171 and 85 µg m^{-3}, while for samples collected three years later (2013) the values were 91, 127 and 56 µg m^{-3}, respectively. These results are consistent with previous studies at the same site by Pereira et al. [43] in 2005 (123 µg m^{-3}; 140 µg m^{-3} and 63 µg m^{-3}), confirming that in 2010 the concentrations were higher than in 2005. This can be explained partly by an increase in the bus fleet; while in 2005 an average of 482 vehicles circulated daily in the bus station, in 2010 this number increased to 519 vehicles. In addition, there were changes in station architecture after 2005, when the air exhaust system was removed for repair, resulting in a lessening of air circulation through the station. The decrease in concentration in 2013 was probably due to new arrangements in the station; after 2012, a new bus platform was added, providing air circulation at the underground floor. Other factors could be associated with this, such as fleet renewal and improvement of fuel quality (diesel) [43].

Table 3 and Figure 2 presented the n-alkane distributions for all sites studied. The C$_{max}$, the most abundant species, was C$_{32}$ for LAP10 and C$_{29}$ for LAP13, which can be related to biogenic contributions [5]. For these samples, it was possible to identify and quantify the homologous series that was distributed between the C$_{16}$ and C$_{33}$. It was observed that the average concentrations did not differ significantly between the two sampling years (1.5–14 ng m^{-3} and 1.3–17 ng m^{-3} in 2010 and 2013, respectively). In addition, there was a greater contribution of n-alkanes of high molecular mass (C$_{24}$ to C$_{33}$). In the same place, a previous analysis by Pereira et al. (2007) with samples collected in 1998 showed concentrations of total n-alkanes ranging between 300 and 580 ng m^{-3}, values above those found in 2010 (range: 72–210 ng m^{-3}) and 2013 (range: 105–202 ng m^{-3}).

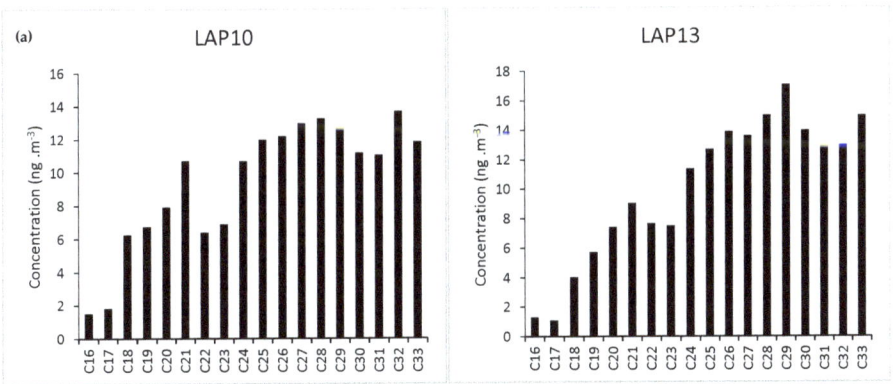

Figure 2. Cont.

(b)

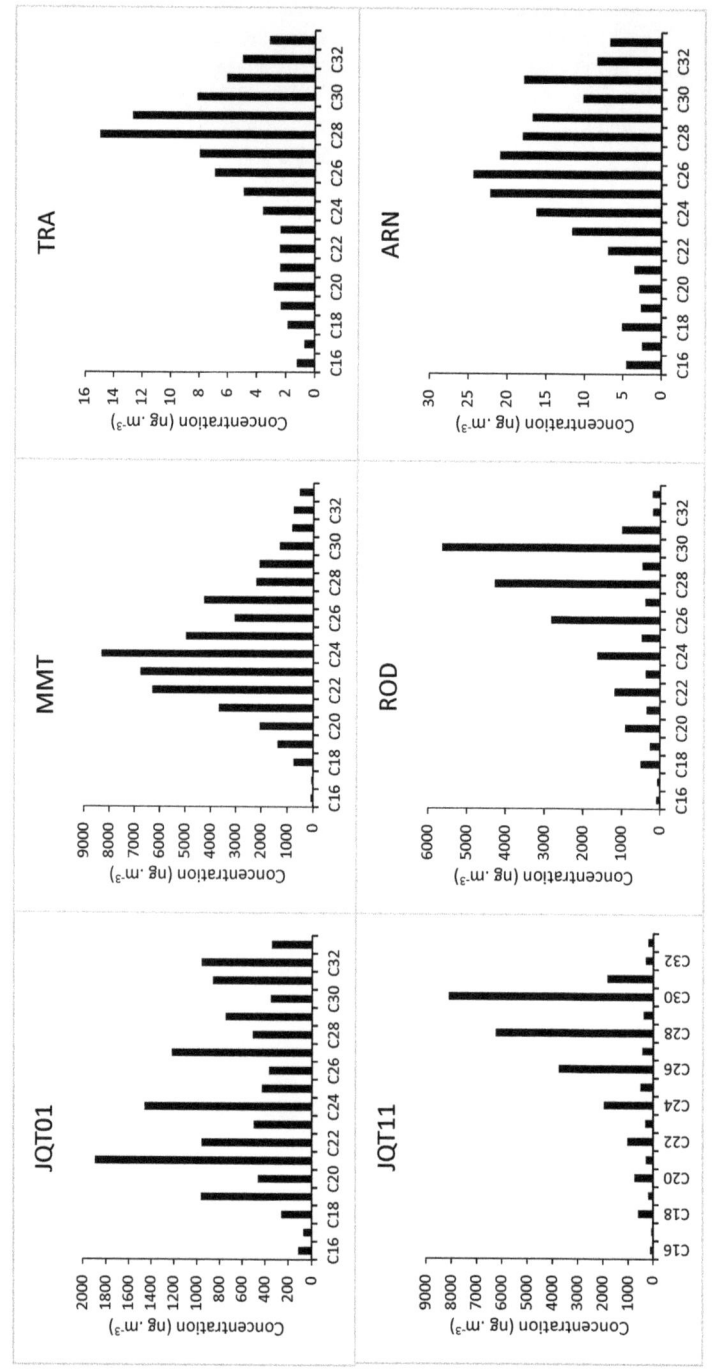

Figure 2. n-alkanes distribution (ng m^{-3}) for (**a**) Salvador and (**b**) São Paulo sites.

The total concentrations of *n*-alkanes found in this study for LAP10 (160 ng m^{-3}) and LAP13 (180 ng m^{-3}), are comparable with values obtained in other studies, which reported concentrations of samples collected in roadway regions and in urban areas, as by Omar et al. [6] (Kuala Lumpur, Malaysian, 103 ng m^{-3}), Wang et al. [44] (Beijing, 137 ng m^{-3}) and Vasconcellos et al. [45] (São Paulo, 106 ng m^{-3}).

The ARN site, an important road close to São Paulo's main avenues, receives a strong influence of bus and car emissions; besides there are many public hospitals around this area. The mean PM$_{2.5}$ concentration at ARN was 79 µg m^{-3}, this value is also three times higher than that recommended for 24 h of exposure by the World Health Organization (WHO) [4]. Regarding *n*-alkanes at ARN, the total concentration was 201 ng m^{-3}, and a predominance of species ranging from C$_{24}$ to C$_{29}$ was observed; the C$_{max}$ was C$_{26}$, which indicates the contribution of mixed sources [40].

Table 2. Means, minima, maxima and standard deviation for PM mass concentrations (µg m^{-3}) of samples collected at each site.

Site	Mean	Min	Max	SD
LAP10 **	141	82	171	37.0
LAP13 **	91	56	127	29.0
JQT01 ***	488	419	571	62.1
MMT ***	253	199	308	44.5
JQT11 *	52	49	59	4.2
ROD *	233	202	282	32.9
TRA *	92	46	145	40.5
ARN *	79	44	124	32.7

* PM$_{2.5}$ mass concentration; ** PM$_{10}$ mass concentration; *** TSP; SD: Standard Deviation.

3.3. Species Contribution

Four chromatograms were chosen to illustrate the difference in profiles according to the site variations (JQT11, ROD, ARN, TRA—Figure 3). The hump areas shown in two chromatograms are chracterized as unresolved complex mixtures (UCM), a term used as one indicator of petroleum or fossil fuel-burning contamination, which represents the area of compounds unable to be resolved by simple gas chromatography. The combination of the two-dimensional gas chromatography technique coupled to time-of-flight mass-spectrometry (GC–TOF/MS) has been reported as a solution to resolve and identify these hydrocarbons individually [46].

Many studies presented similar humps characterized by petroleum contributions in different environmental matrices, such as sediments, rivers, and road dust [20,46]. The hydrocarbons that originate in the UCM hump are predominantly products of biodegraded crude oil and refined fractions, such as lubricating oils. Booth et al. [46] have shown that most of the compounds contained in the UCM area were derived from cyclic hydrocarbons such as alkylbenzenes and tetralins that are potentially toxic to humans.

Comparing the four chromatograms, ROD and JQT exhibited similar distributions and negligible unresolved complex mixture areas while the chromatograms regarding ARN and TRA, two open areas, contain large UCM areas (Figure 3). ARN and TRA showed higher contributions of heavy molecular weight hydrocarbons unresolved by gas chromatography than ROD and JQT.

Another calculation adopted by previous studies [27] is the ratio between the unresolved area of chromatogram to the total resolved area. This ratio describes the degree of contamination by hydrocarbons emitted by heavy fraction of petroleum [36]. ROD and JQT presented ratios equal to 1.7 and 1.6, while ARN and TRA obtained values of 17 and 5, respectively. With these values, it is possible to infer that ARN and TRA sites have different sources of pollution that can contribute significantly to this unresolved mixture.

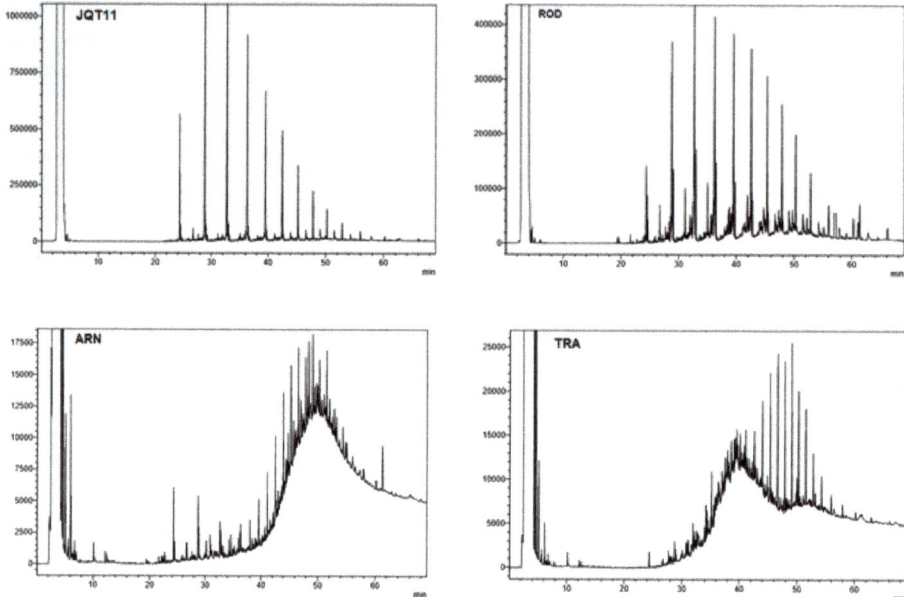

Figure 3. Chromatograms of four different sites (ROD, JQT, ARN and TRA) analyzed by gas chromatography coupled to flame ionization detection (GC-FID).

A large bus corridor is surrounded by five important hospitals at ARN. The municipal legislation requires hospitals to have their own power generator [47] which uses fossil fuels to produce energy. These emissions can add particulate matter containing organic compounds.

At the TRA site, an industrial neighborhood where most of the commercial buildings are for transportation but chemical and plastic industries are also present as well as one of the largest Brazilian airports that is over 4 km away from the sampling site. Masiol and Harrison [48] reported that jet fuel burning emits normal paraffin, iso-paraffin, cycloparaffin, aromatic and alkene classes. The PM emitted by these sources may contain species impacting the atmosphere and air quality.

There are different n-alkane ranges that can be selected to calculate the carbon preference index (CPI), which can show the biogenic versus anthropogenic predominance or petrogenic contribution [49,50]. Considering the CPI_1 (Table 3), which indicates the impact of biogenic and anthropogenic sources [49], all the sites demonstrate a strong influence of anthropogenic sources, since the values are close to 1. These results are similar with the CPI_1 (mean = 1.3) obtained during the winter in 2005 for $PM_{2.5}$ collected at the Chinese megacity Nanjing, at a typical urban site [51].

The CPI_2 shows the contribution of petrogenic hydrocarbons [49]. According to the literature, the smaller values represent a higher contribution of anthropogenic activities [52]. In this study, the smaller values were for JQT11 (mean = 0.3) and ROD (mean = 0.4), indicating again, the high impact of petrogenic emissions.

Table 3 also shows average chain lengths (ACL) at all sites, according to calculation previous presented in Section 2.3 Data Analysis. ACL is a source index relative to lipids, for n-alkanes. The ACL values ranged from 26 (MMT) to 29 (JQT11). According to previous studies, in tropical climates, ACL values between C_{22} and C_{29} were indicative of mixed source contributions [28,53]. Additionally, ACL is a parameter that can be used as an ambient temperature indicator, by the fact that higher values were observed during warm seasons and lower values in cold periods [54]. However, this parameter must be used sparingly, since cloistered places may show different behavior. The samples collected during the warm season were from LAP13 and ARN site, the other samplings were conducted during cold season.

Table 3. Carbon preference index (CPI), average chain length (ACL) values and C_{max} for the n-alkanes at all sites.

Site	CPI_1	CPI_2	ACL	%PNA	C_{max}
LAP10	1.1	1.2	27	97	C_{32}
LAP13	0.9	1.1	28	97	C_{29}
JQT01	1.3	1.2	28	75	C_{21}
MMT	1.0	1.0	26	96	C_{24}
JQT11	0.2	0.3	29	99	C_{30}
ROD	0.2	0.4	28	99	C_{30}
ARN	1.1	1.2	28	93	C_{26}
TRA	0.9	1.4	28	97	C_{30}

The percentage of petrogenic n-alkanes (%PNA) falls between 75% (JQT01) and 99% (ROD). Comparing JQT results obtained in the 2001 and 2011 samples, the %PNA was lower in 2001 (75%) than in 2011 (99%). This difference could be attributed to the increase in the number of vehicles in the city over time. In 2001, there were over 4 million vehicles running in São Paulo, while by 2011 the number increased to 7 million [1], representing an increase of 78%.

It is worth noting that, while the CPI values and C_{max} (Table 3) point to the prevalence of petrogenic sources (LAP10, MMT, JQT11, ROD, ARN, TRA), other sources may have minor roles, such as emissions from activities around the sites, parks (JQT), food commerce (LAP, ARN) and industries (ROD and TRA).

The OEP ratio is an odd-to-even calculation, which is a measure of the local odd-even ratio [29]. It is a useful tool that can provide information about the variety of sources of n-alkanes [55]. The OEP measure was established to eliminate mathematical limitations found in the CPI ratio, as mentioned before in Section 2.3. Data analysis. Although the CPI ratio is the most popular calculation for n-alkanes, high oscillations in the concentration observed for odd/even carbon could appear and the errors of this parameter can increase as the number of carbon chains decreases [56]. In an attempt to use distinguishing running ratios for correlation determinations, the OEP function was established to avoid these fluctuations.

In previous studies, OEP values were plotted against carbon chain length to make the OEP curves. Tunnels sites (JQT01, JQT11, and ROD) showed a similar flat profile, with the largest variation values (ranging from 0.3 to 2.9), and the highest values were below C_{25} (Figure 4), which provides evidence for the predominance of the anthropogenic contribution [19]. The other sites (MMT, TRA, ARN, LAP10, and LAP13) also presented a shape of OEP curves similar to those for urban sites (Figure 4) reported by previous literature studies conducted at sites highly impacted by vehicular emissions [30,55]. Despite this, the range of values was smaller (between 0.1 and 1.6) than the three tunnel samplings, which may indicate one major source for each site [56].

Figure 5 shows the ratio between individual n-alkane concentrations and the sum of n-alkanes for each tunnel ranging from C_{20} to C_{32} in an attempt to evaluate the specific contribution of each compound to the total concentration. It was expected that n-alkanes below C_{25} showed higher contributions relating to the sum of concentrations than the compounds above C_{25}, since the first range ($<C_{25}$) is attributed to anthropogenic sources, which is dominant inside tunnels [5,18]. It was observed that the percentage of even homologs from C_{20} to C_{30} was similar to those of the JQT11 and ROD tunnels. In fact, JQT11 prioritizes LDV during the peak period, prohibiting HDV during this period [56]. For the samples collected 10 years before, MMT and JQT01, at a time when Brazilian legislation allowed HDV circulation at both sites, n-alkanes lower than C_{25} were predominant in these locations, indicating a high contribution of diesel burning [19].

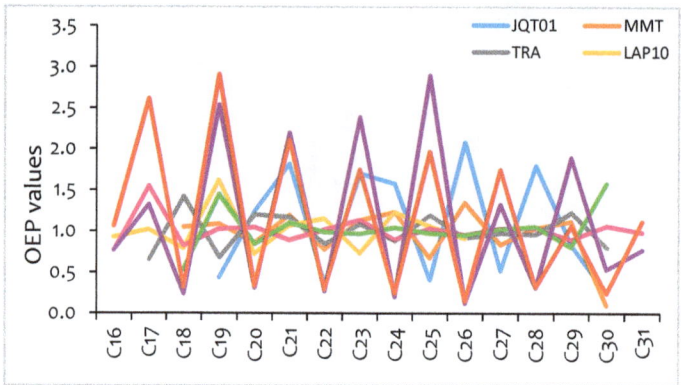

Figure 4. Odd-to-even predominance (OEP) curve of the homologues series of *n*-alkanes.

Since 2008, the percentage of alcohol in Brazilian fuels has increased to 25% [32]. Pacheco and collaborators [57] reported that the addition of ethanol to gasoline may increase the emissions of aldehydes, acetic acid and unburned alcohol, negatively affecting the air quality. For the samples collected at JQT in 2001, C_{21}, C_{24} and C_{27} were the most abundant homologs as opposed the samples collected in 2011, where C_{26}, C_{28} and C_{30} presented the highest concentrations. The change in Brazilian fuels between these years resulted in reducing PM concentrations in the urban areas, but the non-exhaust vehicular emissions such as tire, brake, and road surface wear increased [58], which can be attributed to different profiles observed for JQT01 and JQT11 (Figure 5).

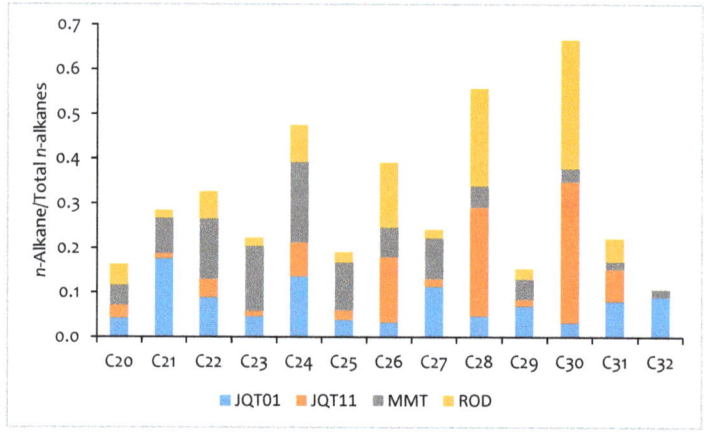

Figure 5. Ratio of *n*-alkanes and sum of total *n*-alkanes concentrations for the tunnels studied.

The homologous distributions presented for the samples collected in tunnels with the predominance of diesel-burning emission, presented distinct shapes. The results show that for MMT samples from 2001, C_{22}, C_{23} and C_{24} were the most abundant homologs, whereas for ROD collected in 2011 the most abundant were C_{26}, C_{28} and C_{30}, showing different contributions.

According to the literature, the most abundant *n*-alkane can be used with respect to the relative source input. C_{max} lower than C_{25} is characteristic of anthropogenic emissions, more specifically by fossil fuel burnings [6,40]. The present study shows different profiles from the studies conducted in other countries; it may be due to the large variety of fuels burned in the Brazilian fleet (C_{max} varied from C_{21} to C_{32}). Besides the fuels commonly used in other countries, in 2004, a governmental program was implemented in Brazil and biofuels were added to diesel [59]. It is expected that this implementation will change the species contribution.

3.4. Statistical Correlations

Pearson's correlations were also calculated using the *n*-alkanes concentrations. The results showed that samples collected at sites with the influence of biodiesel burnt in buses (LAP10, LAP13 and ARN), were strongly correlated with the homologues ranging from C_{22} to C_{32} (R = 0.62–0.99). For samples collected at the ROD and TRA sites where the trucks are fueled with biodiesel, the correlations were also stronger, from C_{15} to C_{32} (R = 0.57–0.99).

On the other hand, low correlations were found for JTQ samples collected during different years. The prohibition of HDV traffic in this tunnel after 2008 may have affected the *n*-alkanes distribution. Results from both tunnels with similar characteristics (ROD and MMT) presented high correlations (R = 0.54–0.99) between C_{15} and C_{25}.

Rogge et al. [18] reported that the profile of *n*-alkanes emitted by fossil fuel burning presented a bimodal distribution with higher emission rates between C_{20} and C_{22} and C_{24} and C_{27}. LAP10 and LAP13 showed a bimodal distribution between C_{19} and C_{21} and from C_{24} to C_{29}. In general, the samples collected at tunnels (TJQ, MMT and ROD) presented a unimodal distribution with even carbon *n*-alkanes higher than odd carbon *n*-alkanes. The TRA and ARN samples exhibited a unimodal profile from C_{20} to C_{29}. At a site with a predominant emission of fossil fuel, an increase of lower molecular weight *n*-alkanes ($\leq C_{22}$) reflects the presence of fossil fuel burning [40].

Hierarchical cluster analysis (HCA) is an unsupervised agglomerative technique that examines inter-point distances between all samples in the data set and represents this information in the form of a two-dimensional graph, the dendrogram. Through the dendrogram, it is possible to visualize the groupings and similarity between the samples and/or variables. Figure 6 outlines grouping tendencies of *n*-alkanes that showed high connections for all sites. In this way, three groups of *n*-alkanes with good correlation were observed, they are: (1) C_{17}, C_{18}, C_{19} and C_{20}; (2) C_{22}, C_{23}, C_{24}, C_{27}, C_{29}, C_{31}, C_{32}, C_{33} and C_{34} and lastly (3) C_{26}, C_{28} and C_{30}.

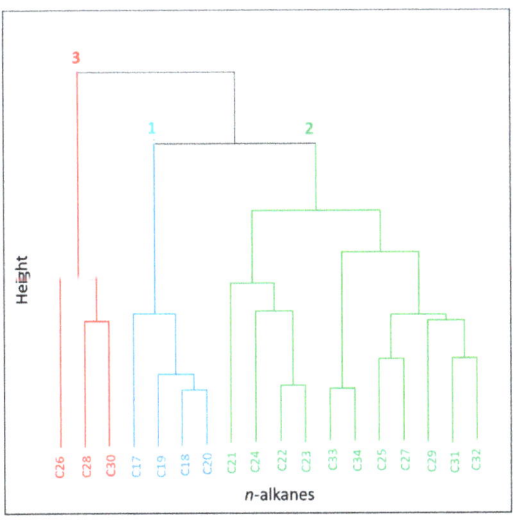

Figure 6. Cluster analysis for all sampling sites.

According to a previous study, 95% of *n*-alkanes emitted by diesel burning have carbon numbers less than C_{19} [18]. By comparison, group 1 might be associated mostly with diesel burning.

Group 2 contains more compounds than the other groups, which is consistent with results obtained previously by Pearson correlation, where compounds from C_{22} to C_{32} showed great associations. Similar results were obtained by independent mathematical methods. This group can be attributed mainly to anthropogenic emissions since the biogenic contribution tends to be small at all sites. According to a previous study, plant wax concentrations were attributed to high-molecular-weight n-alkanes (from C_{24} to C_{35}) [59]; however, the waxes of n-alkanes are presented in the rubber material which produces tires, as antiozonants, to protect it from ozone attack [18]. Therefore, compounds grouped in this cluster with long carbon chains (C_{29}, C_{31}, C_{32}, C_{33}) can be attributed mostly to tire wear particles generated during the rolling shear of the tire tread against the road surface.

The third group contained three even n-alkanes which can be attributed to different sources, since these even compounds have carbon long chains that can be emitted by a source mixture with anthropogenic and biogenic contribution.

3.5. Principal Component Analysis (PCA)

PCA is a useful tool for understanding emission sources. To define the principal sources of n-alkanes, enrichment factors associated with aerosols sources were calculated for all sites. Due to a large range of concentrations, PCA was calculated in two ways: one for large value concentrations (JQT01, MMT, JQT11 and ROD) and the other for small-value concentrations (LAP10, LAP13, ARN and TRA). Table A1 (Appendix A) shows the scores of the two principal components for large and small values. Principal component 1 accounts for 56% of the total data variance, while the principal component 2 explained 32% for the data with large values. In contrast, for small values, PC1 corresponded to 80% and PC2 to 10%.

For large-value concentrations (JQT01, MMT, JQT11 and ROD), most of n-alkanes were explained by factor 1 (Figure 7a). This factor ranged from C_{19} to C_{29}, (except to C_{26} and C_{28}). The range from C_{19}–C_{25} is often attributed to fossil fuel emissions, as this carbon range is predominant in anthropogenic activities [60]. Rogge et al. [18] showed that vehicular tires have higher n-alkanes in their compositions, revealing different compositions with additional pyrolysis products. Therefore, this factor presents mixed sources.

It seems that the samples collected at JQT11 (Figure 7a) presented a higher contribution to factor 2, due to the predominance of even n-alkanes (C_{18}, C_{26}, C_{28} and C_{30}). As described earlier, this tunnel is characterized by light-duty vehicles.

To permit easier graphical visualization, the following nomenclatures were adopted: JQT01 as TQ, MMT as MM, JQ is JQT11 and RD is ROD (Figure 7a).

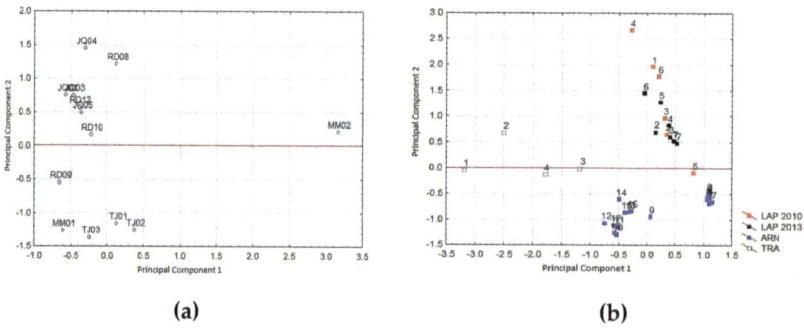

Figure 7. Score graphs of the two principal components for each site with (a) large concentrations and (b) small concentrations.

For the sites with lower concentration values of n-alkanes (LAP10, LAP13, ARN and TRA) factor 1 was attributed predominantly to buses emissions (combustion of diesel plus biodiesel) since this cluster is well correlated (Figure 7b). On the other hand, for TRA samples, different sources besides truck emissions, such as industrial emissions, as expected, can impact this site. Factor 2 with n-alkanes (C_{20} and C_{21}) well correlated can be attributed to unknown sources.

4. Conclusions

The present study reports particulate matter and n-alkane concentrations at different sites impacted by vehicular emission in Brazil. Regarding PM size and concentrations, samples collected at tunnels in 2001 presented higher levels since TSP samples were collected; the addition of biodiesel was implemented years later. The massive difference between the samples collected at JQT in 2001 and 2011 was attributed to PM size and, especially, to the restriction of truck circulation implemented in 2008 by the local government of São Paulo city. The bus station in Salvador (LAP) presented a decrease in PM concentration between 2010 and 2013 probably due to a change in the station's architecture which provided better air circulation at the site.

Despite the fact that PM concentration has decreased in smaller aerosol fractions, the n-alkanes concentrations increased, highlighting the predominance of organic compounds in the fine PM fraction. The n-alkane concentrations were characterized mostly by anthropogenic emissions at all sites, as expected for areas strongly impacted by vehicular emissions. The chromatographic profiles were different for samples collected at tunnels that showed a low contribution of an unresolved complex mixture when compared to the samples collected at urban sites (ARN and TRA). This difference can be attributed to other possible PM local sources, such as power generators and industries.

The CPI values suggest the prevalence of petrogenic sources at LAP10, MMT, JQT11, ROD, ARN and TRA. The OEP values were typical of urban areas and the profiles were similar in spite of different site characteristics. Among the sites, C_{max} varied from C_{21} to C_{32}, showing the complexity of emissions of different fuels.

Pearson's correlations indicated the different site characteristics. The sites affected by buses emissions (LAP10, LAP13 and ARN) were strongly correlated; also well correlated were the samples collected in the ROD and in TRA, impacted by diesel emissions. JTQ01 and JQT11 samples presented poor correlations, probably due to truck restriction after 2008.

PCAs were calculated in two ways: for large and small concentrations. In sites with large values, most of n-alkanes were explained by factor 1, attributed to mixed sources (vehicular and tires emissions), whereas factor 2 was more affected by light-duty vehicles emissions. PCA for small values were attributed to bus emissions (factor 1) and for unknown sources (factor 2).

The use of alternative fuels such as those used in the Brazilian fleet can lead to cleaner emissions, but this should be further studied. It would be interesting if the legislation were more rigid, recommending lower concentrations of particulate material, consequently reducing the concentrations of many pollutants.

Author Contributions: Conceptualization, P.C.V.; data curation, S.C. and P.C.V.; methodology, S.C.; software, S.C. and R.E.B.; validation, S.C. and R.E.B.; formal analysis, S.C.; investigation, S.C.; resources, P.C.V.; writing—original draft preparation, S.C.; writing—review and editing, P.C.V.; supervision, P.C.V.; project administration, P.C.V.; funding acquisition, S.C., and P.C.V. All authors have read and agreed to the published version of the manuscript.

Funding: This research was funded by FAPESP (grant number 2016/23339-1 and grant number 2008/58104-8) and CNPq (grant number 301503/2018-4).

Acknowledgments: The authors thank INCT-Energy and Environment.

Conflicts of Interest: The authors declare no conflict of interest.

Appendix A

Table A1. Factors of principal components analysis.

	Large Values			Small Values	
	PC1	PC2		PC1	PC2
C_{17}	-	-	C_{17}	−0.931	-
C_{18}	0.472	0.781	C_{18}	−0.931	-
C_{19}	0.757	0.444	C_{19}	−0.864	0.324
C_{20}	0.761	0.481	C_{20}	−0.672	0.716
C_{21}	0.870	−0.394	C_{21}	−0.545	0.768
C_{22}	0.964	-	C_{22}	−0.892	-
C_{23}	0.968	-	C_{23}	−0.923	-
C_{24}	0.952	-	C_{24}	−0.926	-
C_{25}	0.964	-	C_{25}	−0.873	−0.303
C_{26}	-	0.961	C_{26}	−0.970	-
C_{27}	0.976	-	C_{27}	−0.984	-
C_{28}	-	0.967	C_{28}	−0.984	-
C_{29}	0.913	-	C_{29}	−0.980	-
C_{30}	−0.328	0.925	C_{30}	−0.910	-
C_{31}	-	0.634	C_{31}	−0.966	-

Values lower than 0.250 in module are omitted; values higher than 0.700 in module are in bold.

References

1. Denatran—Departamento Nacional de Trânsito. Frota de Veículos. 2018. Available online: https://www.denatran.gov.br/estatistica/635-frota-2018 (accessed on 30 June 2018).
2. Brito, J.; Rizzo, L.V.; Herckes, P.; Vasconcellos, P.; Caumo, S.E.S.; Fornaro, A.; Ynoue, R.; Artaxo, P.; Andrade, M.F. Physical–chemical characterisation of the particulate matter inside two road tunnels in the São Paulo Metropolitan Area. *Atmos. Chem. Phys. Discuss.* **2013**, *13*, 12199–12213. [CrossRef]
3. Pereira, G.; Caumo, S.E.S.; Soares, S.; Teinilä, K.; Custódio, D.; Hillamo, R.; Vasconcellos, P.C.; Alves, N.D.O.; Alves, C. Chemical composition of aerosol in São Paulo, Brazil: Influence of the transport of pollutants. *Air Qual. Atmos. Health* **2016**, *10*, 457–468. [CrossRef]
4. WHO. *Air Quality Guidelines for Particulate Matter, Ozone, Nitrogen Dioxide and Sulfur Dioxide*; WHO: Geneva, Switzerland, 2005. Available online: http://www.euro.who.int/Document/E87950.pdf (accessed on 10 December 2019).
5. Simoneit, B.R.; Sheng, G.; Chen, X.; Fu, J.; Zhang, J.; Xu, Y. Molecular marker study of extractable organic matter in aerosols from urban areas of China. *Atmos. Environ. Part A. Gen. Top.* **1991**, *25*, 2111–2129. [CrossRef]
6. Omar, N.Y.M.J.; Bin Abas, M.R.; Rahman, N.A.; Tahir, N.M.; Rushdi, A.; Simoneit, B.R.T. Levels and distributions of organic source tracers in air and roadside dust particles of Kuala Lumpur, Malaysia. *Environ. Earth Sci.* **2007**, *52*, 1485–1500. [CrossRef]
7. Simoneit, B.R.; Kobayashi, M.; Mochida, M.; Kawamura, K.; Lee, M.; Lim, H.; Turpin, B.J.; Komazaki, Y. Composition and major sources of organic compounds of aerosol particulate matter sampled during the ACE-Asia campaign. *J. Geophys. Res. Space Phys.* **2004**, *109*, 1–22. [CrossRef]
8. Duan, F.; He, K.; Liu, X. Characteristics and source identification of fine particulate n-alkanes in Beijing, China. *J. Environ. Sci.* **2010**, *22*, 998–1005. [CrossRef]
9. Finlayson-Pitts, B.J.; Pitts, J.N. Kinetics and Atmospheric Chemistry. In *Chemistry of the Upper and Lower Atmosphere*; Elsevier: Amsterdam, The Netherlands, 2000; pp. 130–178.
10. Atkinson, R. Gas-phase tropospheric chemistry of volatile organic compounds: 1. Alkanes and alkenes. *J. Phys. Chem. Ref. Data* **1997**, *26*, 215–290. [CrossRef]
11. Bidleman, T.F.; Billings, W.N.; Foreman, W.T. Vapor-Particle Partitioning of Semivolatile Organic Compounds: Estimates from Field Collections. *Environ. Sci. Technol.* **1986**, *20*, 1038–1043. [CrossRef]
12. Gill, C.O.; Ratledge, C. Toxicity of n-Alkanes, n-Alk-1-enes, n-Alkan-1-ols and n-Alkyl-1-bromides towards Yests. *J. Gen. Microbiol.* **1972**, *72*, 165–172. [CrossRef]

13. Barnes, I.; Bastian, V.; Becker, K.; Fink, E.; Zabel, F. Reactivity studies of organic substances towards hydroxyl radicals under atmospheric conditions. *Atmos. Environ. (1967)* **1982**, *16*, 545–550. [CrossRef]
14. Pitts, J.N. Nitration of gaseous polycyclic aromatic hydrocarbons in simulated and ambient urban atmospheres: A source of mutagenic nitroarenes. *Atmos. Environ. (1967)* **1987**, *21*, 2531–2547. [CrossRef]
15. Enya, T.; Suzuki, H.; Watanabe, T.; Hirayama, T.; Hisamatsu, Y. 3-Nitrobenzanthrone, a Powerful Bacterial Mutagen and Suspected Human Carcinogen Found in Diesel Exhaust and Airborne Particulates. *Environ. Sci. Technol.* **1997**, *31*, 2772–2776. [CrossRef]
16. Sicre, M.; Marty, J.; Saliot, A.; Aparicio, X.; Grimalt, J.O.; Albaiges, J. Aliphatic and aromatic hydrocarbons in different sized aerosols over the Mediterranean Sea: Occurrence and origin. *Atmos. Environ. (1967)* **1987**, *21*, 2247–2259. [CrossRef]
17. Abu-Allaban, M.; A Gillies, J.; Gertler, A.W.; Clayton, R.; Proffitt, D. Tailpipe, resuspended road dust, and brake-wear emission factors from on-road vehicles. *Atmos. Environ.* **2003**, *37*, 5283–5293. [CrossRef]
18. Rogge, W.F.; Hildemann, L.M.; Mazurek, M.A.; Cass, G.R.; Simoneit, B.R.T. Sources of fine organic aerosol. 3. Road dust, tire debris, and organometallic brake lining dust: Roads as sources and sinks. *Environ. Sci. Technol.* **1993**, *27*, 1892–1904. [CrossRef]
19. Rogge, W.F.; Hildemann, L.M.; Mazurek, M.A.; Cass, G.R.; Simoneit, B.R.T. Sources of fine organic aerosol. 2. Noncatalyst and catalyst-equipped automobiles and heavy-duty diesel trucks. *Environ. Sci. Technol.* **1993**, *27*, 636–651. [CrossRef]
20. Readman, J.W.; Fillmann, G.; Tolosa, I.; Bartocci, J.; Villeneuve, J.-P.; Catinni, C.; Mee, L.D. Petroleum and PAH contamination of the Black Sea. *Mar. Pollut. Bull.* **2002**, *44*, 48–62. [CrossRef]
21. Instituto Brasileiro de Geografia e Estatística. Estatística Populacional Brasileira. 2017. Available online: https://cidades.ibge.gov.br/brasil/sp/sao-paulo/panorama (accessed on 30 June 2018).
22. Vasconcellos, P.; Souza, D.Z.; Magalhães, D.; Da Rocha, G.O. Seasonal Variation of n-Alkanes and Polycyclic Aromatic Hydrocarbon Concentrations in PM10 Samples Collected at Urban Sites of São Paulo State, Brazil. *Water Air Soil Pollut.* **2011**, *222*, 325–336. [CrossRef]
23. Da Silva, R.L. *Hidrocarbonetos Alifáticos (n-alcanos) Associados ao Material Particulado Atmosferico da Estação da Lapa e Regiões no Entorno da Baía de Todos os Santos*; Universidade Federal da Bahia: Salvador, Brazil, 2014.
24. Vasconcellos, P.; Zacarias, D.; Pires, M.A.; Pool, C.S.; Carvalho, L.R. Measurements of polycyclic aromatic hydrocarbons in airborne particles from the metropolitan area of São Paulo City, Brazil. *Atmos. Environ.* **2003**, *37*, 3009–3018. [CrossRef]
25. Caumo, S.; Vicente, A.; Custódio, D.; Alves, C.; Vasconcellos, P. Organic compounds in particulate and gaseous phase collected in the neighbourhood of an industrial complex in São Paulo (Brazil). *Air Qual. Atmos. Health* **2017**, *11*, 271–283. [CrossRef]
26. Wei, S.; Huang, B.; Liu, M.; Bi, X.; Ren, Z.; Sheng, G.; Fu, J. Characterization of PM2.5-bound nitrated and oxygenated PAHs in two industrial sites of South China. *Atmos. Res.* **2012**, *109*, 76–83. [CrossRef]
27. Alves, C.A. Characterisation of solvent extractable organic constituents in atmospheric particulate matter: An overview. *Anais da Academia Brasileira de Ciências* **2008**, *80*, 21–82. [CrossRef]
28. Gupta, S.; Gadi, R.; Mandal, T.; Sharma, S. Seasonal variations and source profile of n-alkanes in particulate matter (PM10) at a heavy traffic site, Delhi. *Environ. Monit. Assess.* **2016**, *189*. [CrossRef]
29. Scala, R.S.; Smith, J.E. An improved measure of the odd-even predominance in the normal alkanes of sediment extracts and petroleum. *Geochem. Cosmochim. Acta* **1970**, *34*, 611–620. [CrossRef]
30. Kavouras, I.G.; Stratigakis, N.; Stephanou, E.G. Iso- and anteiso-alkanes: Specific tracers of environmental tobacco smoke in indoor and outdoor particle-size distributed urban aerosols. *Environ. Sci. Technol.* **1998**, *32*, 1369–1377. [CrossRef]
31. Kavouras, I.G.; Lawrence, J.; Koutrakis, P.; Stephanou, E.G.; Oyola, P. Measurement of particulate aliphatic and polynuclear aromatic hydrocarbons in Santiago de Chile: Source reconciliation and evaluation of sampling artifacts. *Atmos. Environ.* **1999**, *33*, 4977–4986. [CrossRef]
32. Andrade, M.F.; Kumar, P.; Freitas, E.D.; Ynoue, R.Y.; Martins, J.; Martins, L.D.; Nogueira, T.; Perez-Martinez, P.J.; De Miranda, R.M.; Albuquerque, T.T.D.A.; et al. Air quality in the megacity of São Paulo: Evolution over the last 30 years and future perspectives. *Atmos. Environ.* **2017**, *159*, 66–82. [CrossRef]
33. Gaffney, J.S.; Marley, N.A. Atmospheric chemistry and air pollution. *Sci. World J.* **2003**, *3*, 199–234. [CrossRef]

34. CONAMA Conselho Nacional do Meio Ambiente. RESOLUÇÃO/conama/N.° 003 de 28 de junho de 1990. Brasil, 1990. Available online: http://www2.mma.gov.br/port/conama/res/res90/res0390.html (accessed on 25 June 2019).
35. CETESB. Padrões de Qualidade do Ar. 2018. Available online: https://cetesb.sp.gov.br/ar/padroes-de-qualidade-do-ar/ (accessed on 6 June 2019).
36. Alves, C.; Oliveira, C.; Martins, N.C.T.; Mirante, F.; Caseiro, A.; Pio, C.; Matos, M.; Silva, H.; Oliveira, C.; Camões, F. Road tunnel, roadside, and urban background measurements of aliphatic compounds in size-segregated particulate matter. *Atmos. Res.* **2016**, *168*, 139–148. [CrossRef]
37. Liu, Y.; Gao, Y.; Yu, N.; Zhang, C.; Wang, S.; Ma, L.; Zhao, J.; Lohmann, R. Particulate matter, gaseous and particulate polycyclic aromatic hydrocarbons (PAHs) in an urban traffic tunnel of China: Emission from on-road vehicles and gas-particle partitioning. *Chemosphere* **2015**, *134*, 52–59. [CrossRef]
38. Brito, J.; Carbone, S.; Dos Santos, D.A.M.; Dominutti, P.; Alves, N.D.O.; Rizzo, L.V.; Artaxo, P. Disentangling vehicular emission impact on urban air pollution using ethanol as a tracer. *Sci. Rep.* **2018**, *8*, 10679. [CrossRef] [PubMed]
39. Simoneit, B.R.T.; Cardoso, J.N.; Robinson, N. An assessment of the origin and composition of higher molecular weight organic matter in aerosols over Amazonia. *Chemosphere* **1990**, *21*, 1285–1301. [CrossRef]
40. Simoneit, B.R.T. Organic matter of the troposphere-III. Characterization and sources of petroleum and pyrogenic residues in aerosols over the western united states. *Atmos. Environ. (1967)* **1984**, *18*, 51–67. [CrossRef]
41. Vasconcellos, P.C.; Carvalho, L.R.F.; Pool, C.S. Volatile organic compounds inside urban tunnels of São Paulo City, Brazil. *J. Br. Chem. Soc.* **2005**, *16*, 1210–1216. [CrossRef]
42. WHO. *Air Quality Guidelines*; WHO: Geneva, Switzerland, 2000.
43. Pereira, P.A.D.P.; Lopes, W.A.; Carvalho, L.S.; Da Rocha, G.O.; Bahia, N.D.C.; Loyola, J.; Quiterio, S.L.; Escaleira, V.; Arbilla, G.; De Andrade, J.B. Atmospheric concentrations and dry deposition fluxes of particulate trace metals in Salvador, Bahia, Brazil. *Atmos. Environ.* **2007**, *41*, 7837–7850. [CrossRef]
44. Wang, H.; Kawamura, K.; Shooter, D. Wintertime organic aerosols in Christchurch and Auckland, New Zealand: Contributions of residential wood and coal burning and petroleum utilization. *Environ. Sci. Technol.* **2006**, *40*, 5257–5262. [CrossRef] [PubMed]
45. Vasconcellos, P.; Souza, D.Z.; Ávila, S.G.; Araújo, M.P.; Naoto, E.; Nascimento, K.H.; Cavalcante, F.S.; Dos Santos, M.; Smichowski, P.; Behrentz, E.; et al. Comparative study of the atmospheric chemical composition of three South American cities. *Atmos. Environ.* **2011**, *45*, 5770–5777. [CrossRef]
46. Booth, A.M.; Sutton, P.A.; Lewis, C.A.; Lewis, A.C.; Scarlett, A.G.; Chau, W.; Widdows, J.; Rowland, S.J. Unresolved Complex Mixtures of Aromatic Hydrocarbons: Thousands of Overlooked Persistent, Bioaccumulative, and Toxic Contaminants in Mussels. *Environ. Sci. Technol.* **2007**, *41*, 457–464. [CrossRef] [PubMed]
47. Paulo, F. Lei Ordinária 316—2001 São Paulo, SP. Legislacao Municipal do Estado de Sao Paulo; 2003. Available online: https://www.al.sp.gov.br/propositura/?id=107180&tipo=1&ano=2001 (accessed on 15 June 2019).
48. Masiol, M.; Harrison, R.M. Aircraft engine exhaust emissions and other airport-related contributions to ambient air pollution: A review. *Atmos. Environ.* **2014**, *95*, 409–455. [CrossRef]
49. Chen, Y.; Cao, J.; Zhao, J.; Xu, H.; Arimoto, R.; Wang, G.; Han, Y.; Shen, Z.; Li, G. n-Alkanes and polycyclic aromatic hydrocarbons in total suspended particulates from the southeastern Tibetan Plateau: Concentrations, seasonal variations, and sources. *Sci. Total. Environ.* **2014**, *470*, 9–18. [CrossRef] [PubMed]
50. Li, W.; Peng, Y.; Bai, Z. Distributions and sources of n-alkanes in PM 2.5 at urban, industrial and coastal sites in Tianjin, China. *J. Environ. Sci.* **2010**, *22*, 1551–1557. [CrossRef]
51. Wang, G.; Kawamura, K. Molecular characteristics of urban organic aerosols from Nanjing: A case study of a mega-city in China. *Environ. Sci. Technol.* **2005**, *39*, 7430–7438. [CrossRef]
52. Tang, X.L.; Bi, X.H.; Sheng, G.Y.; Tan, J.; Fu, J.M. Seasonal Variation of the Particle Size Distribution of n-Alkanes and Polycyclic Aromatic Hydrocarbons (PAHs) in Urban Aerosol of Guangzhou, China. *Environ. Monit. Assess.* **2006**, *117*, 193–213. [CrossRef]
53. Oros, D.R.; Standley, L.J.; Chen, X.; Simoneit, B.R. Epicuticular Wax Compositions of Predominant Conifers of Western North America. *Zeitschrift für Naturforschung C* **1999**, *54*, 17–24. [CrossRef]

54. Sun, Q.; Xie, M.; Shi, L.; Zhang, Z.; Lin, Y.; Shang, W.; Wang, K.; Li, W.; Liu, J.; Chu, G. Alkanes, compound-specific carbon isotope measures and climate variation during the last millennium from varved sediments of Lake Xiaolongwan, northeast China. *J. Paleolimnol.* **2013**, *50*, 331–344. [CrossRef]
55. Ladji, R.; Yassaa, N.; Balducci, C.; Cecinato, A.; Meklati, B.Y. Annual variation of particulate organic compounds in PM10 in the urban atmosphere of Algiers. *Atmos. Res.* **2009**, *92*, 258–269. [CrossRef]
56. De Miranda, R.M.; Perez-Martinez, P.J.; Andrade, M.D.F.; Ribeiro, F.N.D. Relationship between black carbon (BC) and heavy traffic in São Paulo, Brazil. *Transp. Res. Part D Transp. Environ.* **2019**, *68*, 84–98. [CrossRef]
57. Pacheco, M.T.; Parmigiani, M.M.M.; Andrade, M.D.F.; Morawska, L.; Kumar, P. A review of emissions and concentrations of particulate matter in the three major metropolitan areas of Brazil. *J. Transp. Health* **2017**, *4*, 53–72. [CrossRef]
58. Ministério do Meio Ambiente. *Proconve: Programa de Controle de Poluição do ar Por Veículos Automotores*; Ministério do Meio Ambiente: Brasilia, Brazil, 2018.
59. Bozzetti, C.; El Haddad, I.; Salameh, D.; Daellenbach, K.R.; Fermo, P.; Gonzalez, R.; Minguillon, M.C.; Iinuma, Y.; Poulain, L.; Elser, M.; et al. Organic aerosol source apportionment by offline-AMS over a full year in Marseille. *Atmos. Chem. Phys. Discuss.* **2017**, *17*, 8247–8268. [CrossRef]
60. Xie, M.; Wang, G.; Hu, S.; Han, Q.; Xu, Y.; Gao, Z. Aliphatic alkanes and polycyclic aromatic hydrocarbons in atmospheric PM10 aerosols from Baoji, China: Implications for coal burning. *Atmos. Res.* **2009**, *93*, 840–848. [CrossRef]

© 2020 by the authors. Licensee MDPI, Basel, Switzerland. This article is an open access article distributed under the terms and conditions of the Creative Commons Attribution (CC BY) license (http://creativecommons.org/licenses/by/4.0/).

Article

Air Quality Degradation by Mineral Dust over Beijing, Chengdu and Shanghai Chinese Megacities

Mathieu Lachatre [1,*,†], Gilles Foret [1], Benoit Laurent [1], Guillaume Siour [1], Juan Cuesta [1], Gaëlle Dufour [1], Fan Meng [2], Wei Tang [2], Qijie Zhang [3] and Matthias Beekmann [1]

1. Laboratoire Inter-Universitaire des Systèmes Atmosphériques (LISA), UMR CNRS 7583, CNRS, Université Paris Est Créteil et Université de Paris, Institut Pierre Simon Laplace, 94000 Créteil, France; Gilles.Foret@lisa.u-pec.fr (G.F.); Benoit.Laurent@lisa.u-pec.fr (B.L.); guillaume.siour@lisa.u-pec.fr (G.S.); Juan.Cuesta@lisa.u-pec.fr (J.C.); gaelle.dufour@lisa.u-pec.fr (G.D.); Matthias.Beekmann@lisa.u-pec.fr (M.B.)
2. Chinese Research Academy of Environmental Sciences, Atmospheric Environment Institute, Beijing 100012, China; mengfan@craes.org.cn (F.M.); tangwei@craes.org.cn (W.T.)
3. Yunyiran Kinton Technology Co. Ltd., Nanchang 330000, China; jerry_eshk@163.com
* Correspondence: mathieu.lachatre@lmd.polytechnique.fr
† Current address: LMD/IPSL, École Polytechnique, Institut Polytechnique de Paris, ENS, PSL Université, Sorbonne Université, CNRS, 91128 Palaiseau, France.

Received: 14 May 2020; Accepted: 29 June 2020; Published: 2 July 2020

Abstract: Air pollution in Chinese megacities has reached extremely hazardous levels, and human activities are responsible for the emission or production of large amounts of particulate matter (PM). In addition to PM from anthropogenic sources, natural phenomena, such as dust storms over Asian deserts, may also emit large amounts of PM, which lead episodically to poor air quality over Chinese megacities. In this paper, we quantify the degradation of air quality by dust over Beijing, Chengdu and Shanghai megacities using the three dimensions (3D) chemistry transport model CHIMERE, which simulates dust emission and transport online. In the first part of our work, we evaluate dust emissions using Moderate Resolution Imaging Spectroradiometer (MODIS) and Infrared Atmospheric Sounding Interferometer (IASI) satellite observations of aerosol optical depth, respectively, in the visible and the thermal infrared over source areas. PM simulations were also evaluated compared to surface monitoring stations. Then, mineral dust emissions and their impacts on particle composition of several Chinese megacities were analyzed. Dust emissions and transport over China were simulated during three years (2011, 2013 and 2015). Annual dust contributions to the PM_{10} budget over Beijing, Chengdu and Shanghai were evaluated respectively as 6.6%, 9.5% and 9.3%. Dust outbreaks largely contribute to poor air quality events during springtime. Indeed it was found that dust significantly contribute for 22%, 52% and 43% of spring PM_{10} events (for Beijing, Chengdu and Shanghai respectively).

Keywords: mineral dust; air quality; modeling

1. Introduction

Chinese atmospheric pollution is a major health problem with more than 82% of Chinese population living in environment exceeding particulate matter ($PM_{2.5}$) concentrations of 75 µg m^{-3} [1] (Chinese National Standard GB 3095-2012 for $PM_{2.5}$ in an urban environment, 24H mean: 75 µg m^{-3}; annual mean: 35 µg m^{-3}). During haze pollution events, $PM_{2.5}$ levels have even exceeded 75 µg m^{-3} in 74 China major cities (800 million inhabitants) for 69% of days in January 2013—reaching a daily concentration of 772 µg m^{-3} for $PM_{2.5}$ [2]. A [3] recent study has aggregated chemical speciation from several Chinese cities and report that particle's composition is dominated by organic matter (26.0%), sulfate (17.7%), mineral dust (11.8%), nitrate (9.8%), ammonium (6.6%) and elemental carbon

(6.0%). The People's Republic of China has seen its population exposed to strongly increasing pollution levels until the first decade of the 21st century. As a consequence, daily sunshine duration has been reduced [4] and premature mortality went up from 0.9 million in 2000 to more than 1.2 million (+33%) in 2010 [5], while population increased only by 6% in the same time. Indeed, China has observed an important growth in several economic sectors such as transport increasing from 24 millions vehicles in 2003 to 78 millions vehicles in 2010 [6], energy production with an increase of 250% between 2000 and 2010 [7], construction and other sectors. To lower anthropogenic pollutants emissions, China has developed its environmental policy and tightened air quality standards, leading to a decrease of $PM_{2.5}$ after 2007 [8] or latter for some components, as of NO_x after 2012 [9]. Besides, China comprises extended arid areas, as the Taklimakan desert (270×10^3 km^2), or the Gobi desert (1.3×10^6 km^2) that naturally emit PM. As a result, Chinese air pollution is a mix of pollutants that come from a multitude of sources. Pollution sources will differently affect Chinese cities depending on their locations, their geographical situations or their meteorological situations.

Mineral dust aerosol from deserts represents about 40% of total aerosol mass emitted each year into the atmosphere [10]. Global dust emission estimates range from 1000 Mt.year^{-1} to 5000 Mt.year^{-1} [11,12]. The authors of [11] compile several model studies and estimate a global emission median value of 1572 Mt.year^{-1}. It is thus the most important aerosol component in term of mass for the global atmosphere. Mineral dust aerosol is characterized by a very high spatial and temporal variability, due to the episodic character of emission events associated to meteorological conditions like surface wind speed, and surface characteristics like protrusion elements. Dust aerosols must also be considered and study for their direct and indirect radiative impacts [13,14]. A previous study [15] has estimated that Chinese and Mongolian annual desert dust emissions between 1996 and 2001 are 240 Mt.year^{-1} (+/− 130 Mt.year^{-1}) contributing to between 10 to 25% of global dust emissions. For spring time only, [16] found an emission of 252 Mt.year^{-1}. Mineral dust lifetime in the atmosphere can vary from few hours to several days, [17,18] depending on meteorological conditions, and mineral dust particles can impact air quality of Chinese cities. Thus, dust can be transported over large areas and impact cities PM load: a one-year study, based on 2000, from [19] resulted in an annual dust average in $PM_{2.5}$ fraction of 13 µg m^{-3} in Beijing and 7 µg m^{-3} in Shanghai (corresponding to about 12% of $PM_{2.5}$ load in both cities). It shows high seasonal variability, with higher values in spring. In this study, maximum daily means dust content reaches 40 µg m^{-3} in Beijing and 30 µg m^{-3} in Shanghai. Recent studies have used Lagrangian modeling to determine the dust contribution to $PM_{2.5}$ and PM_{10} in Chinese cities or study dust vertical diffusion and transport pathways [20,21]. Reference [20] has simulated dust plume over East China region and focused on a case study in spring 2015 which shows a large contribution of mineral elements to $PM_{2.5}$ (34% of Ca^{2+}). The authors of [21] consider that dust plumes from Taklimakan are most commonly transported around 1.7 km above sea level.

In this work, we simulated mineral dust aerosol emissions from deserts and their transport across China. From these simulations, mineral dust impact on particulate matter load in several Chinese cities compared to anthropogenic pollution was studied. Three years were simulated (2011, 2013 and 2015) to study inter-annual variability and seasonal variations. In this study, we first evaluated the accuracy of modeled dust emissions, using satellite Aerosol Optical Depth (AOD) observations over source regions, then, we evaluated the regional simulated PM concentration with in situ measurements. Finally, the model capacity to represent PM chemical speciation on a daily basis and during several years allowed us to investigate the part of dust to PM average and peak burdens in several cities. It was also possible to determine dust burden from each desert area for each studied city. The results depend on the model ability to reproduce as correctly as possible for a long-term period the various geophysical processes that control Chinese cities air quality, from mineral dust emissions, transport to anthropogenic pollutants modeling. Thus, the CHIMERE model allows us to propose quantitative results and study the frequencies of pollution events controlled by dust transported to Chinese megacities, which brings additional information to the existing literature.

Beijing, Chengdu and Shanghai are respectively the 2nd, 6th and 1st most populated cities in China, with 22.5, 14.5 and 24.5 million of inhabitants. These three cities also present different geographic situations, Beijing is located in Northern China (39°54′13″ N; 116°23′15″ E), Chengdu in Central China (30°39′00″ N; 104°04′00″ E) and Shanghai in Eastern China (31°13′56″ N; 121°28′09″ E). Beijing and Shanghai are in a similar monsoon system, with strong rainfall within summer season. Chengdu is located in a basin surrounded by plateau and high mountains (Tianshan, Qinling to the north) and climate, especially wind, is more continental and quite different. This induces different exposure to mineral dust pollution from sources mainly located in Western and Northern China (Cities location displayed later in the document). This choice is similar to cities selection made in previous work which reviewed observations of PM composition for representative Chinese megacities [22].

In Section 2, the paper first presents the material and method used in the study. Model and observations are described, then evaluation methods are explained. In Section 3, results are presented, starting with the characterization of mineral dust emissions, and then evaluating its contribution to urban particle pollution for the selected cities. Section 4 gives conclusions.

2. Materials and Method

2.1. The CHIMERE Chemistry Transport Model

In this work, we used the CHIMERE 3D regional chemistry transport model (CTM) ([23,24], 2014b version) run over a 0.25° × 0.25° regular grid. The domain is chosen large enough to include the Taklimakan desert on the West, Japan on the East and China's northern and southern territories (72°30′ E–145° E; 17°30′ N–55° N, simulation domain is displayed Figure 2). The domain is composed by 290 (longitude) × 150 (latitude) grid cells and 17 vertical layers, from the ground to 200 hPa. Vertical layer thicknesses are increasing with altitude, 8 layers lie within the first 2 km of altitude. Advection is resolved using the Van Leer [25] second-order slope-limited transport scheme. Meteorological forcing is generated by ECMWF-IFS meteorological forecasts [26] and then interpolated to hourly resolution.

In our simulations, aerosols were distributed into 10 size-classes (also denoted as bins), from 0.05 µm to 40 µm. 6 bins that correspond to $PM_{2.5}$, 2 bins describe PM between 2.5 µm and 10 µm and finally 2 describe PM coarser than 10 µm. The model's AOD is calculated using a Fast-J photolysis scheme [27] and considering optical properties (Mie scattering, absorption) for each bin and each aerosols species.

2.1.1. Anthropogenic and Biogenic Aerosols Modeling

EDGAR-HTAP V2.2 emission inventories [28] based on 2010 were used to generate anthropogenic emissions (gaseous species and particulate matter). NO_x and SO_2 emissions for 2013 and 2015 have been derived from remote sensing observation (SO_2 and NO_x columns from OMI instrument), with a method used and evaluated in [29], leading to decreases of 37% (SO_2) and 21% (NO_x) between 2011 and 2015, similar to recent inventory trends [30]. Biogenic emissions are generated by the MEGAN-v2 model [31]. Climatological boundary and initial conditions for dust are obtained from the GOCART global model [17] and from the LMDZ-INCA global model for others species [32,33].

Composition and phase state of inorganic aerosol is tabulated by using the ISORROPIA V2006 module [34]. It calculates the partitioning of $NH_{3(g)}/NH_{4(p)}^+$, $HNO_{3(g)}/NO_{3(p)}^-$, and $H_2SO_{4(p)}/HSO_{4(p)}^-/NO_{3(p)}^-$, from the initial gaseous and particulate precursors content and meteorological conditions (temperature, relative humidity). Organic and inorganic species gas-phase chemistry is described with the reduced MELCHIOR2 mechanism [35], and the equilibrium for organic species between gas and particle phase is calculated using [36,37].

In this study, PM chemical speciation is divided into 8 different species (from 13 in our simulations): Dust (only representing the desert and natural fraction), $NH_{4(p)}^+$ (ammonium), SO_4^{2-} (sulfate), $NO_{3(p)}^-$ (nitrate), AMPP (Anthropogenic Mineral Primary Particulate matter—including soil dust emitted from anthropogenic activities such as construction), OM (primary and secondary Organic Matter from biogenic and anthropogenic sources) , BC (Black Carbon) and SALT (sea salt).

2.1.2. Mineral Dust Aerosol Modeling

The CHIMERE model dust modeling was evaluated in a multi model comparison on Northern Chinese areas [38]. In this study, CHIMERE is the only model used and evaluate without need in tuning dust concentrations and which showed good performances.

Mineral dust emissions are a threshold phenomenon, which occurs when the wind friction force exercised on soil particle aggregates is higher than forces maintaining these aggregates on the ground. When a threshold wind friction velocity is exceeded, the wind kinetic energy will mobilize soil aggregates into a horizontal flux, a process called saltation. During saltation, soil aggregates go rolling along the surface, and then, shocks between aggregates and soil will liberate fine soil material from aggregates, in a process called sandblasting. The dust emission calculation requires a soil properties database (roughness length, texture and soil aggregates size distribution) as developed in [39] for China. Saltation fluxes are defined in [40] and calculated as follows (Equation (1)):

$$F_h(D_p) = \frac{K\rho_{air}U^{*3}}{g}\left(1 - \frac{U_t^*}{U^*}\right)\left(1 + \frac{U_t^*}{U^*}\right)^2 \tag{1}$$

where U^* is the friction wind speed, U_t^* is the threshold friction wind speed, calculated following Marticorena and Bergametti (1995). K is constant with value equal to 1, ρ_{air} is the air density and the gravitational acceleration $g = 9.8\,\mathrm{m.s^{-2}}$.

The sandblasting process and vertical emission fluxes were calculated in CHIMERE from a parameterization by [41], which calculates the emission fluxes for a given aerosol size distribution, following Equation (2) [40]:

$$F_{(v,m,i)}(D_p) = \sum_{k=1}^{Nclass} \frac{\pi\rho_p\beta p_i(D_{p,k})d_{m,i}^3}{6e_i} dF_h(D_p) \tag{2}$$

where $Nclass$ corresponds to the intervals of soil size distribution, p_i the relative fraction corresponding to $Nclass$, e_i is individual kinetic energy of aggregate, ρ_p is particle density and β an acceleration constant. We incorporated an optimization of emissions modeling, by adding a criterion on meteorological conditions. When the precipitation rate (inquired by ECMWF) is greater than $0.01\,\mathrm{kg.m^{-2}.h^{-1}}$, we assume an increase of binding energies in the soil that will inhibit the dust emissions for the next two hours.

Three dust size-distribution parameterizations based on a physical description of the emission processes are commonly used in models: [41–44], these parameterizations have been validated on a reduced set of data and the simulation of the dust size distribution remains uncertain mainly due to our ability to document correctly the soil properties at the emissions. The multi model evaluation [38] study included CHIMERE model with [41,43] parameterizations; finally the parameterization [41] was retained for their multi model comparison, due to better results. In our simulations [41,45] was used, about 15% of dust emissions were distributed into $PM_{2.5}$ (similar as [46]), 61% were distributed into the coarse fraction of PM_{10} excluding $PM_{2.5}$ ($PM_{2.5}/PM_{10} = 0.20$), finally, 24% were distributed into a PM fraction with a diameter larger 10 µm, a fraction, which is not always considered [47]. This slightly differs from the 9.8% distribution proposed by [43,47,48] for the mass distributed into the $PM_{2.5}$ which might results to higher AOD at 550 nm close to source regions (AOD at that wavelength being particularly sensitive to particles between 0.5 µm-1 µm [49]). However, even if the chosen parameterization impacts the $PM_{2.5}/PM_{10}$ ratio, the study mainly focus on the PM_{10} fraction (76% of total mass emitted in our case). The main issue to the $PM_{2.5,dust}/PM_{10,dust}$ distribution is finally the difference of transport behavior between the 2 particle range sizes, as the $PM_{2.5,dust}$ fraction is most efficiently transported—the ratio increases to 0.32, 0.25 and 0.34 for Beijing, Chengdu and Shanghai respectively—which might induce an overestimation of $PM_{10,dust}$ mass.

CHIMERE model and modules are described in details in the model documentation.

In our study, simulations were performed for 2011, 2013 and 2015, to study inter annual variability of dust emissions. The years 2011 and 2013 were chosen because numerous dust events have been observed for those years [50,51]. However, these years were not abnormally high in total dust emissions from deserts [52] or in mineral dust detected in cities [53].

2.1.3. Method to Determine Dust Origins in Cities

To determine the contribution of each area to dust load in target cities, we successively inhibited dust emissions from the various arid areas to tag dust desert origin (Equation (3)).

$$Contribution_{(n,i,t)} = \frac{(dust)_{(0,i,t)} - (dust)_{(n,i,t)}}{(dust)_{(0,i,t)}} \tag{3}$$

where $(dust)_{(0)}$ represents the baseline simulation in which emissions are included for all areas, n indicates computed simulations while inhibiting arid areas emissions, i pixel locations and t timestep. Simulations have been performed inhibiting one area at a time, and an additional simulation has been conducted by removing dust coming from outside the domain (i.e., boundary conditions) to evaluate their impact. This method will lead to some error in deposition process modeling, because deposition processes are non linear, depend on multiple factors such as meteorology, interactions between particles and soils with different types and properties [54,55]. The resulting underestimation for the dust concentration of 9% for Beijing, 8% for Chengdu and 13% for Shanghai and is considered as still acceptable.

2.2. AOD Data Set and Its Use for Model Evaluation

Satellite based information from the Moderate Resolution Imaging Spectroradiometer (MODIS) instrument, based on the NASA Terra satellite, was used to evaluate simulated dust emissions, as in [56]. The MODIS AOD product was largely used to study particulate pollution and decently compared to other AOD products as AERONET or MISR [52]. The MODIS instrument provides a global sampling of the Earth once a day, passing around 10:30 a.m. at local time. It provides information on atmospheric column content as a combined Dark Target and Deep Blue 550 nm AOD product, on a $1° \times 1°$ resolution grid. Additionally, we compared AODs derived at 10 µm from thermal infrared measurements of the Infrared Atmospheric Sounding Interferometer (IASI) onboard the MetOp-A satellite (overpass at 09:30 a.m. local time) with those calculated from simulated dust distributions by CHIMERE ([57], using a Mie code and Asian dust refractive indexes from). IASI-derived AODs in the thermal infrared are estimated with the newly-developed AEROIASI retrieval approach ([58,59], version 2). We used CHIMERE simulations at 10:00 a.m. local time for the comparisons of AODs derived by both MODIS and IASI.

We first focused on dust emissions areas, to evaluate the simulated emissions. We verified that changes brought by dust emission inhibition by rainfall have improved correspondences between MODIS and CHIMERE AODs. For the 4 arid sub regions, AOD is mostly controlled by mineral dust, as mineral dust emissions represent more than 90% of the mass of total pollutant emissions for these regions.

The large dust emitting areas were divided into 4 regions (Figure 1) and compared to MODIS observations: the so-called Taklimakan desert area, which includes Taklimakan, the Kumtaq desert and the Qaidam desert; the Gobi desert, mainly situated over Mongolia; the Gurban desert, situated between the Taklimakan and Gobi desert; the last and smallest area, which is also the closest to China's biggest cities, contains several Chinese deserts (Tengger desert, Ulan Buh desert, Qubqi desert and Mu Us sandy land) and is called hereafter the Northern China desert. This separation of arid areas into smaller areas allows for a finer comparison between model and observations and a better analysis of dust variability; indeed the heterogeneity (especially in terms of emission frequencies) between each arid sub region is very important. The delimited sub-regions contain most of dust emissions, in our case, ~98% of total dust emissions. The sub regions are different with respect to their emission/surface ratio (Table 1). Taklimakan is the main dust source region with $1\,460 \times 10^3$ km^2. The Northern China desert region is the smallest area (260×10^3 km^2), but is located closer to China East coast, and then to densely populated areas. The Mongolian Gobi desert is the largest area ($1\,630 \times 10^3$ km^2), but is split into two areas, the Gurban Tunggut desert (211×10^3 km^2; Called Gurban desert hereafter) and the main part of the Mongolian Gobi desert ($1\,104 \times 10^3$ km^2). It was necessary to split up the Mongolian Gobi Desert, as the temporality in dust emissions for the considered sub domains was different. Statistical results are presented in Table 2.

The AERONET network (AErosol RObotic NETwork, aeronet.gsfc.nasa.gov last consulted 09/30/2018) distributes observations of AOD for several wavelengths, and Ångström coefficients, with an hourly resolution for 2011, 2013 and 2015 in Beijing. Ångström coefficients provide information on particle size distribution in the atmospheric column, and can be calculated from AOD measurements at several wavelengths. Larger particles are characterized by lower Ångström coefficients and vice-versa. A value below 0.4 characterizes coarse particles such as dust [60].

Table 1. Mineral dust emissions from the different source regions, annual average calculated from 2011, 2013 ad 2015 simulations. Last column indicates the proportion of mineral dust emitted during the 5% and 20% strongest emitting days.

Areas	Mean Emissions (Mt.y^{-1})/ Contribution	Standard dev. (Mt)	Emis/Surf (10^{-3}t.km^{-2})	% of Mass Emitted in 5%/20% Strongest Days
Taklimakan desert	198 Mt/70%	15 Mt	135	62%/94%
Mongolian G. desert	65 Mt/23%	7 Mt	50	53%/90%
Northern C. desert	18 Mt/ 7%	6 Mt	71	82%/99%
Total domain	283 Mt/-	28 Mt	-	54%/87%

2.3. Surface Measurements Data Set and Comparison Methods

CHIMERE PM modeling in populated areas was also evaluated against PM$_{2.5}$ and PM$_{10}$ measurements from background stations close to cities, during spring periods and full years. We first wanted to assure that CHIMERE correctly models particulate pollution levels and variability in Beijing, Chengdu and Shanghai.

A comparison between model and hourly measurements from Chinese monitoring was performed from PM$_{2.5}$ and PM$_{10}$ spring season for 2013 (March–April) and 2015 (March–April–May). We focused our comparisons on these periods, because of data availability and because the dust impact was expected to be the strongest in spring. PM measurements are from TEOM instruments [61].

We first used a method to evaluate the representativeness of monitoring stations, from [62] which uses the relative diurnal variability of SO_2 or O_3, in order to evaluate if the considered stations were representative of rural, suburban, urban or traffic environment. This was motivated by the need to compare model simulations to monitoring stations with a coherent spatial resolution representativeness and in our case, stations near sources are not eligible. Based on [62] method results, the Beijing, Chengdu and Shanghai stations are located in a rural environment.

Additionally, similar information is also available for an other 30 stations located in a rural environment and 3 located in a suburban environment, located in 21 different cities. These stations were used to evaluate model performances on various areas, but the results are not detailed. Station locations can be found in Figure S5 (in the supplementary section).

Model versus measurement comparisons were conducted on an hourly basis, sampling model output depending on measurements' availability. Simulated model concentrations are bi-linearly interpolated at stations' coordinates.

We also used data from the U.S. Embassy and consulates over 3 cities (e.g., in Beijing, Chengdu and Shanghai, in urban background environment) to evaluate the $PM_{2.5}$ correspondences over a larger period (2011, 2013 and 2015). $PM_{2.5}$ is measured on an hourly basis with a MetOne BAM 1020 instrument [63]. It has to be noted that 2011 measurements are available only for Beijing. Comparison methods applied are the same as for previous comparisons, but daily values are calculated from hourly data depending on measurements availability, and compared.

Statistical results of comparisons, such as normalized bias, Normalized Root Mean Square Error (NRMSE) and correlation mean are presented in Tables 3 and 4 for spring seasons in 2011, 2013 and 2015 and in Figure S2 (in supplement).

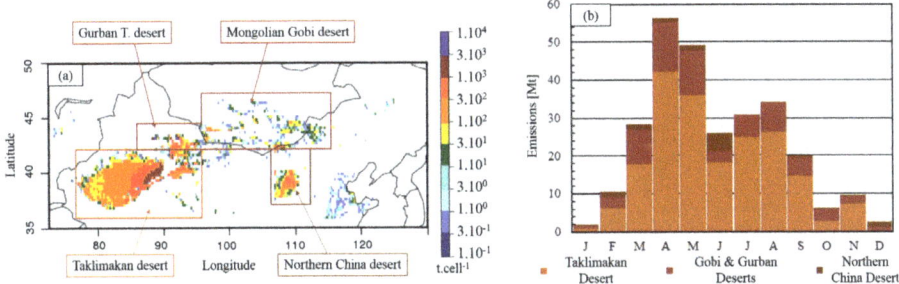

Figure 1. (a) Mean annual dust emissions from the main arid areas in Asia, simulated with CHIMERE for 2011, 2013 and 2015, in t.cell^{-1}. (b) Monthly emissions for the main arid areas in Mt.

3. Results and Discussion

3.1. Dust Emissions and Evaluation

3.1.1. Mineral Dust Emissions and Seasonality

Dust emissions yearly means for each desert areas are displayed in Table 1 and Figure 1a. The Taklimakan desert is the main dust source region with 198 Mt.year^{-1} emissions. We simulated 18.5 Mt.year^{-1} dust emissions from the Northern China Desert (71×10^{-3} t.km^{-2}). From the Mongolian Gobi desert 65.2 Mt.year^{-1} were emitted (50×10^{-3} t.km^{-2}). On average, total emissions reach about 283 Mt.year^{-1}, with 51% of mass emitted within spring season, 31% emitted in summer, 12% emitted in fall and about 6% emitted in winter (Figure 1b). A result close to [11], who indicate a median value of 294 Mt.year^{-1} for Asia (standard deviation: 253 Mt.year^{-1}), obtained from an ensemble with 15 models. Lower results were obtained in [15,64] with values respectively of 242 Mt (1996–2001 average; $\sigma = 131$ Mt) and 213 Mt (2006 and 2010 average). The CHIMERE dust spatial distribution also fits results found in more recent study as [65].

Dust emissions can strongly vary from one year to another: for 2011, 2013 and 2015 respectively 321, 255 and 274 Mt were emitted each year, which makes a 66 Mt difference between the highest and lowest value (23% of annual mean emissions). Compared to yearly emissions from (294 Mt, 242 Mt, 213 Mt [11,15,65]), we concluded that 2011 emissions are slightly higher than mean values for emissions, but 2013 and 2015 represent average years.

Mean spring emissions for the three selected years are 145 Mt, slightly higher than in [66] with 120 Mt (1960–2003 average), and lower than in [15] with values respectively of 182 Mt (1996–2001 average).

Table 1 presents the percent of dust emitted from the deserts during the 5% highest emitting days. For the Mongolian Gobi desert and the Northern China desert more than 80% of total dust mass is emitted during these days, which indicates intensive, but infrequent events. For the Taklimakan desert and the Gurban desert, more than 50% of mass is emitted during this subset of high emitting days, which still implies intense emissions events but with also more regular mineral dust emissions according to study of [15].

Figure 2 presents CHIMERE dust related 550 nm AOD over China (from ground to 200 hPa), for each season and, calculated from an 2011, 2013 and 2015 average. The largest AOD_{dust} values near emissions areas occur during spring, and are efficiently transported over China and overseas (Figure 2). During summer season (Figure 2) AOD values over deserts are still high, but transport to Eastern China is not pronounced. This is due to particular meteorological conditions during the summer monsoon season, with high rainfall and southerly winds unfavorable for advection of dust to Eastern China. For winter and fall season, modeled dust emissions are low, and thus dust related AODs are low over emissions areas.

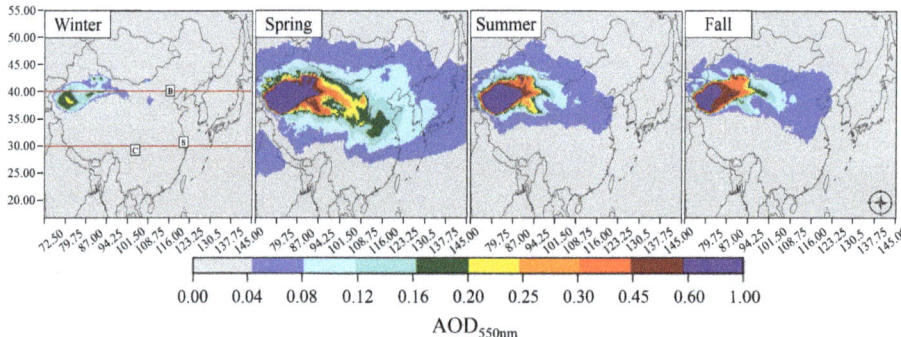

Figure 2. CHIMERE seasonal 550 nm Aerosol Optical Depth (AOD) associated to dust aerosols for Winter, Spring, Summer and Fall. Beijing (B), Chengdu (C) and Shanghai (S) locations are indicated on Winter map. Red lines on Winter map indicate 39.5° and 31.0° latitudes.

3.1.2. CHIMERE AOD Evaluation in Dust Emission Source Areas

Comparisons of CHIMERE AOD against MODIS AOD are gathered in Table 2, the table shows the temporal correspondences (day per day comparison of AOD values). We proceed to a comparison over the different areas for 2011, 2013 and 2015. Daily comparisons were conducted, calculating daily mean AOD for each sub domain with MODIS and CHIMERE outputs, filtering model data depending on MODIS pixels data availability. A daily sub-domain average value was accepted if at least 40% of MODIS data were available. Statistical information was calculated on time series (Bias(%), NRMSE(%) and Pearson correlation coefficient (r)) to estimate correspondences between model and satellite observations.

Table 2. Temporal correspondences between CHIMERE 550 nm AOD and MODIS 550 nm AOD, for 2011, 2013 and 2015 over main arid areas. NRMSE is the Normalized Root Mean Square Error. r is the correlation coefficient. n indicates number of daily mean value for sub domain.

Areas	Bias (%)	NRMSE (%)	r	n
Taklimakan desert	+50%	159%	0.74	1014
Mongolian Gobi desert	−31%	60%	0.51	897
Gurban desert	+64%	161%	0.54	937
Northern China desert	−10%	65%	0.66	862

Figure S1 (in supplements) represents the 550 nm AOD spatial distributions in a $1° \times 1°$ resolution for 2011, 2013 and 2015, as measured by the MODIS instrument and modeled by CHIMERE. Statistical results for AODs spatial distribution comparisons are resumed in Table S1 (in the supplementary section). A similar spatial pattern between simulation and observations can be noted corresponding to a Pearson correlation coefficient of 0.68. For most of the grid cells, CHIMERE AOD underestimates MODIS values (bias $= -40\%$), excepted over Taklimakan and Gurban desert areas, where AOD are overestimated (bias $= +50\%$ and $+64\%$ respectively). Thus, this overestimation could partly be induced by an excessive distribution of mineral dust mass into $PM_{2.5}$.

Indeed, CHIMERE AOD mainly overestimates MODIS AOD for the two most dust emitting areas, but a good daily correlation is obtained (Pearson correlation coefficient $r = 0.74$ for Taklimakan and $r = 0.54$ for the Gurban desert). This statement is also verified for the Mongolian Gobi desert and the Northern Desert ($r = 0.51$ and $r = 0.66$ respectively). These correlations show that the model reproduces at least part of the dust emissions events in a correct timing.

Comparisons for 2011, 2013 and 2015 separately have shown similar results for 2011 and 2013. As for Taklimakan example Pearson correlation coefficients equal $r_{2011} = 0.76$; $r_{2013} = 0.77$ bias equal $+59\%$ and $+51\%$. In 2015, Pearson correlation coefficient is 0.69 and bias equals $+39\%$. The same variability is found for Northern China desert area. For Mongolian Gobi desert, slightly more variability is found between the three years, as for Mongolian Gobi desert area, Pearson correlation coefficient equal $r_{2011} = 0.60$, $r_{2013} = 0.41$ and $r_{2015} = 0.50$, nonetheless bias is quite stable (-35%, -27% and -30%). A similar variability is found for the Gurban desert area.

Additional information is gained from AEROIASI retrieval, deriving AOD at 10 µm. CHIMERE AOD for dust at 10 µm was calculated to compare with an additional independent instrument particularly sensitive to the coarse fraction of aerosols. In this analysis only focusing on dust, we have evaluated the model ability to simulate the occurrence of dust events (with AODs larger than 0.2). Frequencies (in %) of daily dust 10 µm exceeding 0.2 in AEROIASI and CHIMERE are displayed on Figure 3, covering 2011, 2013 and 2015 period (used here for model evaluation, but not as a climatology). The AOD threshold is most likely to be exceeded in desert areas and both retrieval and model highlighted the same patterns. Similarly, as in the MODIS comparison, the Mongolian Gobi desert frequencies are underestimated and Taklimakan frequencies are slightly overestimated but well reproduced.

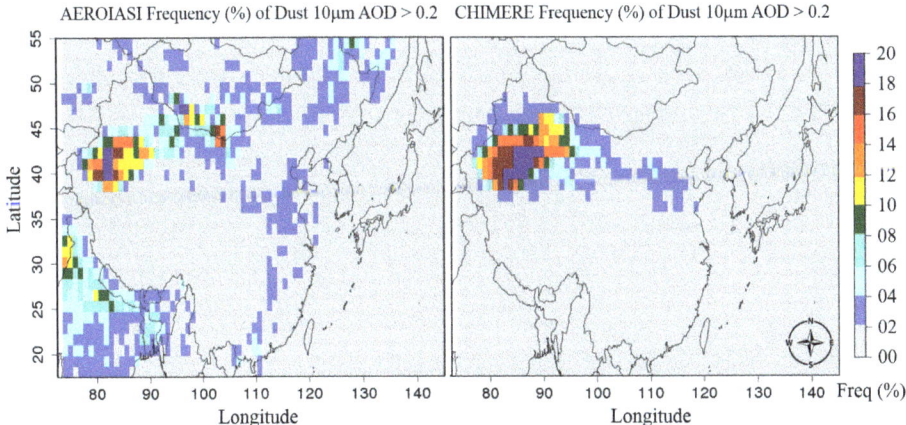

Figure 3. (**left**) AEROIASI Frequency (%) of daily dust 10 µm AOD > 0.2. (**right**) CHIMERE Frequency (%) of daily dust 10 µm AOD > 0.2. Frequencies are calculated over daily value for 2011, 2013 and 2015.

An additional comparison has been done between AEROIASI and CHIMERE AOD with the same criteria as done for the MODIS vs CHIMERE comparison. Similar results have been found for the Taklimakan area ($r = 0.75$, $n = 945$, $NRMSE = 134\%$) with the notable exception for the bias ($bias = 9\%$), which is rather small and much lower than the bias with respect to MODIS AOD at 550 nm. Distinctly of the 550 nm AOD, the 10 µm AOD is mostly controlled by the coarser fraction of dust aerosol. This aspect provides reliability on the amounts of dust simulated by CHIMERE, since most of its mass corresponds to the coarse fraction of aerosols (85% of dust mass correspond to particle larger than 2.5 µm, q.v. Section 2.1).

3.1.3. Dust Vertical Dispersion

Over the Taklimakan area (39.5° of latitude), CHIMERE simulations shows desert dust vertically transported up 4 and 8 km of altitude depending on the season, with a a marked vertical gradient between the surface and the upper layers (see Figure 4 1st row). Such vertical distribution of dust with a marked gradient over several kilometers is consistent with a climatology (2007–2015) of desert dust vertical distribution over this area [67], done with measurements form the CALIOP instrument ([68,69], Cloud Aerosol Lidar with Orthogonal Polarization). However, dust plume transport appears to be more vertically spread in CHIMERE simulations than in the CALIOP climatology in which dust do not reach altitude higher than 6 km, while in CHIMERE simulations they can reach 10 km of altitude. This excessive spreading in vertical transport modeling has already been observed for simulation of volcanoes plumes [24,70,71], and it is a point which requires improvement (e.g., vertical transport scheme or vertical resolution). As a consequence, mineral dust plume is transported over the Himalayas (2nd row, 31.0° of latitude) a path that is not observed in [67] and pollutant in plume can be excessively diffused, leading to low concentrations.

3.2. Dust Contribution to Urban Chinese Particle Pollution

3.2.1. Evaluation of PM Concentration Modeling

CHIMERE simulation results compared to measurements for PM_{10} and $PM_{2.5}$ during 2013 and 2015 springs are displayed in Table 3. A model evaluation with surface measurements is necessary to quantify model skills and limits. As China is a large country with various pollution sources (anthropogenic or naturals), evaluation of surface concentrations is necessary to determine which cities present lower uncertainties in the available database to afterwards evaluate the impact of dust on PM load. PM content is underestimated in eastern and north—western areas, but slightly overestimated over central and southern China compared to measurements. Also, in these areas, NRMSE is smaller than over northern areas. The $PM_{2.5}$ daily variation is better represented than PM_{10}, with higher Pearson correlation coefficients.

Mean, normalized bias, NRMSE and correlation are presented in Table 3, for three selected cities. Normalized mean errors were calculated for $PM_{2.5}$ and PM_{10}, estimated respectively to 21% and 31%. The authors of [72] suggest to evaluate PM modeling performances based on mean fractional bias and error statistical indicators, expressed as in Equations (4) and (5) respectively:

$$MFB = \frac{1}{N} \sum_{i=1}^{N} \frac{(C_m - C_o)}{(C_o + C_m/2)} \quad (4)$$

$$MFE = \frac{1}{N} \sum_{i=1}^{N} \frac{|C_m - C_o|}{(C_o + C_m/2)} \quad (5)$$

where C_m is the CHIMERE estimated concentration at i station, C_o the measured concentration at station i, and N the number of available stations for the considered time period. The authors of [72] suggest that model performances for PM is fairly good for the considered period if MFB is lower than

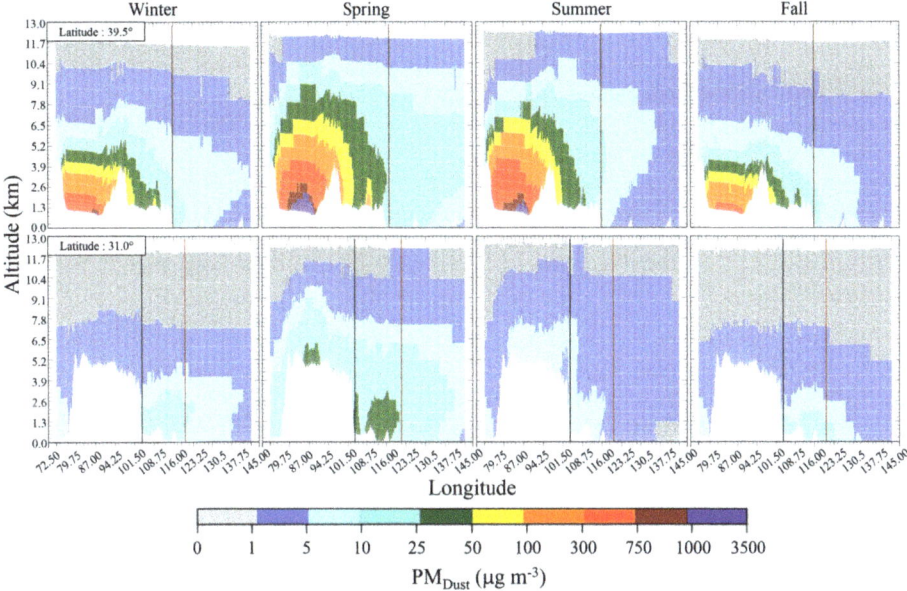

Figure 4. CHIMERE dust concentration vertical profiles longitude cut for winter (1st column), spring (2nd column), summer (3rd column) and fall (4th column). 1st row shows profiles for 39.5° latitude, 2nd row shows profiles for 31.0° latitude. Altitudes are calculated compared to sea level. All figures are produced with 2011, 2013 and 2015 data. Beijing is identified on 1st row by the brown vertical line, Chengdu is identified on 2nd row by the black vertical line and Shanghai is identified on 2nd row by the red vertical line. Transect latitudes are displayed on Figure 2.

0.3 and MFE lower than 0.5. Both criteria are met for $PM_{2.5}$ (−0.002 and 0.14) and PM_{10} (−0.21 and 0.24) for MFB and MFE respectively (MFE and MFB values are calculated on all stations available in the dataset see Section 2.3).

Table 3. CHIMERE Model PM_{10} and $PM_{2.5}$ comparisons to hourly measurements from monitoring stations for springs 2013 and 2015. n indicates number of hourly measurements.

Stations	Meas Mean	Bias (%)	NRMSE(%)	r	n
Beijing PM_{10}	120.7 µg m^{-3}	−26%	73%	0.47	2921
Chengdu PM_{10}	165.1 µg m^{-3}	−10%	67%	0.69	3533
Shanghai PM_{10}	85.9 µg m^{-3}	+05%	47%	0.69	2704
Beijing $PM_{2.5}$	77.3 µg m^{-3}	−06%	57%	0.77	3556
Chengdu $PM_{2.5}$	82.5 µg m^{-3}	+23%	64%	0.69	3648
Shanghai $PM_{2.5}$	54.6 µg m^{-3}	+19%	55%	0.69	3341

We then evaluated $PM_{2.5}$ over 2011, 2013 and 2015 for the three selected cities, from US embassy data. CHIMERE vs US embassy $PM_{2.5}$ comparisons results for 2011, 2013 and 2015 are shown in Table 4. Daily time series for Beijing, Chengdu and Shanghai are displayed in Figure S2 (in supplement). 2011 measurements are available only for the Beijing station. The highest annual mean $PM_{2.5}$ is observed in Beijing (92.2 µg m^{-3}), and the lowest in Shanghai (55.3 µg m^{-3}). Measurements show for all of the three stations a seasonal cycle, with highest pollution levels in winter and lowest pollution levels in summer. It was observed in available time series that Beijing, Chengdu and Shanghai exceed

Chinese standard for daily $PM_{2.5}$ (75 µg m^{-3}; Ambient Air Quality Standards—National Standard GB 3095-2012) respectively on 47%, 43% and 21% of available days. The daily $PM_{2.5}$ variability is correctly represented in CHIMERE with Pearson correlation coefficients of 0.75, 0.72 and 0.76 in Beijing, Chengdu and Shanghai respectively. CHIMERE simulated $PM_{2.5}$ overestimates measurements values, particularly in Chengdu (+52%). Differences between measurements and simulations are larger in 2015, maybe because of a decreasing pollution trend not completely reproduced by 2010 based inventories [29]. CHIMERE also satisfies $PM_{2.5}$ modeling performance goals as suggested by [72], with a MFB and MFE of 0.19 for the three stations for the considered period.

Table 4. CHIMERE model $PM_{2.5}$ comparison to daily measurements from monitoring stations for 2011, 2013 and 2015. n indicates number of daily mean measurements.

Stations	Meas Mean	Bias (%)	NRMSE(%)	r	n
Beijing $PM_{2.5}$	94.2 µg m^{-3}	+19%	61%	0.75	1085
Chengdu $PM_{2.5}$	83.8 µg m^{-3}	+52%	69%	0.72	687
Shanghai $PM_{2.5}$	55.3 µg m^{-3}	+24%	57%	0.76	723

3.2.2. Dust Contribution to Cities' Air Pollution and Dust Origin

Figure S4 (in the supplementary section) presents PM_{10} and respective dust contributions for 2011, 2013 and 2015 spring seasons to PM_{10}. As dust emissions present high variability (2015 emissions are 25% lower than 2011 emissions), dust impact to populated areas is also varying causing a significant change of dust contribution in Beijing (from 25% in 2011 to 16% in 2015), Chengdu (24% to 18%) and Shanghai (21% to 14%). It also can be noted that absolute PM_{10} values decrease between 2011 and 2015.

Figure 5 shows the monthly variability of dust contributions to PM_{10} in three Chinese cities, Beijing, Chengdu and Shanghai. The specific desert mineral dust is originating from is also displayed on Figure 5.

It can be observed for the three cities that the highest daily PM_{10} levels are reached during winter, and the lowest ones during spring (for Beijing) or summer (Shanghai and Chengdu). Amplitudes for these daily values are smaller for Shanghai, ranging from 60 µg m^{-3} to 160 µg m^{-3} than for Beijing and Chengdu, ranging between 90 µg m^{-3} to 250 µg m^{-3}. These values are particularly high compared to Chinese National Standard GB 3095-2012 for PM_{10} in an urban environment, i.e., 24 h mean: 150 µg m^{-3} and annual mean: 70 µg m^{-3}.

Seasonal differences can be explained by meteorological parameters, as winter presents low temperature, low atmospheric dispersion and low rainfall frequencies and on the contrary, summer presents higher temperature, thicker boundary layer and mainly, during monsoon, higher rainfall. Anthropogenic pollutant emissions slightly increase during winter, but not with as much amplitude as PM concentrations.

The mineral dust contribution to PM_{10} mean computed for 2011, 2013 and 2015 is 6.6%, 9.5% and 9.3% respectively for Beijing, Chengdu and Shanghai. It can be observed that mineral dust has its highest impact on PM concentrations during spring (with contribution to PM_{10} for Beijing, Chengdu and Shanghai of 18.9%, 24.1% and 18.3% respectively). The authors of [73], using source apportionment for spring 2009 calculated a dust contribution to $PM_{2.5}$ up to 15% in Chengdu, which is similar to CHIMERE dust contribution to $PM_{2.5}$, of 12.5% (and 24.1% for PM_{10}). For all cities, a smaller contribution of dust during fall and almost no contribution during winter and summer were simulated. As little emissions are observed during winter from desert areas (5% of annual total), it was expected to have limited contribution to PM_{10} for this season. The summer period differs, as even if dust emissions are large, only small impact from dust is observed, as already observed in Figure 4 on the distribution of dust related AOD. Summer in China is a particular season, with high rainfall frequencies over East China, because of monsoon. It is likely that atmospheric aerosols will be deposited because of rainfall,

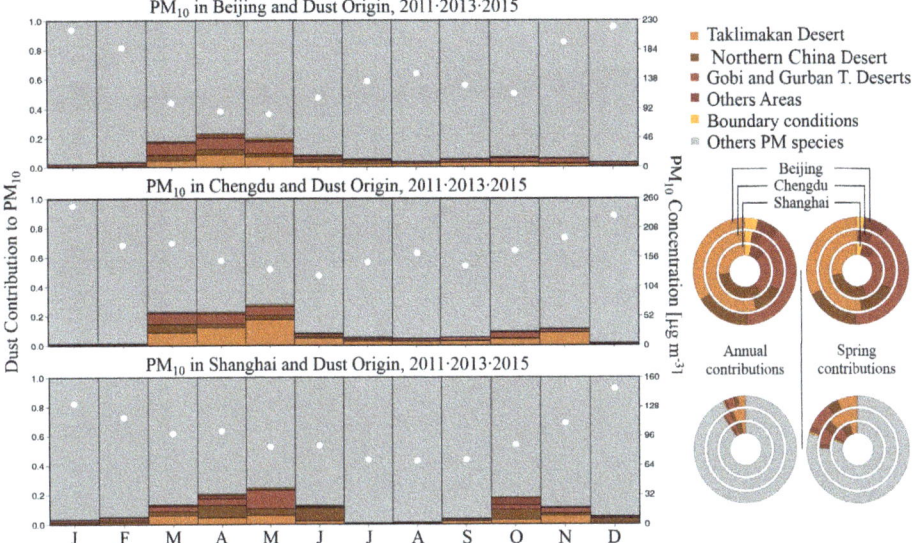

Figure 5. Dust contribution to PM$_{10}$ pollution in Beijing, Chengdu and Shanghai—Monthly variation calculated from 2011, 2013 and 2015 simulations. Ring plots show annual and spring contributors to PM$_{10}$ (bottom) and dust origin area (top) for the three cities. White dots variations in monthly PM$_{10}$ concentrations for the three cities.

before reaching populated areas. Also, southerly monsoon winds are not prone for advecting dust from western to eastern China.

Also, the Ångström median for spring decreases of 10% compared to annual Ångström at the Beijing AERONET station, from 1.19 to 1.07, showing a higher contribution of coarse particulate to pollution during spring. On the opposite, we note highest Ångström values (1.27) in winter which indicate a larger contribution to finer particulate matter in the atmospheric column. This result is consistent with CHIMERE simulations, with Ångström equal to 0.91 during spring and Ångström equal to 1.32 for winter.

It can also be observed that dust origins are similar for Shanghai and Beijing, with mineral dust coming from all of the main dust areas in equivalent proportions. Results are different for Chengdu, localized in central China, with more than 50% of mineral dust coming from Taklimakan, and less impact from Gobi and Northern China deserts than for Beijing and Shanghai. These results show that all considered desert areas have an impact on cities pollution: even if 70% of dust in our model is emitted from Taklimakan, it will represents less than 30% of the dust in Shanghai and Beijing, and about 50% in Chengdu. For these first two cities, the Gobi and Gurban deserts are the largest contributors (about 40%).

Dust load in Chinese cities origin has already been investigated, mostly using retro-trajectories [74, 75], the use of a tagging method from numerical computing provide an additional way to determine dust origins. Using retro-trajectories, the study [75] estimates for Chengdu's 2013 winter that for 2 days out 90, air pollution is controlled by air masses coming from Xinjiang region (mineral dust from Taklimakan source area)—a statement similar to the very low contribution found for this season in our results as a consequence of mineral dust low emissions.

3.2.3. PM Chemical Composition and Comparison to Observations

Table S2 presents the CHIMERE simulated PM_{10} composition in Beijing, Chengdu and Shanghai. For the three cities, the main contributor is anthropogenic mineral primary particular ($AMPP$). Nitrates ($NO_{3(p)}^-$) are second highest contributor, in Beijing and Shanghai, and 4th in Chengdu. In the three cities, ammonium ($NH_{4(p)}^+$) contributions are similar, but acid-base balances for nitrates and sulfate ($SO_{4(p)}^{2-}$) are different for Chengdu, compared to Shanghai and Beijing, with more sulfate and acidity in Chengdu. Organic matter (OM, includes POA and SOA) is the third contributor in Beijing and Chengdu, and the fourth contributor in Shanghai to PM_{10}. BC in the three cities corresponds to around 6% of PM_{10} pollution. In addition, sea salt presents a little contribution of 3.3% to Shanghai PM_{10}. It can be noted that evaluating particles components in CHIMERE over China would be of great interest but it is beyond the scope of this paper.

As a consequence, the previous analysis has to be considered as a preliminary semi-quantitative analysis. Here, we will briefly discuss, how our model results correspond to results from observational studies [76–78] even if measurement periods do not exactly correspond. Contributions comparisons are condensed in Table 5 for an easier reading.

Table 5. CHIMERE contribution compared to bibliography.

Considered Species	CHIMERE	Reference	City
$BC \subset PM_{2.5}$	6.4%	5%; [76]	Beijing
$OM \subset PM_{2.5}$	16.0%	20%; [76]	Beijing
$NH_{4(p)}^+ \subset PM_{2.5}$	10.5%	10%; [76]	Beijing
$NO_{3(p)}^- \subset PM_{2.5}$	23.8%	15%; [76]	Beijing
$SO_{4(p)}^{2-} \subset PM_{2.5}$	11.2%	15%; [76]	Beijing
$Dust \subset PM_{2.5}$	2.9%	7.5%; [76]	Beijing
$SO_{4(p)}^{2-} \subset PM_{2.5}$	21.3%	17%; [77]	Chengdu
$SO_{4(p)}^{2-} \subset PM_{10}$	17.5%	17%; [77]	Chengdu
$NO_{3(p)}^- \subset PM_{2.5}$	12.9%	10%; [77]	Chengdu
$NO_{3(p)}^- \subset PM_{10}$	10.0%	10%; [77]	Chengdu
$SO_{4(p)}^{2-} \subset PM_{2.5}$	14.6%	21.7%; [78]	Shanghai
$NO_{3(p)}^- \subset PM_{2.5}$	19.6%	19.6%; [78]	Shanghai
$NH_{4(p)}^+ \subset PM_{2.5}$	10.4%	12.7%; [78]	Shanghai
$OM \subset PM_{2.5}$	11.1%	20.2%; [78]	Shanghai

The authors of [76] study Beijing's $PM_{2.5}$ annual trends and chemical speciation between 2000 and 2015. Considering 2011, 2013 and 2015 observations, the study presents close results to CHIMERE in Beijing for $PM_{2.5}$ speciation, with 5% of BC (against 6.4%), 20% of OM (against 16%), 10% of $NH_{4(p)}^+$ (against 10.5%). Results differ slightly between $NO_{3(p)}^-/SO_{4(p)}^{2-}$ balance, with about 15% of each in the measures, against 23.8%/11.2% in CHIMERE. Dust represents 7.5% of $PM_{2.5}$ load but is measured as "Soil dust" also includes road and construction dust in addition to mineral desert dust (2.9% in model).

The authors of [77] present Chengdu PM pollution chemical speciation between 2007 and 2013, with its seasonal variability. An increase of $NO_{3(p)}^-$ between 2007 and 2013 when a decrease of $SO_{4(p)}^{2-}$ is observed which is also observed in our simulations (2011 to 2015) and better described in [29]. $SO_{4(p)}^{2-}$ abundances for both $PM_{2.5}$ and PM_{10} range around 17% (for 2011 and 2013) which is quite similar to modeled abundances (respectively 21.3% and 17.4%). $NO_{3(p)}^-$ contributions for $PM_{2.5}$ and PM_{10} range around 10%, when CHIMERE modeled, respectively, 12.9% and 10.0%. Al, Si and Ca represent together about 20% of total PM, with maximum contribution in spring, but this does include anthropogenic and natural crustal elements and cannot be directly compared to our modeling results of desert dust.

The authors of [78] sampled Shanghai PM$_{2.5}$ pollution composition from 2011 to 2013. Results for Shanghai are quite similar to those observed in Beijing, with an underestimation of $SO_{4(p)}^{2-}$ (14.6% against 21.7%) and OM (11.1% against 20.2%) but good correspondences for $NO_{3(p)}^-$ (19.6% against 19.6%) and $NH_{4(p)}^+$ (10.4% against 12.7%).

3.2.4. Daily Variability of PM$_{10}$ Component Concentrations

PM$_{10}$ components daily concentration distributions for Beijing, Chengdu and Shanghai are displayed in Figure 6. Median values for dust concentrations correspond to few micrograms only, in the three cities, which rank it as one of lowest contributors to PM$_{10}$ pollution in our simulations, as it has been also observed in Table S2, with mean concentration contributions. The additional information provided in the Figure 6 is the daily variability of each component, and it appears that dust presents the largest variability and in the three cities, the highest daily concentration among PM$_{10}$ components is due to dust (i.e., reaching 400 µg m^{-3}, 540 µg m^{-3} and 350 µg m^{-3} respectively for Beijing, Chengdu and Shanghai). This particular aspect follows dust emissions dynamics, very localized in time (q.v. Table 1, 5th column), and responsible of very strong pollution events. Thus, it is judicious to focus on dust contribution to PM pollution events in cities rather than the average impact of dust, and this is the key point of our work, exploiting the most original CHIMERE model abilities.

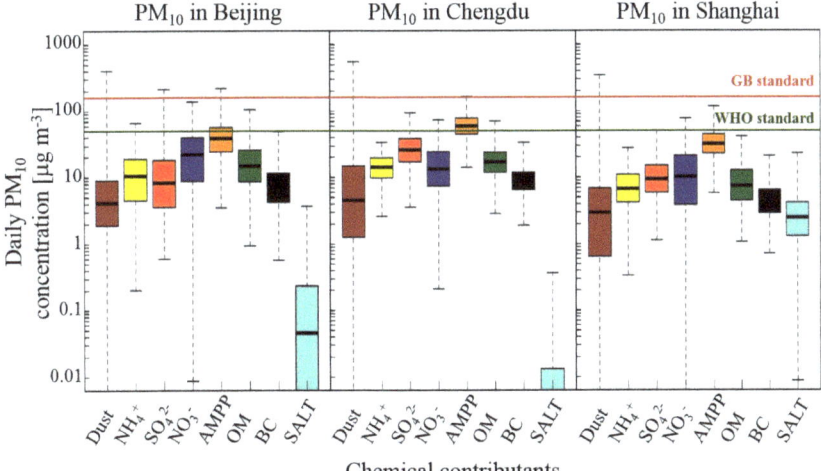

Figure 6. Boxplot distribution for simulated daily concentrations (1095 days considered) of species contributing to PM$_{10}$. (**left**) Beijing (**center**) Chengdu and (**right**) Shanghai. Model species correspond to: Dust (only representing the desert and natural fraction), $NH_{4(p)}^+$ (ammonium), SO_4^{2-} (sulfate), $NO_{3(p)}^-$ (nitrate), AMPP (Anthropogenic Mineral Primary Particulate matter—including soil dust emitted from anthropogenic activities such as construction), OM (primary and secondary Organic Matter, BC (Black Carbon), from biogenic and anthropogenic sources) and SALT (sea salt). Chinese and WHO standards for daily PM$_{10}$ have been added to evaluate the frequency of overshoot for each species.

3.2.5. Dust Contribution during High Pollution Episodes

From Figure 7 and also Table 6 it can be observed for the whole period, that mineral dust is a minor contributor to PM$_{10}$ and even more to PM$_{2.5}$ concentrations. For instance, for most of the days (between 70% to 75%) for the three cities, the dust contribution to PM$_{10}$ is below 10% (Figure 7), only for 7.2% to 11.0% of days, dust it is more than 25 %. When studying the 25% most strongly polluted days, dust have slightly larger contributions for Chengdu and Shanghai, but not for Beijing. For the first two cities, this means that the dust presence is correlated with pollution events, although it is generally not dominant. On the contrary, for Beijing, PM pollution events occur during winter due to anthropogenic sources, and dust has only little impact on air pollution for this season.

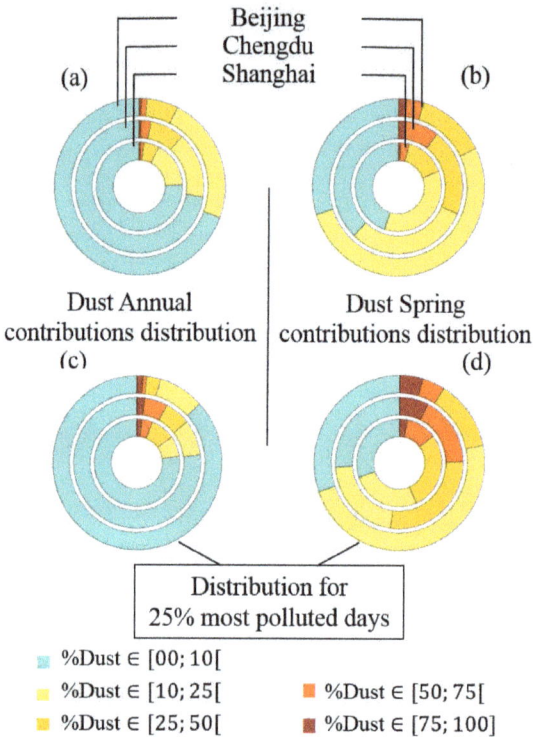

Figure 7. (a) Percentage of days when dust contributes between 0–10%, 10–25%, 25–50%, 50–75% and 75–100% to daily PM_{10} averages; (b) same as (a), but for spring only; (c) Same as (a), but for the 25% of most polluted days only (in terms of PM_{10} level); (d) Same as for (c), but for spring only.

If we focus now on spring season, we can identify a larger impact from mineral dust to cities PM pollution, an impact which gets even stronger when selecting the 25% most polluted PM_{10} days. During spring, PM_{10} pollution events can frequently be affected by dust in addition to anthropogenic PM pollution sources: the dust contribution is higher than 25% for 22% of days for Beijing, as much as 52% for Chengdu and 43% for Shanghai. To a lesser extent, mineral dust is implied to $PM_{2.5}$ events (Fraction of 9% for Beijing, 19% for Chengdu and 14% for Shanghai).

Table 6. Fraction of days when the dust contributions to daily $PM_{2.5}$ or PM_{10} are at least 25%. These values were calculated for Beijing, Chengdu and Shanghai, and for various samples: all days in 2011, 2013 and 2015, only for the spring season, for days with the largest pollution (in terms of $PM_{2.5}$ and PM_{10} respectively), for these latter days, but only during spring. As an example the table reads as follows: For Chengdu, during the 25% most polluted days in spring 2011, 2013 and 2015, 52% of days displayed dust contributions larger than 25%. $P_{75,Year}/P_{75,Spring}$ indicates the PM_{10} 75th percentiles value. They are threshold values, above which a day is considered as one of the 25% most polluted days. pol. days stands for "polluted days", with $[PM_x] > P_{75,PM_x}$. F.o.d stands for "Fraction of days".

Cities and Species	days year^{-1}/ days spring^{-1} with Dust contrib.> 25%	(Year/Spring) F.o.d with Dust contrib.> 25%	$P_{75,Year}$/ $P_{75,Spring}$ ($\mu g\,m^{-3}$)	(Year/Spring)$_{pol.days}$ F.o.d with Dust contrib.> 25%
Beijing $PM_{2.5}$	7.3/7.3	0.02/0.08	144/97	0.02/0.09
Beijing PM_{10}	25.5/16.4	0.07/0.18	185/123	0.04/0.22
Chengdu $PM_{2.5}$	14.6/10.1	0.04/0.11	150/128	0.05/0.19
Chengdu PM_{10}	40.1/28.3	0.11/0.31	206/183	0.14/0.52
Shanghai $PM_{2.5}$	10.9/04.5	0.03/0.05	94/84	0.06/0.14
Shanghai PM_{10}	29.2/17.3	0.08/0.19	122/112	0.15/0.43

Additional information from the AERONET station located in Beijing (see Figure S3 in supplements) shows that the fraction of days with Ångström coefficients below 0.4 (which characterizes a dominant dust contribution to the PM column) is nearly a factor of 2 larger for spring than for the rest of the year. The same increase is observed from simulated Ångström coefficients. In addition, for the 25% of days with the largest AOD values, the Beijing AERONET station shows a larger fraction of days with an Ångström coefficient below 0.4: 0.04 for all seasons, and 0.08 for spring only. These values are somewhat higher than in CHIMERE (respectively 0.02 and 0.06). As a conclusion, also for AOD, the dust contribution is for most of the days not dominant.

4. Conclusions

In this study, we aimed at evaluating the impact of mineral dust on air quality in three Chinese megacities with the regional CHIMERE CTM. Modeled dust emissions average 283 Mt.year^{-1}, with about 70% emitted from Taklimakan area, 23% emitted from the Gobi desert and 6.5% emitted from Northern China desert. MODIS information was used to validate correspondences between CHIMERE AOD and satellite AOD variation, and results appeared to be satisfying, which allows us to then study the dust contribution to cities' air pollution. It has been chosen to focus on Beijing, Chengdu and Shanghai, because of their large population and different geographical locations with respect to emission zones. We have estimated the impact of dust on cities' air pollution: we found average contributions of 10 $\mu g\,m^{-3}$ (6.6%), 17 $\mu g\,m^{-3}$ (9.5%) and 9 $\mu g\,m^{-3}$ (9.3%) to PM_{10} in Beijing, Chengdu and Shanghai respectively. These values values must be taken into account while evaluating Chinese cities air quality, as they represent a significant burden for the cities air quality in regards to the annual WHO threshold of 20 $\mu g\,m^{-3}$. The dust contribution is highly seasonally dependent, with highest contributions during spring, when contributions to PM_{10} reach 22 $\mu g\,m^{-3}$ (18.9%), 37.0 $\mu g\,m^{-3}$ (24.1%) and 12 $\mu g\,m^{-3}$ (18.3%). In Beijing and Shanghai, dust is advected rather equally from all main Chinese deserts, while in Chengdu, it originates mainly from the Taklimakan desert. Analysis for days with enhanced PM_{10} pollution (above the 75th percentile) especially in the spring season shows that dust is often a main contributor (dust contribution larger than 25%) during PM_{10} polluted days especially in the spring season: this holds at 22% of the days for Beijing, 52% for Chengdu and 43% for Shanghai. Considering all seasons, the impact on polluted events is smaller, especially during winter, when particulate matter pollution mostly originates from anthropogenic sources. Finally, if mineral dust do not represent a predominant source compared to anthropogenic pollutant on average, mineral dust alone leads to an exceedance of the daily WHO threshold for PM_{10} (50 µg m^{-3}) in Bejing for 10 days per year, 29 days per year in Chengdu and 13 days per year in Shanghai.

In the future, dust vertical transport modeling should be investigated more deeply and compared to products which contains information on vertical dispersion (CALIOP [69], AEROIASI [59]). In addition, updated emissions inventories should be used and *SOA* modeling should be improved

Supplementary Materials: The following are available at http://www.mdpi.com/2073-4433/11/7/708/s1: Figure S1: 550 nm AOD spatial distribution to 1° × 1° resolution for 2011, 2013 and 2015. Table S1: Spatial correspondences between MODIS 550 nm AOD and CHIMERE 550 nm AOD. Table S2: CHIMERE Model PM_{10} annual composition for Bejing, Chengdu and Shanghai. Figure S2: Daily $PM_{2.5}$ time series for monitoring stations and CHIMERE simulations. Figure S3: Daily Ångström coefficient distribution at Beijing station. Figure S4: PM_{10} simulated surface concentration. Figure S5: PM station locations.

Author Contributions: M.L., G.F. and M.B. designed the experiments and M.L. carried them out. G.S. prepared meteorological and emission data. F.M. and W.T. provided data surface measurements from CRAES institution. M.L. adapted the model code and performed the simulations. B.L. developed the database of surface characteristics for Asian deserts. Q.Z., J.C. and G.D. help to the analysis. M.L. prepared the manuscript and all authors contributed to the text, interpretation of the results and reviewed the manuscript. All authors have read and agreed to the published version of the manuscript.

Funding: M. Lachatre was funded by the Sorbonne Université, l'Ecole Doctorale des Sciences de l'Environnement d'Ile de France (ED129) and PolEASIA ANR project under the allocation ANR-15-CE04-0005.

Acknowledgments: This work was granted access to the HPC resources of TGCC under the allocation A0030107232 made by GENCI. We acknowledge the free use of MODIS AOD data from the Terra from (https://terra.nasa.gov/, last consulted 25/09/2018). Surface measurements provided by U.S. Department of State Air Quality Monitoring Program, Mission China (www.stateair.net, last consulted 07/10/2018). We also acknowledge the Chinese Research Academy of Environmental Sciences (CRAES) in Beijing, for providing us $PM_{2.5}$ and PM_{10} surface concentrations data. We also acknowledge the NASA's AERONET network. The development and production of AEROIASI retrievals are supported by the Centre National des Études Spatiales (CNES, the French Space Agency, IASI project/Terre, Océan, Surfaces continentales, Atmosphère) and the Programme National de Télédétection Spatiale (PNTS, http://programmes.insu.cnrs.fr/pnts, grant n° PNTS-2013-05, project "SYNAEROZON"). IASI is a joint mission of EUMETSAT and CNES.

Conflicts of Interest: The authors declare no conflict of interest.

References

1. Geng, G.; Zhang, Q.; Martin, R.V.; van Donkelaar, A.; Huo, H.; Che, H.; Lin, J.; He, K. Estimating long-term $PM_{2.5}$ concentrations in China using satellite-based aerosol optical depth and a chemical transport model. *Remote Sens. Environ.* **2015**, *166*, 262–270, doi:10.1016/j.rse.2015.05.016.
2. Huang, R.J.; Zhang, Y.; Bozzetti, C.; Ho, K.F.; Cao, J.J.; Han, Y.; Daellenbach, K.R.; Slowik, J.G.; Platt, S.M.; Canonaco, F.; et al. High secondary aerosol contribution to particulate pollution during haze events in China. *Nature* **2014**, *514*, 2018–222, doi:10.1038/nature13774.
3. Liu, Z.; Gao, W.; Yu, Y.; Hu, B.; Xin, J.; Sun, Y.; Wang, L.; Wang, G.; Bi, X.; Zhang, G.; et al. Characteristics of $PM_{2.5}$ mass concentrations and chemical species in urban and background areas of China: Emerging results from the CARE-China network. *Atmos. Chem. Phys.* **2018**, *18*, 8849–8871, doi:10.5194/acp-18-8849-2018.
4. Kaiser, D.P.; Qian, Y. Decreasing trends in sunshine duration over China for 1954–1998: Indication of increased haze pollution? *Geophys. Res. Lett.* **2002**, *29*, 38-1–38-4, doi:10.1029/2002GL016057.
5. Xie, R.; Sabel, C.E.; Lu, X.; Zhu, W.; Kan, H.; Nielsen, C.P.; Wang, H. Long-term trend and spatial pattern of PM 2.5 induced premature mortality in China. *Environ. Int.* **2016**, *97*, 180–186, doi:10.1016/j.envint.2016.09.003.
6. Wang, Y.; Teter, J.; Sperling, D. China's soaring vehicle population : Even greater than forecasted ? *Energy Policy* **2011**, *39*, 3296–3306, doi:10.1016/j.enpol.2011.03.020.
7. Liu, J.; Niu, D.; Song, X. The energy supply and demand pattern of China: A review of evolution and sustainable development. *Renew. Sustain. Energy Rev.* **2013**, *25*, 220–228, doi:10.1016/j.rser.2013.01.061.
8. Ma, Z.; Hu, X.; Sayer, A.; Levy, R.; Zhang, Q.; Xue, Y.; Tong, S.; Bi, J.; Huang, L.; Liu, Y. Satellite-Based Spatiotemporal Trends in PM2.5 Concentrations: China, 2004–2013. *Environ. Health Perspect.* **2016**, *124*, 184–192, doi:10.1289/ehp.1409481.

9. Liu, F.; Beirle, S.; Zhang, Q.; Van Der A, R.J.; Zheng, B.; Tong, D.; He, K. NOxemission trends over Chinese cities estimated from OMI observations during 2005 to 2015. *Atmos. Chem. Phys.* **2017**, *17*, 9261–9275, doi:10.5194/acp-17-9261-2017.
10. IPCC. *Climate Change 2001: The Scientific Basis. Contribution of Working Group I to the Third Assessment Report of theIntergovernmental Panel on Climate Change*; Houghton, J.T., Ding, Y., Griggs, D.J., Noguer, M., van der Linden, P.J., Dai, X., Maskell, K., Johnson, C.A., Eds.; Cambridge University Press: Cambridge, UK; New York, NY, USA, 2011; 881pp.
11. Huneeus, N.; Schulz, M.; Balkanski, Y.; Griesfeller, J.; Prospero, J.; Kinne, S.; Bauer, S.; Boucher, O.; Chin, M.; Dentener, F.; et al. Global dust model intercomparison in AeroCom phase I. *Atmos. Chem. Phys.* **2011**, *11*, 7781–7816, doi:10.5194/acp-11-7781-2011.
12. Hamilton, D.S.; Scanza, R.A.; Feng, Y.; Guinness, J.; Kok, J.F.; Li, L.; Liu, X.; Rathod, S.D.; Wan, J.S.; Wu, M.; et al. Improved methodologies for Earth system modelling of atmospheric soluble iron and observation comparisons using the Mechanism of Intermediate complexity for Modelling Iron (MIMI v1.0). *Geosci. Model Dev.* **2019**, *12*, 3835–3862, doi:10.5194/gmd-12-3835-2019.
13. Sokolik, I.N.; Toon, O.B. Incorporation of mineralogical composition into models of the radiative properties of mineral aerosol from UV to IR wavelengths. *J. Geophys. Res. Atmos.* **1999**, *104*, 9423–9444, doi:10.1029/1998JD200048.
14. Li, L.; Sokolik, I.N. The Dust Direct Radiative Impact and Its Sensitivity to the Land Surface State and Key Minerals in the WRF-Chem-DuMo Model: A Case Study of Dust Storms in Central Asia. *J. Geophys. Res. Atmos.* **2018**, *123*, 4564–4582, doi:10.1029/2017JD027667.
15. Laurent, B.; Marticorena, B.; Bergametti, G.; Mei, F. Modeling mineral dust emissions from Chinese and Mongolian deserts. *Glob. Planet. Chang.* **2006**, *52*, 121–141, doi:10.1016/j.gloplacha.2006.02.012.
16. Gong, S.L.; Zhang, X.Y.; Zhao, T.L.; McKendry, I.G.; Jaffe, D.A.; Lu, N.M. Characterization of soil dust aerosol in China and its transport and distribution during 2001 ACE-Asia: 2. Model simulation and validation. *J. Geophys. Res. Atmos.* **2003**, *108*, doi:10.1029/2002JD002633.
17. Ginoux, P.; Chin, M.; Tegen, I.; Prospero, J.M.; Holben, B. adn Dubovik, O.; Lin, S. Sources and distributions of dust aerosols simulated with the GOCART model. *J. Geophys. Res. Atmos.* **2001**, *106*, 255–273.
18. Mahowald, N.; Kohfeld, K.; Hansson, M.; Balkanski, Y.; Harrison, S.P.; Prentice, I.C.; Schulz, M.; Rodhe, H. Dust sources and deposition during the last glacial maximum and current climate: A comparison of model results with paleodata from ice cores and marine sediments. *J. Geophys. Res. Atmos.* **1999**, doi:10.1029/1999JD900084.
19. Yang, F.; Ye, B.; He, K.; Ma, Y.; Cadle, S.H.; Chan, T.; Mulawa, P.A. Characterization of atmospheric mineral components of PM 2 . 5 in Beijing and Shanghai , China. *Sci. Total Environ.* **2005**, *343*, 221–230, doi:10.1016/j.scitotenv.2004.10.017.
20. Pan, X.; Uno, I.; Zhe, W.; Nishizawa, T.; Sugimoto, N.; Yamamoto, S.; Kobayashi, H.; Sun, Y.; Fu, P.; Tang, X.; et al. Real-time observational evidence of changing Asian dustmorphology with the mixing of heavy anthropogenic pollution. *Sci. Rep.* **2017**, *7*, 335, doi:10.1038/s41598-017-00444-w.
21. Yu, Y.; Kalashnikova, O.V.; Garay, M.J.; Notaro, M. Climatology of Asian dust activation and transport potential based on MISR satellite observations and trajectory analysis. *Atmos. Chem. Phys.* **2019**, *19*, 363–378, doi:10.5194/acp 19-363-2019.
22. Yang, F.; Tan, J.; Zhao, Q.; Du, Z.; He, K.; Ma, Y.; Duan, F.; Chen, G.; Zhao, Q. Characteristics of $PM_{2.5}$ speciation in representative megacities and across China. *Atmos. Chem. Phys.* **2011**, *11*, 5207–5219, doi:10.5194/acp-11-5207-2011.
23. Menut, L.; Bessagnet, B.; Khvorostyanov, D.; Beekmann, M.; Blond, N.; Colette, a.; Coll, I.; Curci, G.; Foret, G.; Hodzic, A.; et al. CHIMERE 2013: A model for regional atmospheric composition modelling. *Geosci. Model Dev.* **2013**, *6*, 981–1028, doi:10.5194/gmd-6-981-2013.
24. Mailler, S.; Menut, L.; Khvorostyanov, D.; Valari, M.; Couvidat, F.; Siour, G.; Turquety, S.; Briant, R.; Tuccella, P.; Bessagnet, B.; et al. CHIMERE-2017: From urban to hemispheric chemistry-transport modeling. *Geosci. Model Dev.* **2017**, *10*, 2397–2423, doi:10.5194/gmd-10-2397-2017.
25. Van Leer, B. Towards the ultimate conservative difference scheme. IV. A new approach to numerical convection. *J. Comput. Phys.* **1977**, *23*, 276–299, doi:10.1016/0021-9991(77)90095-X.
26. Owens, R.G.; Hewson, T. *ECMWF Forecast User Guide*; ECMWF, Reading, UK, 2018. doi:10.21957/m1cs7h.

27. Olivier, W.; Xin, Z.; Michael, J.P. Fast-J: Accurate Simulation of In- and Below-Cloud Photolysis in Tropospheric Chemical Models. *J. Atmos. Chem.* **2000**, doi:10.1023/A:1006415919030.
28. Janssens-Maenhout, G.; Crippa, M.; Guizzardi, D.; Dentener, F.; Muntean, M.; Pouliot, G.; Keating, T.; Zhang, Q.; Kurokawa, J.; Wankmüller, R.; et al. HTAP-v2.2: A mosaic of regional and global emission grid maps for 2008 and 2010 to study hemispheric transport of air pollution. *Atmos. Chem. Phys.* **2015**, *15*, 11411–11432, doi:10.5194/acp-15-11411-2015.
29. Lachatre, M.; Fortems-Cheiney, A.; Foret, G.; Siour, G.; Dufour, G.; Clarisse, L.; Clerbaux, C.; Coheur, P.F.; Van Damme, M.; Beekmann, M. The unintended consequence of SO_2 and NO_2 regulations over China: Increase of ammonia levels and impact on $PM_{2.5}$ concentrations. *Atmos. Chem. Phys.* **2019**, *19*, 6701–6716, doi:10.5194/acp-19-6701-2019.
30. Zheng, B.; Tong, D.; Li, M.; Liu, F.; Hong, C.; Geng, G.; Li, H.; Li, X. Trends in China's anthropogenic emissions since 2010 as the consequence of clean air actions. *Atmos. Chem. Phys.* **2018**, *18*, 14095–14111, doi:doi.org/10.5194/acp-2018-374.
31. Guenther, A.B.; Jiang, X.; Heald, C.L.; Sakulyanontvittaya, T.; Duhl, T.; Emmons, L.K.; Wang, X. Model Development The Model of Emissions of Gases and Aerosols from Nature version 2.1 (MEGAN2.1): An extended and updated framework for modeling biogenic emissions. *Geosci. Model Dev.* **2012**, 1471–1492, doi:10.5194/gmd-5-1471-2012.
32. Hauglustaine, D.A.; Hourdin, F.; Jourdain, L.; Filiberti, M.A.; Walters, S.; Lamarque, J.F.; Holland, E.A. Interactive chemistry in the Laboratoire de Météorologie Dynamique general circulation model: Description and background tropospheric chemistry evaluation. *J. Geophys. Res. Atmos.* **2004**, *109*, n/a–n/a, doi:10.1029/2003JD003957.
33. Hourdin, F.; Musat, I.; Bony, S.; Braconnot, P.; Codron, F.; Dufresne, J.L.; Fairhead, L.; Filiberti, M.A.; Friedlingstein, P.; Grandpeix, J.Y.; et al. The LMDZ4 general circulation model: Climate performance and sensitivity to parametrized physics with emphasis on tropical convection. *Clim. Dyn.* **2006**, *27*, 787–813, doi:10.1007/s00382-006-0158-0.
34. Nenes, A.; Pilinis, C.; Pandis, S. ISORROPIA: A new thermodynamic model for inorganic multicomponent atmospheric aerosols. *Aquatic. Geochem.* **1998**, pp. 4:123–152.
35. Derognat, C.; Beekmann, M.; Baeumle, M.; Martin, D.; Schmidt, H. Effect of biogenic volatile organic compound emissions on tropospheric chemistry during the Atmospheric Pollution Over the Paris Area (ESQUIF) campaign in the Ile-de-France region. *J. Geophys. Res. Atmos.* **2003**, *108*, doi:10.1029/2001JD001421.
36. Pankow, J.F. AN ABSORPTION MODEL OF GAS/PARTICLE PARTITIONING OF ORGANIC COMPOUNDS IN THE ATMOSPHERE. *Atmos. Environ.* **1994**, *28*, 185–188.
37. Kaupp, H.; Umlauf, G. ATMOSPHERIC GAS-PARTICLE PARTITIONING OF ORGANIC COMPOUNDS: COMPARISON OF SAMPLING METHODS. *Atmos. Environ.* **1992**, *26*, 2259–2267, doi:10.1016/0960-1686(92)90357-Q.
38. Ma, S.; Zhang, X.; Gao, C.; Tong, D.Q.; Xiu, A.; Wu, G.; Cao, X.; Huang, L.; Zhao, H.; Zhang, S.; et al. Multimodel simulations of a springtime dust storm over northeastern China: Implications of an evaluation of four commonly used air quality models (CMAQ v5.2.1, CAMx v6.50, CHIMERE v2017r4, and WRF-Chem v3.9.1). *Geosci. Model Dev.* **2019**, *12*, 4603–4625, doi:10.5194/gmd-12-4603-2019.
39. Laurent, B.; Marticorena, B.; Bergametti, G. Simulation of the mineral dust emission frequencies from desert areas of China and Mongolia using an aerodynamic roughness length map derived from the POLDER/ADEOS 1 surface products. *J. Geophys. Res. Atmos.* **2005**, *110*, 1–21, doi:10.1029/2004JD005013.
40. Menut, L.; Schmechtig, C.; Marticorena, B. Sensitivity of the Sandblasting Flux Calculations to the Soil Size Distribution Accuracy. *J. Atmos. Ocean. Technol.* **2005**, *22*, 1875–1884.
41. Alfaro, S.; Gomes, L. Modeling mineral aerosol production by wind erosion: Emission intensities and aerosol size distributions in source areas. *J. Geophys. Res. Atmos.* **2001**, *106*, 18075–18084.
42. Shao, Y. A model for mineral dust emission. *J. Geophys. Res. Atmos.* **2001**, *106*, 20239–20254, doi:10.1029/2001JD900171.
43. Kok, J.F. A scaling theory for the size distribution of emitted dust aerosols suggests climate models underestimate the size of the global dust cycle. *Proc. Natl. Acad. Sci. USA* **2011**, *108*, 1016–1021, doi:10.1073/pnas.1014798108.

44. Albani, S.; Mahowald, N.M.; Perry, A.T.; Scanza, R.A.; Zender, C.S.; Heavens, N.G.; Maggi, V.; Kok, J.F.; Otto-Bliesner, B.L. Improved dust representation in the Community Atmosphere Model. *J. Adv. Model. Earth Syst.* **2014**, *6*, 541–570, doi:10.1002/2013MS000279.
45. Alfaro, S.; Gaudichet, A.; Gomes, L.; Maillé, M. Modeling the size distribution of soil aerosol product by sandblasting. *J. Geophys. Res. Atmos.* **1997**, *102*, 11239–11249.
46. Mahowald, N.M.; Muhs, D.R.; Levis, S.; Rasch, P.J.; Yoshioka, M.; Zender, C.S.; Luo, C. Change in atmospheric mineral aerosols in response to climate: Last glacial period, preindustrial, modern, and doubled carbon dioxide climates. *J. Geophys. Res. Atmos.* **2006**, *111*, doi:10.1029/2005JD006653.
47. Mahowald, N.; Albani, S.; Kok, J.F.; Engelstaeder, S.; Scanza, R.; Ward, D.S.; Flanner, M.G. The size distribution of desert dust aerosols and its impact on the Earth system. *Aeolian Res.* **2014**, *15*, 53–71, doi:10.1016/j.aeolia.2013.09.002.
48. Kok, J.F.; Mahowald, N.M.; Fratini, G.; Gillies, J.A.; Ishizuka, M.; Leys, J.F.; Mikami, M.; Park, M.S.; Park, S.U.; Van Pelt, R.S.; et al. An improved dust emission model – Part 1: Model description and comparison against measurements. *Atmos. Chem. Phys.* **2014**, *14*, 13023–13041, doi:10.5194/acp-14-13023-2014.
49. Foret, G.; Bergametti, G.; Dulac, F.; Menut, L. An optimized particle size bin scheme for modeling mineral dust aerosol. *J. Geophys. Res. Atmos.* **2006**, *111*, doi:10.1029/2005JD006797.
50. Wang, G.H.; Cheng, C.L.; Huang, Y.; Tao, J.; Ren, Y.Q.; Wu, F.; Meng, J.J.; Li, J.J.; Cheng, Y.T.; Cao, J.J.; et al. Evolution of aerosol chemistry in Xi'an, inland China, during the dust storm period of 2013—Part 1: Sources, chemical forms and formation mechanisms of nitrate and sulfate. *Atmos. Chem. Phys.* **2014**, *14*, 11571–11585, doi:10.5194/acp-14-11571-2014.
51. Fu, X.; Wang, S.X.; Cheng, Z.; Xing, J.; Zhao, B.; Wang, J.D.; Hao, J.M. Source, transport and impacts of a heavy dust event in the Yangtze River Delta, China, in 2011. *Atmos. Chem. Phys.* **2014**, *14*, 1239–1254, doi:10.5194/acp-14-1239-2014.
52. Mikalai, F.; Haowen, Y.; Zhongrong, Z.; Shuwen, Y.; Wei, L.; Yanming, L. Author Correction: Combined use of satellite and surface observations to study aerosol optical depth in different regions of China. *Sci. Rep.* **2019**, doi:10.1038/s41598-019-54734-6.
53. Xin, W.; Jun, L.; Huizheng, C.; Fei, J.; Jingjing, L. Spatial and temporal evolution of natural and anthropogenic dust events over northern China. *Sci. Rep.* **2018**, doi:10.1038/s41598-018-20382-5.
54. Seinfeld, J.H.; Pandis, S.N. *ATMOSPHERIC From Air Pollution to Climate Change SECOND EDITION*; Wiley-Interscience: Hoboken, NJ, USA, 2006; pp. 628–674.
55. Zhang, L.; Gong, S.; Padro, J.; Barrie, L. A size-segregated particle dry deposition scheme for an atmospheric aerosol module. *Atmos. Environ.* **2001**, *35*, 549–560, doi:10.1016/S1352-2310(00)00326-5.
56. Zhang, Q.; Laurent, B.; Velay-Lasry, F.; Ngo, R.; Derognat, C.; Marticorena, B.; Albergel, A. An air quality forecasting system in Beijing—Application to the study of dust storm events in China in May 2008. *J. Environ. Sci.* **2012**, *24*, 102–111, doi:10.1016/S1001-0742(11)60733-X.
57. Di Biagio, C.; Formenti, P.; Balkanski, Y.; Caponi, L.; Cazaunau, M.; Pangui, E.; Journet, E.; Nowak, S.; Caquineau, S.; Andreae, M.O.; et al. Global scale variability of the mineral dust long-wave refractive index: A new dataset of in situ measurements for climate modeling and remote sensing. *Atmos. Chem. Phys.* **2017**, *17*, 1901–1929, doi:10.5194/acp-17-1901-2017.
58. Cuesta, J.; Eremenko, M.; Flamant, C.; Dufou, G.; Laurent, B.; Bergametta, G.; Höpfner, M.; Orphal, J.; Zhou, D. Three-dimensional distribution of a major desert dust outbreak over East Asia in March 2008 derived from IASI satellite observations Juan. *J. Geophys. Res. Atmos.* **2015**, 7099–7127, doi:10.1002/2014JD022406.
59. Cuesta, J.; Flamant, C.; Gaetani, M.; Knippertz, P.; Fink, A.H.; Chazette, P.; Eremenko, M.; Dufour, G.; Di Biagio, C.; Formenti, P. Three-dimensional pathways of dust over the Sahara during summertime 2011 as revealed by new IASI observations. *Q. J. R. Meteorol. Soc.* **2020**, doi:10.1002/qj.3814.
60. Eck, T.F.; Holben, B.N.; Reid, J.S.; Dubovik, O.; Smirnov, A.; O'Neill, N.T.; Slutsker, I.; Kinne, S. Wavelength dependence of the optical depth of biomass burning, urban, and desert dust aerosols. *J. Geophys. Res. Atmos.* **1999**, *104*, 31333–31349, doi:10.1029/1999JD900923.
61. Weilin, W.; Suli, Z.; Limin, J.; Michael, T.; Boen, Z.; Gang, X.; Haobo, H. Estimation of PM2.5 Concentrations in China Using a Spatial Back Propagation Neural Network. *Sci. Rep.* **2019**, doi:10.1038/s41598-019-50177-1.
62. Flemming, J.; Stern, R.; Yamartino, R.J. A new air quality regime classification scheme for O3, NO2, SO2 and PM10 observations sites. *Atmos. Environ.* **2005**, *39*, 6121–6129, doi:10.1016/j.atmosenv.2005.06.039.

63. Martini, F.M.S.; Hasenkopf, C.A.; Roberts, D.C. Statistical analysis of PM2.5 observations from diplomatic facilities in China. *Atmos. Environ.* **2015**, *110*, 174–185, doi:10.1016/j.atmosenv.2015.03.060.
64. Tan, S.C.; Li, J.; Che, H.; Chen, B.; Wang, H. Transport of East Asian dust storms to the marginal seas of China and the southern North Pacific in spring 2010. *Atmos. Environ.* **2017**, *148*, 316–328, doi:10.1016/j.atmosenv.2016.10.054.
65. Chen, S.; Huang, J.; Qian, Y.; Zhao, C.; Kang, L.; Yang, B.; Wang, Y.; Liu, Y.; Yuan, T.; Wang, T.; et al. An Overview of Mineral Dust Modeling over East Asia. *J. Meteorol. Res.* **2017**, *31*, 633–653.
66. Hou, Z.J.Z. A Simulated Climatology of Asian Dust Aerosol and Its Trans-Pacific Transport. Part I : Mean Climate and Validation. *J. Clim.* **2006**, *19*, 88–104.
67. Proestakis, E.; Amiridis, V.; Marinou, E.; Georgoulias, A.K.; Solomos, S.; Kazadzis, S.; Chimot, J.; Che, H.; Alexandri, G.; Binietoglou, I.; et al. Nine-year spatial and temporal evolution of desert dust aerosols over South and East Asia as revealed by CALIOP. *Atmos. Chem. Phys.* **2018**, *18*, 1337–1362, doi:10.5194/acp-18-1337-2018.
68. Ansmann, A.; Bösenberg, J.; Chaikovsky, A.; Comerón, A.; Eckhardt, S.; Eixmann, R.; Freudenthaler, V.; Ginoux, P.; Komguem, L.; Linné, H.; et al. Long-range transport of Saharan dust to northern Europe: The 11–16 October 2001 outbreak observed with EARLINET. *J. Geophys. Res. Atmos.* **2003**, *108*, doi:10.1029/2003JD003757.
69. Liu, Z.; Omar, A.; Vaughan, M.; Hair, J.; Kittaka, C.; Hu, Y.; Powell, K.; Trepte, C.; Winker, D.; Hostetler, C.; et al. CALIPSO lidar observations of the optical properties of Saharan dust: A case study of long-range transport. *J. Geophys. Res. Atmos.* **2008**, *113*, doi:10.1029/2007JD008878.
70. Colette, A.; Favez, O.; Meleux, F.; Chiappini, L.; Haeffelin, M.; Morille, Y.; Malherbe, L.; Papin, A.; Bessagnet, B.; Menut, L.; et al. Assessing in near real time the impact of the April 2010 Eyjafjallajokull ash plume on air quality. *Atmos. Environ.* **2011**, *45*, 1217–1221, doi:10.1016/j.atmosenv.2010.09.064.
71. Boichu, M.; Clarisse, L.; Péré, J.C.; Herbin, H.; Goloub, P.; Thieuleux, F.; Ducos, F.; Clerbaux, C.; Tanré, D. Temporal variations of flux and altitude of sulfur dioxide emissions during volcanic eruptions: Implications for long-range dispersal of volcanic clouds. *Atmos. Chem. Phys.* **2015**, *15*, 8381–8400, doi:10.5194/acp-15-8381-2015.
72. Boylan, J.W.; Russell, A.G. PM and light extinction model performance metrics, goals, and criteria for three-dimensional air quality models. *Atmos. Environ.* **2006**, *40*, 4946–4959, doi:10.1016/j.atmosenv.2005.09.087.
73. Tao, J.; Zhang, L.; Engling, G.; Zhang, R.; Yang, Y.; Cao, J. Chemical composition of PM2.5 in an urban environment in Chengdu , China : Importance of springtime dust storms and biomass burning. *Atmos. Res.* **2013**, *122*, 270–283, doi:10.1016/j.atmosres.2012.11.004.
74. Wang, Y.; Zhang, X.; Arimoto, R. The contribution from distant dust sources to the atmospheric particulate matter loadings at XiAn, China during spring. *Sci. Total Environ.* **2006**, *368*, 875—883, doi:10.1016/j.scitotenv.2006.03.040.
75. Liao, T.; Wang, S.; Ai, J.; Gui, K.; Duan, B.; Zhao, Q.; Zhang, X.; Jiang, W.; Sun, Y. Heavy pollution episodes, transport pathways and potential sources of PM2.5 during the winter of 2013 in Chengdu (China). *Sci. Total Environ.* **2017**, *584-585*, 1056–1065, doi:10.1016/j.scitotenv.2017.01.160.
76. Lang, J.; Zhang, Y.; Zhou, Y.; Cheng, S.; Chen, D.; Guo, X.; Chen, S.; Li, X.; Xing, X.; Wang, H. Trends of PM2.5 and Chemical Composition in Beijing, 2000–2015. *Aerosol Air Qual. Res.* **2017**, *17*, 412–425, doi:10.4209/aaqr.2016.07.0307.
77. Shi, G.L.; Tian, Y.Z.; Ma, T.; Song, D.L.; Zhou, L.D.; Han, B.; Feng, Y.C.; Russell, A.G. Size distribution, directional source contributions and pollution status of PM from Chengdu, China during a long-term sampling campaign. *J. Environ. Sci.* **2017**, *56*, 1–11, doi:10.1016/j.jes.2016.08.017.
78. Wang, H.; Qiao, L.; Lou, S.; Zhou, M.; Ding, A.; Huang, H.; Chen, J.; Wang, Q.; Tao, S.; Chen, C.; et al. Chemical composition of PM2.5 and meteorological impact among three years in urban Shanghai, China. *J. Clean. Prod.* **2016**, *112*, 1302–1311. doi:10.1016/j.jclepro.2015.04.099.

© 2020 by the authors. Licensee MDPI, Basel, Switzerland. This article is an open access article distributed under the terms and conditions of the Creative Commons Attribution (CC BY) license (http://creativecommons.org/licenses/by/4.0/).

Article

Application of Positive Matrix Factorization Receptor Model for Source Identification of PM10 in the City of Sofia, Bulgaria

Elena Hristova *, Blagorodka Veleva, Emilia Georgieva and Hristomir Branzov

National Institute of Meteorology and Hydrology, 1784 Sofia, Bulgaria; blagorodka.veleva@meteo.bg (B.V.); emilia.georgieva@meteo.bg (E.G.); hristomir.branzov@meteo.bg (H.B.)
* Correspondence: elena.hristova@meteo.bg

Received: 31 July 2020; Accepted: 20 August 2020; Published: 23 August 2020

Abstract: The Positive Matrix Factorization (PMF) receptor model is used for identification of source contributions to PM10 sampled during the period January 2019–January 2020 in Sofia. More than 200 filters were analyzed by X-Ray Fluorescence (XRF), Inductively Coupled Plasma Mass Spectrometry (ICP-MS), and Ion chromatography for chemical elements and soluble ions. Seasonal patterns of PM10 mass and elements' concentration are observed with minimum in the summer months and maximum in the cold period. The results from source apportionment (SAP) study showed that the resuspension factor is the main contributor to the total PM10 mass (25%), followed by Biomass burning (BB) (23%), Mixed SO_4^{2-} (19%), Sec (16%), Traffic (TR) (9%), Industry (IND) (4%), Nitrate rich (4%), and Fuel oil burning (FUEL) (0.4%) in Sofia. There are some similarities in relative contribution of the main factors compared to the years 2012–2013. The differences are in identification of the new factor described as mixed sulphate as well as the decrease of the FUEL factor. The results of comparing SAP with EPA PMF 5.0 and chemical transport models (CTM), given by Copernicus Atmosphere Monitoring Service, are presented and discussed for the first time for Bulgaria.

Keywords: PM10; chemical characterization; source apportionment; PMF

1. Introduction

Urban air pollution is the 10th most important risk factor for human health in the middle and high-developed countries according to the World Health Organization [1]. In a number of European countries, the main problem with air pollution in cities is related to exceedances of limit values for nitrogen oxides and fine particulate matter (PM) [2].

Air particulates are emitted into the atmosphere by a number of anthropogenic sources such as energy, industry, road transport, the burning of various solid and liquid fuels, and waste incineration. Natural sources, such as resuspension and erosion of soils, marine aerosols and volcanic eruptions, and formation of secondary particles by biogenic emissions, can also be significant depending on the region [3]. The harmful effect of air particulate on human health depends mainly on its size (Total Suspended Particulate, PM10, PM2.5, PM1), its concentration, and its chemical composition. Fine PM is easily deposited in the respiratory tract where they cause inflammations and diseases of the respiratory and cardiovascular system [1,4].

Southeast Europe is one of the hot spots on the continent with exceedances of PM10 limit values [2,5,6]. In Bulgaria many towns face problems with high PM10 concentrations, especially during the winter [7–9]. It is reported that the highest number of exceedances of daily permitted PM10 concentration (50 µg m^{-3}) for 2017 are registered in Bulgaria: Plovdiv—127, Burgas—116, and Sofia—96 µg m^{-3}.

Sofia is the most densely populated area in the country with a number of different air pollution sources from the industrial, traffic, and domestic sector. The city is located in a semi closed valley because of which unfavorable meteorological conditions contribute to the accumulation of pollutants [10–12]. Recent studies on PM elemental composition [13–15] indicate a distinguished seasonal pattern in the macro- and microelements concentration with maximums during the cold period of the year. This is due to the temperature inversions and stable stratification in the atmospheric boundary layer that creates conditions for limited dispersion of airborne particulates. The emissions of PM and precursor gases in winter increase because of the domestic heating [11,13,15].

The regulatory air quality monitoring network in Sofia is comprised of five stations in the city and one at a nearby mountain site (about 800 m above the Sofia valley). The data from the regulatory network is not sufficient to understand the distribution of pollutants in the complex environment and the impact of potential sources of emissions. Chemical transport models (CTM) have proven useful to fill in gaps in observations. They are also widely used to forecast pollutant concentrations and to simulate effects of different emission scenarios. The Bulgarian chemical weather forecasting system, running operationally since 2012, provides surface concentrations for the region of Sofia [16] but it underestimates observed PM mass concentrations mainly due to lack of detailed emission inventories on a small and regional scale [17]. Other important factors, like soil and road dust resuspension, are also not treated in the modelling system. For this reason, it was not used for source apportionment studies in Bulgaria. Other methods for apportioning of PM mass concentrations sources are receptor-oriented methods, which are based on observations for the PM chemical composition at a specific site.

Receptor oriented models are applied for many tasks in studying urban air pollution in many countries and regions not only in Europe but in Asia, North Africa, North and South America [18–24]. One of the most promising approaches is the advanced factor analysis technique (Positive matrix factorization—PMF), developed by Paatero and Tapper [25], further improved with the Multilinear Engine algorithm [26] and the error estimation procedures [27].

The success of applying receptor methods depends on the availability of data for the PM chemical composition and their quality. For Sofia, such data was obtained during the short time experimental campaigns during 2012–2013 [13,15]. Some of this data was used for a source apportionment study for some cities in the Danube region [28], which was first of a kind for Bulgaria.

In this work, we present results from newly obtained data on the chemical composition of PM10 in Sofia, covering one year (January 2019–January 2020). The data was analyzed using the Positive Matrix Factorization model recommended by the US Environmental Protection Agency (EPA PMF 5.0) following the recommendations given in [29,30]. The results are compared to previous analysis for Sofia [28].

2. Materials and Methods

2.1. Sampling Site

Sofia is the capital and the largest city in Bulgaria, located in the western part of the country (Figure 1). According to 2018 data, the city has a population of 1.3 million inhabitants [31]. Sofia is placed in the semi-closed Sofia valley surrounded by many mountains: Vitosha Mountain to the South with highest peak Cherni Vrah of 2290 m; Ljulin to South-West; Balkan and Murgash Mountains (1687 m) to the North and Northeast, and parts of Sredna gora; Lozen Mountain to the South-East (1226 m height), and Vakarel mountains to the East (700–900 m). Sofia valley is 75 km long (in direction NW to SE) and from 5 to 20 km wide. This topography of the city prevents dispersion of the pollutants and results in unfavorable air quality conditions [15]. Sofia has a humid continental climate with an average annual temperature of 10.4 °C. The regime of precipitation has a well-expressed continental character—the quantity of precipitation in the winter is substantially lower than in the summer. The complex relief of the Sofia valley also influences the wind regime. The winds are usually weak with a high number of days with calm conditions (about 40%). The prevailing wind directions are from west (more than 26%) followed from east, south-east, and south-west. The inversions are frequent in more than 50% of days in the year, mainly in the autumn and the winter [10,11].

Figure 1. Maps with main point source of SO2/SOx (a) and PM10 (b).

The main point emission sources of SO_2 in the country (from west to east—about 400 km) and PM10 from industrial facilities in Sofia region are shown in Figure 1a,b. The main SO_2 sources are coal fired thermal power plants (TPP). The biggest TPP in the Balkan Peninsula is located in south-east Bulgaria. Two other TPPs are located south-ward of Sofia at 30–60 km distance. The biggest industrial plant in the region of Sofia is a copper production plant (60 km to the east of the city) and cement plants (60–70 km to the north, north-east of the city).

PM sampling was carried out at the Central Meteorological Observatory (CMO) in Sofia (42.655 N, 23.384 E, at 586 m a.s.l.) (Figure 2). This is an urban background (UB) site located within the NIMH campus (south-east part of the city) not far away from the main traffic roads—Tsarigradsko shose and Malinov str. Being a populated area, it is also influenced by domestic emissions, including residential wood burning in the suburban and districts in the outskirt of the Vitosha mountain.

Figure 2. Map with Automatic monitoring stations (AMS), NIMH site, and sampling instruments.

2.2. PM10 Sampling, Chemical Analysis, and Data Quality Control

PM10 samples were collected on a daily basis (24 h) with 3 different standardized low volume samplers (2.3 m^3 h^{-1}), according to EN-12341 standard: Tecora Echo PM Sampler, LVS—Sven Leckel

GmbH, and automatic sequential sampler SQ1 (Giano and Gemini—Dado Lab company). All samplers were placed in CMO on a grass filed and the sampling heads were at 2 m. a.g.l (Figure 2). The samples were collected every day at 9:00 LST (local standard time). A total of 289 PM10 samples were collected on a quartz fiber filter (47 mm, Whatman QMA) during the period from 7 January 2019 to 2 February 2020 (67 samples with Tecora, 63 by SVL and 151 by SQ1). The samplers were in operation as follow: the Tecora from 7 January to 22 April 2019, and from 11 December to 15 December 2019; the LVS sampler from 13 February to 25 April 2019, and the SQ1 from 30 March 2019 to 18 April, from 24 June to 31 July 2019, from 11 September 2019 to 2 February 2020. The PM mass concentrations were obtained by gravimetric analysis (EN12341) with analytical balance (Mettler Toledo, MS105DU/M).

To ensure compatible results and as a part of quality control simultaneous, sampling was performed in 42 days with Tecora and SVL. PM10 samples with SQ1 were also simultaneously collected 20 days in April and 3 days in December. The results of PM10 mass concentration in these samples are summarized in Table S1 (in the Supplementary Materials). The difference in the concentration between Tecora and SVL samplers vary from 0% up to 15% in fewer than 41 days. The regression can be written as PM10 (Tecora) = 1.046 × PM10 (SVL)—0.0574 with R^2 = 0.985. The mass concentrations obtained with SQ1 are slightly higher. It was decided to make no corrections in the derived PM10 mass concentrations but to increase the uncertainty when PM10 mass is analyzed with the EPA PMF 5.0 software.

The analyses for PM10 elemental composition were performed in the Institute for Medical Research and Occupational Health, Zagreb by Energy Dispersive X-Ray Fluorescence (ED XRF) technique (PANalytical Epsilon 5 Instrument). This technique was successfully tested and applied in previous studies [28,32,33].

Soluble ions from the punch of $\frac{1}{4}$ of the PM10 filters were analyzed in the certified laboratory "Aquateratest" Sofia by the Ion Chromatograph (ICS 1100, DIONEX), ICP OES (Vista MPX CCD Simultaneous, Varian) and the Spectrophotometer S-20, following standards: EN ISO 10304-1:2009, EN ISO11885:2009, and ISO 7150-1:2002. The elements and the number of samples above the Detection Limit (DL) are given in Section 3.1. The results for elemental concentration were compared to the measured ion concentration and high correlation was observed for Ca/Ca^{2+} Cl/Cl^-, K/K^+ (with few outliers). In case of Sulfur and SO_4^{2-} time series, the agreement between measured concentrations is acceptable (Figure S1 in the Supplementary Materials).

The concentrations of the analyzed elements and ions were measured in some simultaneously collected samples. The comparison of some elements in the filters collected with different samplers is presented in Figure S2 (in the Supplementary Materials). The concentrations of macro-elements like S, Cl, K, Ca, Fe are very close in both PM10 samples with R^2 about 0.90. Some of the microelements with concentration close to the DL show higher variability. The Scandium, detected mainly in Tecora PM10 samples, was not taken into account in further analyses.

2.3. Source Apportionment by PMF

The analysis was performed on 231 samples using the free software EPA PMF 5.0, implementing the ME-2 algorithm developed by Paaero et al., 2014 [27]. PMF is an advanced factor analysis technique that uses measured concentrations and their uncertainties to solve the mass balance equation X = G × F + E, where: X is the chemical composition matrix; G is the source contributions; F the factor profiles; and E the residual.

Strong, weak, and bad variables were selected according to their S/N (signal to noise ratio), as defined by Paatero and Taaper, 1994 [25]. Error estimation (EE), Bootstrap (BS), and Displacement (DISP) methods were used for analyzing factor analytical solutions. The uncertainty of elements is based on an estimate of the uncertainty for each species in a sample. The values below the detection limit were substituted with 1/2 of the DL and the corresponding uncertainty with 5/6 of the DL [34]. The analytical uncertainty for all soluble ions was set to 10%. Extra modeling uncertainty of 10% was used. One sample (25 April 2019) was excluded from the PMF analysis because of the presence of many outliers, most of which were crustal elements.

The solutions with different number of factors (from 4 up to 9) were tested. As it is recommended, one hundred simulations were performed and the differences in scaled residuals between the different simulations were very low. The summary of input data, EPA PMF settings, output data, and error estimation (EE) are presented in Section 3.2.

2.4. Source Apportionment by Copernicus Atmosphere Monitoring Service (CAMS) CTM

The source receptor (SR) analysis for Sofia, available from the Copernicus Atmosphere Monitoring Service (CAMS) on Policy Support [35], gives insight to the origin of PM10—the contribution by local/non-local and natural sources, the geographical origin, and the speciation. The SR analysis for 2019 and the four seasons, provided by CAMS, is based on two state-of-the-art chemical transport modelling systems. It is running operationally to forecast daily speciation for major EU cities—EMEP/MSC-W rv4.15, further on EMEP [36], and LOTOS-EUROSv2.0 [37]. Both models are run by one meteorological driver (ECMWF-IFS) and emissions input (TNO-MACC-III for 2011), but have different approach in the additional simulations for SR analysis—perturbation of emissions (EMEP) and labeling approach (LOTOS-EUROS). Further details on the models set up for SR analysis is given by [38,39]. The comparison of these two modelling systems for Sofia is useful to understand model uncertainty and to identify long-range transport effects on the chemical composition of PM10.

The results for Sofia refer to an area of 60 × 60 km (interpreted as "local") and, considering the characteristics of the region, it is rather coarse and inhomogeneous, including the city, a part of Vitosha Mountain, and another town (Pernik) in southward direction. Thus, the SR analysis from the two modelling systems is used here to highlight the chemical composition of background PM in the region of Sofia.

3. Results and Discussion

3.1. PM10 Mass Concentrations

The mean annual PM10 concentrations and the number of exceedances measured by Automatic Monitoring Stations (AMS) of the national Executive Environmental Agency, Ministry of Environment and Waters presents a clear tendency of decrease in Sofia municipality since 2011 (Figure 3a). The number of days with mean daily concentration above 50 $\mu g\,m^{-3}$ (EU daily limit value) exceeds 35 at 4 of 5 urban background AMS in 2019. There are 65 days with exceedance in AMS Nadejda (NW part of the city) [8]. The mean PM10 quarterly concentrations (I, II, III, and IV quarter) for air quality monitoring stations in Sofia are presented in Figure 3b. Clear seasonal pattern of PM10 concentration with maximum in the cold period of the year (from October until March) is due to variations in meteorological conditions and an increase in the number and strength of the sources. Time variation of PM10 data in this experimental study presented on Figure 4 is similar to that described for previous years based on mean quarterly concentrations with maximums in the January, February, November, December 2019, and January 2020. Minimums are observed in summer months when intensive turbulent mixing and a high atmospheric boundary layer (ABL) are present, resulting in lower measured concentrations. It is important to note that on a daily scale the pollutants and PM10 in particular are well mixed in the urban boundary layer. This is proved with high correlation of PM10 values between the 5 AMS stations. During 2019, the coefficient of determination R^2 varies from the lowest 0.79 (Pavlovo-Drujba) to $R^2 = 0.91$ between the AMS Hipodruma, the most eastern AMS Mladost, and the west-northern AMS Hadejda stations.

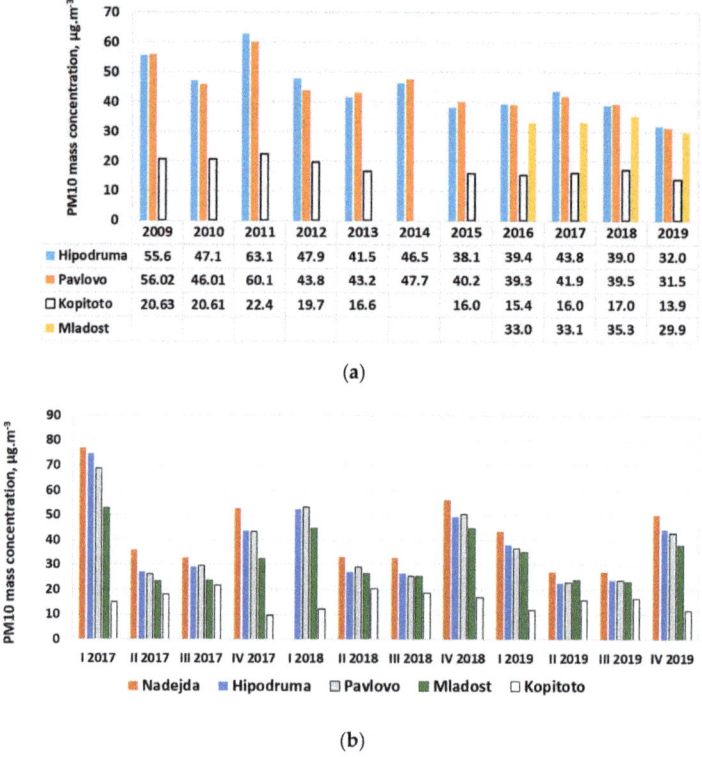

Figure 3. Mean annual (**a**) and mean quarterly (**b**) PM10 concentrations in AMS.

The average PM10 concentration in this study (30.9 µg m^{-3}) is close to the reported mean value (29.9 µg m^{-3}) in the AMS Mladost, placed in the NIMH yard at a distance of 150 m from sampling site.

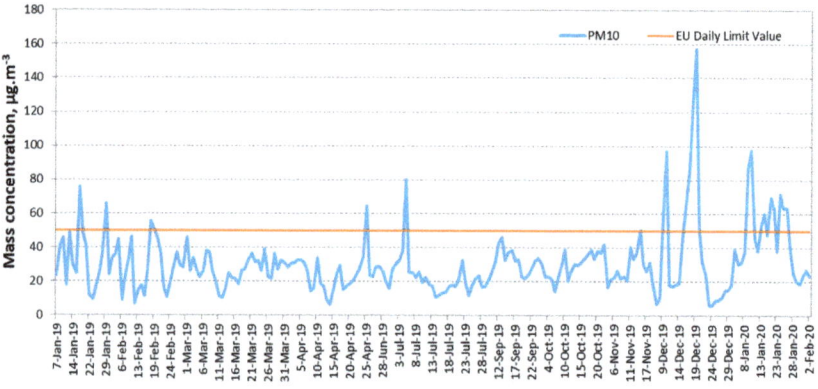

Figure 4. Mass PM10 concentration and daily limit value.

The percentage of the highest daily concentrations (>50 µg m^{-3}) is 10%. In the neighboring station, the days with exceedance were 30. The summarized results of chemical composition in PM10 samples are given in Table 1.

Table 1. Statistical results for the measured PM10 and element concentrations in ng m^{-3}.

Species	Mean	Median	Max	Min	STDEV	Count	% of Conc >DL
PM10	30,861	26,582	157,512	6063	18,974	234	100.0%
Al	992	911	2195	604	363	20	8.5%
S	537	468	2070	65	305	234	100.0%
Cl	244	94	2282	2	380	233	99.6%
K	261	209	1606	28	213	234	100.0%
Ca	872	752	4370	15	686	234	100.0%
Sc	3.37	3.51	5.32	1.99	0.95	29	12.4%
Ti	24.1	19.5	157.5	1.7	19.6	227	97.0%
V	1.69	1.55	2.50	1.30	0.46	6	2.6%
Cr	2.13	1.92	4.45	0.91	0.86	66	28.2%
Mn	17.6	13.9	87.6	2.7	13.0	228	97.4%
Fe	542	425	4110	65	485	234	100.0%
Co	2.4	2.4	2.8	2.1	0.2	8	3.4%
Ni	2.6	1.8	5.7	1.2	1.4	107	45.7%
Cu	36	25	235	3	34	234	100.0%
Zn	101	56	960	1	131	233	99.6%
Br	4.3	3.7	21.1	2.1	2.5	132	56.4%
Sr	4.4	3.6	29.2	2.0	3.3	117	50.0%
Zr	0.004	0.003	0.010	0.002	0.001	88	37.6%
Mo	0.010	0.006	0.023	0.003	0.007	84	35.9%
Cd	5.3	4.5	8.9	3.8	1.5	17	7.3%
Sn	12.5	11.3	26.2	6.7	5.9	12	5.1%
Sb	10.3	9.5	20.7	6.3	3.3	72	30.8%
Ba	29.1	20.0	368.1	12.6	35.2	130	55.6%
Pb	13.1	10.2	65.0	4.9	9.5	200	85.5%
Cl-	314	134	3087	28	489	217	92.7%
NO_3^-	1868	1319	10,196	194	1509	221	94.4%
SO_4^{2-}	3026	2658	10,399	415	1706	234	100.0%
Ca^{2+}	766	663	4246	21	613	234	100.0%
K^+	361	262	2025	24	317	203	86.8%
Mg^{2+}	126	110	647	8	90	234	100.0%
Na^+	375	309	1800	9	277	220	94.0%
NH_4^+	754	533	4018	48	719	189	80.8%

The data for Sodium (Na), measured by ED XRF, is not included in Table 2 because of its high concentration in the second batch of the blank filters. Some elements were above DL in only 2 or 4 days (Ag of about 5 ng.m^{-3} Rb of 2.6 ng.m^{-3} and Y of 0.002 ng.m^{-3}).

The concentration of sulfate ions (SO_4^{2-}) is the highest one, with a mean value of 3.03 µg m^{-3}, followed by nitrate (NO_3^-) of 1.87 µg m^{-3}, and then in decreasing order by Ca^{2+}, Na^+, K^+, Cl^-, and Mg^{2+}. The mean SO_4^{2-} concentration is comparable with the measured in eastern, sought eastern, and part of central sites of EMEP network in Europe [40]. It is lower than reported for the years 2012–2013 [28] as it can be expected from a decrease of SO_2 emissions on the continent [2]. The concentration of NO_3^- and other ions are comparable to those for some European countries [25,41] and lower than reported for sites in Asia [42,43]. PM10 in Sofia is enriched with Fe, Cu, Mn, and Zn in comparison to other European sites [44–46], which is a distinguished signal of industrial activities. Their concentrations are near to those observed in Barcelona, Porto, Milano, Firenze, and Athens [24] and lower in comparison with denser and more industrialized as Turkey [46] and Asian regions [47].

3.2. Source Apportionment Results

Several factors were tested (4–9) in order to obtain the optimum PMF solution. The solution with eight factors was chosen with regard to the highest number factors with physical meaning. The important summary of input and output data are summarized in Table 2. The identification of the factors is based on the key elements in the source fingerprint and the correlations between them.

The diagnostic tests indicated acceptable Q/Qexp values (<2) for the majority of the elements, with the exception of Pb. The obtained ratio between the Qtrue and Qrobust values is lower than the value of 1.5 [48]. The PM mass reconstruction was satisfactory with R^2 between modeled and real PM mass higher than 0.89 (Figure S3). The unaccounted PM10 mass is 3%. In the BS test, the obtained factors were mapped in >99%, with the exception of Traffic and Mixed SO_4^{2-} (that were mapped 86% of the time) (Table 2).

Table 2. Summary of the Positive Matrix Factorization (PMF) input and output data.

Input Data	
Data	PM$_{10}$, Al, S, Cl, K, Ca, Sc, Ti, Cr, Mn, Fe, Ni, Cu, Zn, Br, Sr, Zr, Mo, Sb, I, Ba, W, Pb, Cl$^-$, NO$_3^-$, SO$_4^{2-}$, Ca^{2+}, K$^+$, Mg^{2+}, NH$_4^+$, Na$^+$
Number of factors	4–9 (8 final)
Total variable	PM$_{10}$—weak
	24
No. of used species	Strong: Cl, Ti, Mn, Fe, Cu, Zn, Br, Sr, Pb, NO$_3^-$, SO$_4^{2-}$, Ca^{2+}, K$^+$, Mg^{2+}, NH$_4^+$
	Weak: Cr, Ni, Zr, Mo, Sb, Ba
№ of BS runs	100
BS random seed	88
Min. Correlation R-Value	0.6
FPEAK test	Yes (Fpeak 0)
Output Data	
Q(Robust)	2522.17
Q(True)	2524.69
Q$_{exp}$	1886
Q (Robust)/Q (true)	1.001
Q (Robust)/Q$_{exp}$	1.338
Species with Q/Q$_{exp}$ >2	Pb,
DISP %dQ	0
DISP swaps	0
Mapping of bootstrap factors to base factors:	Secondary, Industry, Nitrate rich—100%, Resuspension, BB, Fuel oil burning—99%, Traffic, Mixed SO$_4^{2-}$—86%

Sources of ambient PM10 have been grouped into 8 categories: Resuspension (RES), Secondary (SEC), Biomass burning (BB), Traffic (TR), Industry (IND), Nitrate rich, Fuel oil burning (FUEL), and Mixed SO_4^{2-}. In order to be sure in factors/source definition, they were compared with chemical profiles from the European database SPECIEUROPE [49]. The chemical factor profiles (the relative mass contribution of each chemical species to PM mass, $\mu g \mu g^{-1}$) and factor contribution (% of each species apportioned to the factors) are presented in Figure 5. The temporal variations of all factors/sources identified by PMF for Sofia are presented in Figure S4 in the Supplementary Materials. The obtained source contribution to the PM10 mass concentration on annual and seasonal bases as pie chart and daily source contributions ($\mu g\ m^{-3}$) are presented in Figure 6.

The Resuspension (RES) factor is dominated by crustal elements such as Mg, Ca, Ti, Sr, Mn, and Fe [25,28]. The name "resuspension" includes contribution from soil and road dust resuspended into the surface air layer from wind blow and mechanically induced resuspension of air particulate from the vehicles. The presence of Zn, Cu, Ni, Sb, and Mo in this factor also suggested a slightly mixed with non-exhaust traffic source [24]. The minimum contribution is observed in the beginning of 2019 (January–February) and the maximum in the middle of December 2019, when stagnant weather conditions with prolonged inversion and presence of fog led to high PM10 concentrations (Figure S4 in the Supplementary Materials). The relatively high impact of RES (33%) is connected to the dryer periods (Figure 6d). In 2019, significantly lower monthly precipitation amounts were measured. After June and in December 2019 and January 2020, the monthly sum of precipitation was below 60% of the climatic norm. The RES source on annual base is the main contributor to the total PM10 mass in Sofia (25%) and is close to the one obtained on a smaller number of samples at the same site in 2012–2013 [28].

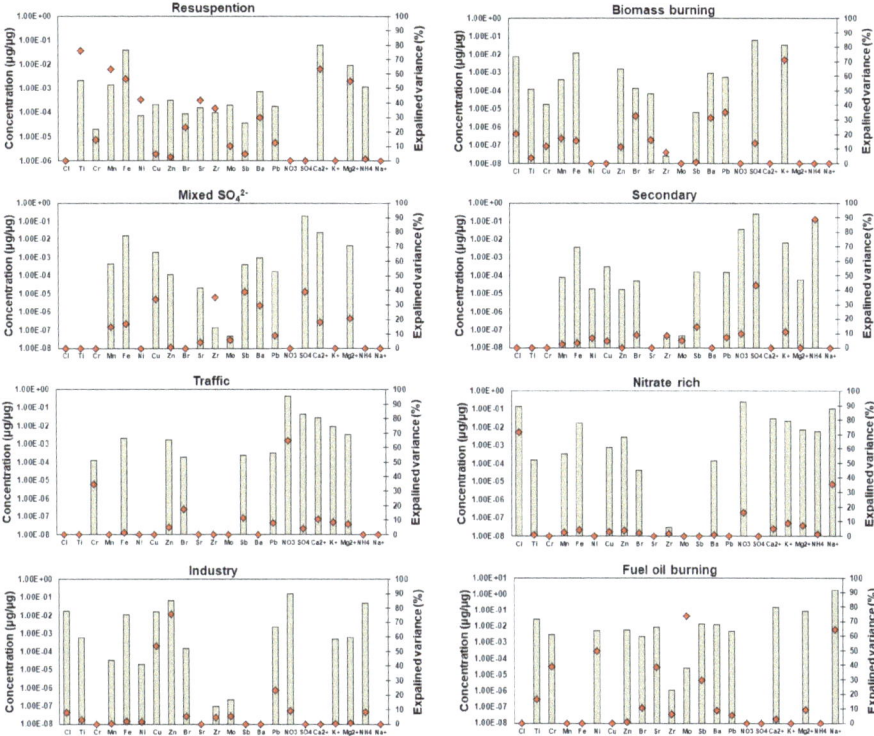

Figure 5. Factor contribution and source chemical profiles.

The biomass burning (BB) factor contains high concentration of the main tracer K^+. Other chemical species, such as Cl^-, SO_4^{2-}, Zn, Pb, and Br, were also associated with this source [21,24,41,50]. The BB presents a typical seasonal trend with higher contributions in winter that gradually decrease to minimum levels in the summer (Figure S4 in the Supplementary Materials). This factor has a small contribution in the summer, probably due to forest fires [51]. The contribution of BB source to PM10 mass on an annual base is 24% (Figure 6a). As expected, the maximum is observed in the cold period with 32% (January and February 2019) (Figure 6b) and 29% (December 2019, January 2020) (Figure 6f).

The chemical profile of "Mixed SO_4^{2-}" suggested a complex origin characterized by high concentrations of SO_4^{2-}, Fe, Ca, Mg, Cu, and Pb. The concentration of this factor was rather stable during the study period with a maximum in the summer. The presence of elements such as Ca, Fe, Zr, and Mo suggested a combination of coal and waste combustion in power plants or cement factories [29,52,53]. Recently, in Bulgaria, cement plants use coal, tiers, and RDF as alternative fuel (Bulgarian National Emission Register, annual reports for 2018/2019) [53,54]. The back-trajectory analysis for some days with high contribution of this factor showed prevailing directions from N and S-SW (Figure S5 in the Supplementary Materials). In both directions there are industrial plants with official permission for RDF burning (Bulgarian National Emission Register), confirming regional influence.

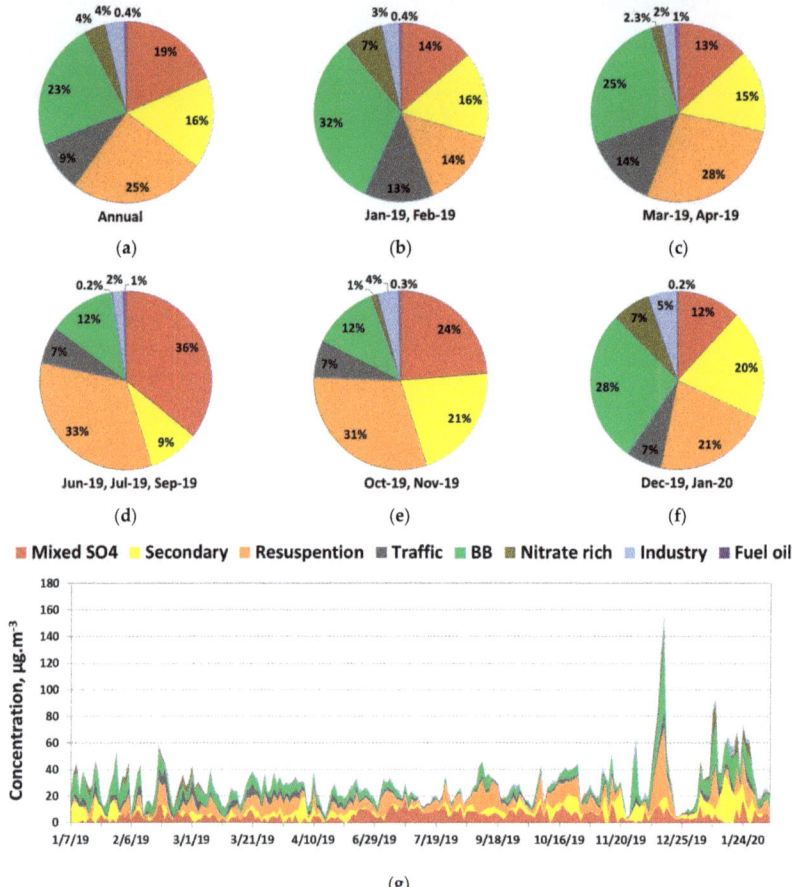

Figure 6. Source profiles (%) on annual (**a**) and seasonal bases (**b**) Jan and Feb 2019; (**c**) Mar and Apr 2019; (**d**) Jun, Jul and Sep 2019; (**e**) Oct and Nov 2019; (**f**) Dec 2019 and Jan 2020), and daily source contributions to PM10 (**g**).

Secondary aerosol (SEC) was dominated by inorganic ions (ammonium sulphate and ammonium nitrate). Secondary inorganic aerosols may be associated with "long range transport". The residence time of sulphates and nitrates in the atmosphere is between 3 and 9 days [3] and is frequently associated with "aged air masses" due to the slow oxidation of SO_2 to SO_4^{2-} [55]. The presence of anthropogenic elements (e.g., Br, Pb) indicated aged primary emissions mixed with secondary sulphate. A contribution of this source was 16% of the total mass and expressed a seasonal trend [28]. The maximum contribution of SEC factor was observed in January 2020 (35 µg m^{-3}), when the height of ABL was low and fog occurred. (Figure S4 in the Supplementary Materials).

The traffic (TR) factor is primarily composed by NO_3^- and elements as Ca, Zn, Sb, Cr, Mg, Fe, and K$^+$ (markers for non-exhaust traffic emissions). NO_3^- is deriving from the ageing of exhaust gases [56,57]. Several studies have shown emissions of Zn, Sb, and Cr, which are associated with brake wear, tires, and lubricants [58–61]. According to Diapouli et. al, 2017 [62], the concentration of Zn is approximately 15 times higher in tires compared to brakes, while concentrations of other heavy metals such as Fe, Sb, Cu, and Ba were higher for brake materials. The elements Ca, Fe, K, and Mg are mostly attributed to road dust [24,28,63]. The lack of information about BC or EC/OC concentration in this analysis probably leads to some mixing of this source with others, like biomass burning. The TR accounts for 9% of the measured PM10. The TR source has a slightly seasonal pattern with maximums in the cold periods in Sofia (13 µg m^{-3}).

The Industry (IND) factor is characterized by high levels of Cu, Zn, and Pb, markers of industrial emissions [24,41]. This factor is associated with local copper alloys and glass production (the Glass factory Drujba is located in the city at about 15 km and the Copper alloy factory is at about 4 km distance from the sampling site) (Bulgarian National Emission Register). The IND source is permanent during the year with some maximums during the cold period when the prolonged inversions occurred (Figure S4 in the Supplementary Materials). The IND factor contributes with 4% to the total PM10 mass.

The Nitrates rich factor is composed mainly by NO_3^-, Cl, and Na and a mixture of Ca, K^+, and Fe. According to Kocak et al., 2011 and Eleftheriadis et al., 2014, [42,63], the presence of Na^+ indicates the production of $NaNO_3$ due to the interaction of sea salt aerosol with anthropogenic gaseous precursors (NO_x). Nitrate is more often related to local rather than regional sources. The mixture of pollutants emitted from both anthropogenic and natural sources, and their interaction and transportation to the receptor site is reflected by the presence of mixed source profiles [64]. The contribution of this factor to the PM10 mass is 4% on annual bases. The Nitrates rich source has seasonality in its contribution to PM10 mass, with a maximum in the cold periods (7%) and a minimum in the summer (1%). The Nitrates rich factor is also related to oxidized local gaseous NOx emissions from thermal power plants (TPP) working on natural gas, which supply thermal energy in more than 60% of households in Sofia. This source can be called Secondary nitrate following [24] or Nitrates/Aged Sea Salt [42,62,63]. They explain "as oxidation product of NO_x emissions (road traffic and industrial plants)" in the five AIRUSE cities Barcelona, Florence, Athens, Porto, and Milan.

The Fuel oil burning (FUEL) factor is characterized by high concentration of Ni, key tracers for oil fuel, and Na, Ti, Sb, Cr, Ba, and Mo [42]. This factor represented 0.4% of the total PM10 (Figure 6a) mass (0.14 µg m^{-3}) which is lower than observed for other sampling sites in Europe (0.8–1.0 µg m^{-3}) [24]. One of the known minor sources is fuel oil burn up, when the gas boilers of TPPs begin to operate. The higher contributions observed in the spring (0.24 µg m^{-3}) might be due to some regional transport when there is more intensive atmospheric air circulation, which favors their movement and distribution. The contribution of FUEL source is less in comparison to a previous study for Sofia (6%) [28]. It is much lower, reflecting the change from fuel oil to gas in small installations for local districts heating.

3.3. Contribution of Outside Sources and Chemical Composition of Background PM Concentrations

The CTM model results for Sofia in 2019 available at CAMS [51] give insight to the contributions from different geographical areas and chemical species to the PM10 background concentrations in the region of Sofia (Figure 7a). The contributions from Bulgarian sources ("domestic") are estimated as 51.4% (EMEP) and 36.8% (LOTOS-EUROS). For the natural contribution, mainly from Saharan dust, there are significant differences—10.8% (EMEP) and 29.4% (LOTOS-EUROS). The rather high share for "others" (representing non-EU countries and model boundary conditions) are comparable for both systems (26.4% and 21.7%). This is due to the position of Sofia in the western part of the country and the prevailing synoptic flow (from west and north-west), suggesting contributions from emission sources in the Western Balkans.

Figure 7b shows the chemical composition from the two model systems along with the results from this study on yearly basis. The comparison is not straightforward as the model results refer to background concentrations for a large area around Sofia, while our results are based on measurements at a fixed point in a residential area. The models use emission inventory for 2011 but emissions certainly have changed from 2011 to 2019, especially reduction in SO_2 emissions and increase of NO_2 emissions in the Balkan region [2]. Nevertheless, there are some common findings using the two CTM SR analysis and the results from the PMF in this study. The contribution of SO_4^{2-} is significant—the models' average is 23%, whereas this study indicates 18%. This is related to processes of secondary aerosol formation and regional contribution from SO_2 emissions in the country and SE Europe [28]. To note the high contribution of the group "rest" that includes secondary organic aerosols, particle-bound water, and unaccounted primary aerosols: models indicate 25% on average, this study—30%. Further studies, including considerations of detailed emissions on local and regional levels, are needed to understand these results. The dust contribution to PM10 concentrations in Sofia seems to be important for the year 2019—with a share varying from 16% on average by the models to about 25% in this study. This suggests that the impact of natural factors on the PM10 concentrations should be further studied.

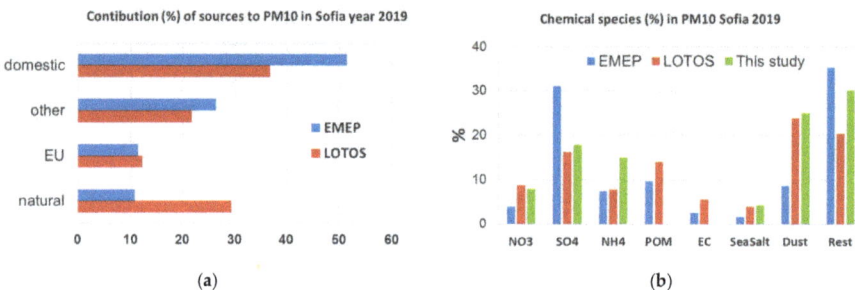

Figure 7. Contribution (a) and chemical species (b) in PM10 for Sofia, 2019.

4. Conclusions

This study has focused on the PM levels, chemical composition, and sources apportionment for a one-year period in the capital of Bulgaria, Sofia. Seasonal patterns of PM10 mass concentration with a minimum in the summer months and a maximum in the cold period is accompanied by the similar time variation of the prevailing macroelements (Ca, Fe), some of the microelements (Pb, Zn, Cu, Mn, Cr, Br, Sr), and ions (SO_4^{2-}, NO_3^-, Cl^-, K^+, NH_4^+, Na^+). This is due to the different meteorological conditions and in some extent to the contribution of additional sources of primary and secondary PM10.

The application of the positive matrix factorization method with EPA PMF 5.0 software led to the identification of eight factors, six of which are with the most significant contribution to the PM10 mass concentration in Sofia (RES, SEC, Mixed SO_4^{2-}, BB, TR, and IND). In comparison with a previous study (2012–2013), a new source appeared (Mixed SO_4^{2-}) with a relatively high contribution during the summer. The RES source is the main contributor to the total PM10 mass in Sofia (25%) and is close to the one obtained in 2012–2013. Second is the BB followed by Mixed SO_4^{2-}, Sec, TR, IND, Nitrate rich, and FUEL.

For the first time the obtained results for contribution of sulphates, nitrates, ammonium ions, and dust (RES) are compared with modeled ones, derived by CAMS models. The variation in SO_4^{2-} and NO_3^- contribution in PM10 mass obtained by two CTM models is seen. The relative contribution to the PM10 in this study is closer to the derived by LOTOS model.

The results from source apportionment (SAP) are of high importance to the development of action plans for improving the air quality based on PM characterization and source apportionment. The authorities can focus on measures to reduce emissions of the sources which they can control. This study provides important information for policy design with specific measures to improve urban air quality in Sofia. Future work on more in-depth analysis for long range transport and Probability Source Contribution Function is planned.

Supplementary Materials: The following are available online at http://www.mdpi.com/2073-4433/11/9/890/s1, Table S1: Comparison of PM10 concentration obtained by tree different samplers; Figure S1: Comparison of obtained results with different analytical techniques; Figure S2: Comparison of results obtained with different samplers; Figure S3: Correlation between experimental PM10 and modeled PM10; Figure S4: Daily source contributions to PM10 levels for the study period in Sofia; Figure S:. HYSPLIT [65] back trajectories for two selected days with high contribution of "Mixed SO_4^{2-}" factor.

Author Contributions: Conceptualization, E.H. and B.V.; Methodology, B.V. and E.H.; Formal Analysis, E.H., B.V., E.G. and H.B.; Writing—Original Draft Preparation, E.H., B.V. and E.G.; Visualization, E.H.; Funding Acquisition, H.B. All authors have read and agreed to the published version of the manuscript.

Funding: This research received no external funding.

Acknowledgments: The authors acknowledge the International Atomic Energy Agency for technical support through project TN-RER7011-1804467 and to the Institute for Medical Research and Occupational Health, Zagreb for the elemental analysis.

Conflicts of Interest: The authors declare no conflict of interest.

References

1. World Health Organization. Available online: https://www.who.int/en/news-room/fact-sheets/detail/ambient-(outdoor)-air-quality-and-health (accessed on 20 June 2020).
2. European Environment Agency. *Air Quality in Europe—2019 Report*; Publications Office of the European Union: Luxembourg, 2019; ISBN 978-92-9480-088-6. [CrossRef]
3. Seinfeld, J.H.; Pandis, N.S. *Atmospheric Chemistry and Physics. From Air Pollution to Climate Change*, 2nd ed.; John Wiley & Sons: Hoboken, NJ, USA, 2006; p. 1225.
4. Lim, S.S.; Vos, T.; Flaxman, A.D.; Danaei, G.; Shibuya, K.; Adair-Rohani, H.; AlMazroa, M.A.; Amann, M.; Anderson, H.R.; Andrews, K.G.; et al. A comparative risk assessment of burden of disease and injury attributable to 67 risk factor clusters in 21 regions, 1990–2010: A systematic analysis for the Global Burden of Disease Study 2010. *Lancet* **2012**, *380*, 2224–2260. [CrossRef]
5. CAMS71_2016SC2. Annual Air Quality Assessment Report for 2015. Available online: https://policy.atmosphere.copernicus.eu/reports/CAMS-71_SC22016_D71.1.3_201801_V2.pdf (accessed on 27 June 2020).
6. Putaud, J.-P.; van Dingenen, R.; Alastuey, A.; Bauer, H.; Birmili, W.; Cyrys, J.; Flentje, H.; Fuzzi, S.; Gehrig, R.; Hansson, H.C.; et al. A European aerosol phenomenology 3: Physical and chemical characteristics of particulate matter from 60 rural, urban, and kerbside sites across Europe. *Atmos. Environ.* **2010**, *44*, 1308–1320. [CrossRef]
7. MOEW Rep. National Report on the State and Protection of the Environment in the Republic of Bulgaria in 2017. Issued 2019 from Ministry of Environment and Waters (MEW). Available online: http://eea.government.bg/bg/soer/2017 (accessed on 16 June 2020).
8. MOEW. *Regional Situation Report of the Environment in 2019*; Regional Inspectorate of Environment and Water: Sofia, Bulgaria, 2020.
9. MOEW. National Report on the State and Protection of The Environment, MoEW, Sofia. Available online: http://eea.government.bg/bg/soer/2014. (accessed on 12 June 2020).
10. Andreev, V.; Branzov, C.; Koleva, E.; Tzenkova, A.; Ivancheva, J.; Videnov, P. *Climate and Human Comfort of Sofia. Ecology of the City of Sofia, Species and Communities in an Urban Environment*; Pensoft Publishers: Sofia, Bulgaria, 2004; pp. 25–54.
11. Batchvarova, E.; Syrakov, D.; Tzenkova, A. Air Pollution Characteristics of a Region of Sofia and Data from Field Experiments (1992–1993). In *Urban Air Pollution*; Allegrini, I., De Santis, F., Eds.; Springer: Berlin, Germany, 1996; Volume 8, pp. 235–242.
12. Naydenova, I.; Petrova, T.; Velichkova, R.; Simova, I. PM10 exceedance in Bulgaria. In Proceedings of the CBU International Conference on Innovations in Science and Education, Prague, Czech Republic, 21–23 March 2018; Volume 6.
13. Hristova, E.; Veleva, B. Variation of air particulate concentration in Sofia, 2005–2012. *Bulg. J. Meteo. Hydr.* **2013**, *18*, 47–56.
14. Veleva, B.; Hristova, E.; Nikolova, E.; Kolarova, M.; Valcheva, R. Elemental composition of air particulate (PM10) in Sofia by EDXRF techniques. *J. Chem. Techno. Metallu.* **2014**, *49*, 163–169.
15. Veleva, B.; Hristova, E.; Nikolova, E.; Kolarova, M.; Valcheva, R. Statistical evaluation of elemental composition data of PM10 air particulate in Sofia. *Int. J. Environ. Pollut.* **2015**, *57*, 175–188. [CrossRef]
16. Syrakov, D.; Prodanova, M.; Etropolska, I.; Slavov, K.; Ganev, K.; Miloshev, N.; Ljubenov, T. A Multy- Domain Operational Chemical Weather Forecast System. In *International Conference on Large-Scale Scientific Computing*; Lirkov, I., Ed.; LSSC 2013, LNCS 8353; Springer: Berlin, Germany, 2014; pp. 413–420. [CrossRef]

17. Georgieva, E.; Syrakov, D.; Prodanova, M.; Etropolska, I.; Slavov, K. Evaluating the performance of WRF-CMAQ air quality modelling system in Bulgaria by means of the DELTA tool. *Int. J. Environ. Pollut.* **2015**, *57*, 272–284. [CrossRef]
18. Viana, M.; Kuhlbusch, T.; Querol, X.; Alastuey, A.; Harrison, R.M.; Hopke, P.K.; Winiwarter, W.; Vallius, M.; Szidat, S.; Prévôt, A.S.H.; et al. Source apportionment of particulate matter in Europe: A review of methods results. *J. Aerosol Sci.* **2008**, *39*, 827–849. [CrossRef]
19. Hopke, P.K. The application of receptor modeling to air quality data. *Pollut. Atmos.* **2010**, *91*, 91–109.
20. Amato, F.; Pandolfi, M.; Escrig, A.; Querol, X.; Alastuey, A.; Pey, J.; Perez, N.; Hopke, P.K. Quantifying Road Dust Resuspension in Urban Environment by Multilinear Engine: A Comparison with PMF2. *Atmos. Environ.* **2009**, *43*, 2770–2780. [CrossRef]
21. Belis, C.A.; Cancelinha, J.; Duane, M.; Forcina, V.; Pedroni, V.; Passarella, R.; Tanet, G.; Douglas, K.; Piazzalunga, A.; Bolzacchini, E.; et al. Sources for PM air pollution in the Po Plain, Italy: I. Critical comparison of methods for estimating biomass burning contributions to benzo(a)pyrene. *Atmos. Environ.* **2011**, *45*, 7266–7275. [CrossRef]
22. Johnson, T.M.; Guttikunda, S.; Wells, G.J.; Artaxo, P.; Bond, T.C.; Russell, A.G.; Watson, J.G.; West, J. *Tools for Improving Air Quality Management: A Review of Top-down Source Apportionment Techniques and Their Application in Developing Countries*; The World Bank Group: Washington, DC, USA, 2011; p. 220.
23. Hwang, I.-J.; Hopke, P.K. Comparison of source apportionment of PM25 using, P.M.F.2.; EPAPMF version 2 Asian. *J. Atmos. Environ.* **2011**, *5*, 86–96. [CrossRef]
24. Amato, F.; Alastuey, A.; Karanasiou, A.; Lucarelli, F.; Nava, S.; Calzolai, G.; Severi, M.; Becagli, S.; Gianelle, V.L.; Colombi, C.; et al. AIRUSE-LIFE +: A harmonized PM speciation and source apportionment in five southern European cities. *Atmos. Chem. Phys.* **2016**, *16*, 3289–3309. [CrossRef]
25. Paatero, P.; Tapper, U. Positive matrix factorization: A non-negative factor model with optimal utilization of error estimates of data values. *Environmetrics* **1994**, *5*, 111–126. [CrossRef]
26. Paatero, P. The multilinear engine a table-driven, least squares program for solving multilinear problems, including the n-way parallel factor analysis model. *J. Comput. Graph. Stat.* **1999**, *8*, 854–888.
27. Paatero, P.; Eberly, S.; Brown, S.G.; Norris, G.A. Methods for estimating uncertainty in factor analytic solutions. *Atmos. Meas. Tech.* **2014**, *7*, 781–797. [CrossRef]
28. Perrone, M.G.; Vratolis, S.; Georgieva, E.; Török, S.; Šega, K.; Veleva, B.; Osánd, J.; Bešlić, I.; Kertész, Z.; Pernigotti, D.; et al. Sources and geographic origin of particulate matter in urban areas of the Danube macro-region: The cases of Zagreb (Croatia), Budapest (Hungary) and Sofia (Bulgaria). *Sci. Tot. Environ.* **2018**, *619–620*, 1515–1529. [CrossRef]
29. Belis, C.A.; Larsen Bo, R.; Amato, F.; El Haddad, I.; Favez, O.; Harrison, R.M.; Hopke, P.K.; Nava, S.; Paatero, P.; Prevot, A.; et al. *European Guide on Air Pollution Source Apportionment with Receptor Models*; JRC Reference Reports, 9789279325144. <10.2788/9307>; Publications Office of European Union: Luxembourg, 2014.
30. Belis, C.A.; Favez, O.; Mircea, M.; Diapouli, E.; Manousakas, M.-I.; Vratolis, S.; Gilardoni, S.; Paglione, M.; Decesari, S.; Mocnik, G.; et al. *European Guide on Air Pollution Source Apportionment with Receptor Models—Revised Version 2019*; EUR 29816 EN; JRC117306; Publications Office of the European Union: Luxembourg, 2019; ISBN 978-92-76-09001-4. [CrossRef]
31. National Statistical Institute. Available online: https://www.nsi.bg/en/content/6710/population-towns-and-sex (accessed on 20 June 2020).
32. Davila, S.; Bešlić, I.; Šega, K. Use of ED-XRF instruments to monitor air quality, 244–249. In Proceedings of the 11th Symposium of the Croatian Radiation Protection Association, Zagreb, Croatia, 5–7 April 2017.
33. Bešlić, I.; Burger, J.; Cadoni, F.; Centioli, D.; Kranjc, I.; van den Bril, B.; Rinkovec, J.; Šega, K.; Zang, T.; Žužul, S.; et al. Determination of As, Cd, Ni and Pb in PM10—comparison of different sample work-up and analysis methods. *Gefahrstoffe - Reinhaltung der Luft* **2020**, *81*, 227–233.
34. Polissar, A.V.; Hopke, P.K.; Paatero, P.; Malm, W.C.; Sisler, J.F. Atmospheric aerosol over Alaska: 2 Elemental composition sources. *J. Geoph. Res. Atmos.* **1998**, *103*, 19,045–19,057. [CrossRef]
35. Atmosphere monitoring service. Available online: https://policy.atmosphere.copernicus.eu/ (accessed on 29 June 2020).
36. Simpson, D.; Benedictow, A.; Berge, H.; Bergström, R.; Emberson, L.D.; Fagerli, H.; Flechard, C.R.; Hayman, G.D.; Gauss, M.; Jonson, J.E.; et al. The EMEP MSC-W chemical transport model—technical description. *Atmos. Chem. Phys.* **2012**, *12*, 7825–7865. [CrossRef]

37. Manders, A.M.M.; Builtjes, P.J.H.; Curier, L.; Denier van der Gon, H.A.C.; Hendriks, C.; Jonkers, S.; Kranenburg, R.; Kuenen, J.J.P.; Segers, A.J.; Timmermans, R.M.A.; et al. Curriculum vitae of the LOTOS–EUROS (v2.0) chemistry transport model. *Geosci. Model Dev.* **2017**, *10*, 4145–4173. [CrossRef]
38. Pommier, M.; Fagerli, H.; Schulz, M.; Valdebenito, A.; Kranenburg, R.; Schaap, M. Prediction of source contributions to urban background PM10 concentrations in European cities: A case study for an episode in December 2016 using EMEP/MSC-W rv4.15 and LOTOS-EUROS v2.0—Part 1: The country contributions. *Geosci. Model Dev.* **2020**, *13*, 1787–1807. [CrossRef]
39. Schulz, M.; Mortier, A.; Tsyro, S.V. Annual Source-Receptor Major European Cities—2019. Issued by: Met Norway, 06/07/2020 Ref: CAMS71_2019SC1_D3.1.5-2020_202007_AnnualSR2019_v2. 2020. Available online: https://policy.atmosphere.copernicus.eu/reports/CAMS71_D3.1.5-2020_v2.pdf (accessed on 21 June 2020).
40. Alastuey, A.; Querol, X.; Aas, W.; Lucarelli, F.; Moreno, T.; Cavalli, F.; Areskoug, H.; Balan, V.; Catrambone, M.; Ceburnis, D.; et al. Geochemistry of PM10 over Europe. *Atmos. Chem. Phys.* **2016**, *16*, 6107–6129. [CrossRef]
41. Taiwo, A.M. Source Apportionment of Urban Background Particulate Matter in Birmingham, United Kingdom Using a Mass Closure Model. *Aerosol Air Qual. Res.* **2016**, *16*, 1244–1252. [CrossRef]
42. Koçak, M.; Theodosi, C.; Zarmpas, P.; Ima, U.; Bougiatioti, A.; Yenigun, O.; Mihalopoulos, N. Particulate matter (PM10) in Istanbul: Origin, source areas and potential impact on surrounding regions. *Atmos. Environ.* **2011**, *45*, 6891–6900. [CrossRef]
43. Sharma, S.K.; Mandal, T.K.; Saxena, M.; Rashmi; Rohtash; Sharma, A.; Gautam, R. Source apportionment of PM10 by using positive matrix factorization at an urban site of Delhi, India. *Urban Clim.* **2014**, *10*, 656–670. [CrossRef]
44. Manousakas, M.; Diapouli, E.; Papaefthymiou, H.; Migliori, A.; Karydas, A.G.; Padilla-Alvarez, R.; Bogovac, M.; Kaiser, R.B.; Jaksic, M.; Bogdanovic-Radovic, I.; et al. Source apportionment by PMF on elemental concentrations obtained by PIXE analysis of PM10 samples collected at the vicinity of lignite power plants and mines in Megalopolis, Greece. *Nucl. Instrum. Methods Phys. Res.* **2015**, *349*, 114–124. [CrossRef]
45. Samek, L.; Stegowski, Z.; Furman, L.; Fiedor, J. Chemical content and estimated sources of fine fraction of particulate matter collected in Krakow. *Air Qual. Atmos. Health* **2017**, *10*, 47–52. [CrossRef]
46. Bozkurt, Z.; Gaga, E.O.; Taşpınar, F.; Arı, A.; Pekey, B.; Pekey, H.; Döğeroğlu, T.; Üzmez, Ö.Ö. Atmospheric ambient trace element concentrations of PM10 at urban and sub-urban sites: Source apportionment and health risk estimation. *Environ. Monit. Assess* **2018**, *190*, 2–17. [CrossRef]
47. Karagulian, F.; Belis, C.A.; Francisco, C.; Dora, C.; Prüss-Ustün, A.M.; Bonjour, S.; Adair-Rohani, H.; Amann, M. Contributions to cities' ambient particulate matter (PM): A systematic review of local source contributions at global level. *Atmos. Environ.* **2015**, *120*, 475–483. [CrossRef]
48. Norris, G.; Duvall, R. *EPA Positive Matrix Factorization (PMF) 5.0 Fundamentals and User Guide. United States*; (EPA/600/R-14/2018); Environmental Protection Agency (EPA); Office of Research and Development: Washington, DC, USA, 2014.
49. Pernigotti, D.; Belis, C.A.; Span, L. SPECIEUROPE: The European data base for PM source profiles. *Atmos. Pollu. Res.* **2016**, *7*, 307–314. [CrossRef]
50. Gunchin, G.; Manousakas, M.; Osan, J.; Karydas, A.G.; Eleftheriadıs, K.; Lodoysamba, S.; Shagjjamba, D.; Migliori, A.; Padilla-Alvarez, R.; Streli, C.; et al. Three-year Long Source Apportionment Study of Airborne Particles in Ulaanbaatar Using X-ray Fluorescence and Positive Matrix Factorization. *Aerosol Air Qual. Res.* **2019**, *19*, 1056–1067. [CrossRef]
51. Copernicus Emergency Management Service. Available online: https://effis.jrc.ec.europa.eu/static/effis_current_situation/public/index.html (accessed on 15 June 2020).
52. SPECISUROPE. Available online: https://source-apportionment.jrc.ec.europa.eu/Specieurope/profiles.aspx?specie=779] (accessed on 21 May 2019).
53. Bulgarian National Emission Register. Available online: http://pdbase.government.bg/forms/public_eprtr.jsp (accessed on 21 May 2019).
54. Annual Reports. Available online: http://eea.government.bg/bg/r-r/r-kpkz/godishni-dokladi-14/index (accessed on 7 May 2019).
55. Lazaridis, M.; Eleftheriadis, K.; Smolik, J.; Colbeck, I.; Kallos, G.; Drossinos, Y.; Zdimal, V.; Mihalopoulos, N.; Mikuska, P.; Bryant, C.; et al. Dynamics of fine particles and photo-oxidants in the Eastern Mediterranean (SUB-AERO). *Atmos. Environ.* **2006**, *40*, 6214–6228. [CrossRef]

56. Raman, R.S.; Hopke, P.K. Source apportionment of fine particles utilizing partially speciated carbonaceous aerosol data at two rural locations in New York State. *Atmos. Environ.* **2007**, *41*, 7923–7939. [CrossRef]
57. Lim, J.M.; Lee, J.H.; Moon, J.H.; Chung, Y.S.; Kim, K.H. Source apportionment of PM10 at a small industrial area using positive matrix factorization. *Atmos. Res.* **2010**, *95*, 88–100. [CrossRef]
58. Wahlin, P.; Berkowicz, R.; Palmgren, F. Characterisation of traffic-generated particulate matter in Copenhagen. *Atmos. Environ.* **2006**, *40*, 2151–2159. [CrossRef]
59. Song, F.; Gao, Y. Size distributions of trace elements associated with ambient particular matter in the affinity of a major highway in the New Jerseye, New York metropolitan area. *Atmos. Environ.* **2011**, *45*, 6714–6723. [CrossRef]
60. Amato, F.; Pandolfi, M.; Moreno, T.; Furger, M.; Pey, J.; Alastuey, A.; Bukowiecki, N.; Prevot, A.S.H.; Baltensperger, U.; Querol, X. Sources and variability of inhalable road dust particles in three European cities. *Atmos. Environ.* **2011**, *45*, 6777–6787. [CrossRef]
61. Pant, P.; Harrison, R.M. Estimation of the contribution of road traffic emissions to particulate matter concentrations from field measurements: A review. *Atmos. Environ.* **2013**, *77*, 78–97. [CrossRef]
62. Diapouli, E.; Manousakas, M.I.; Vratolis, S.; Vasilatou, V.; Pateraki, S.; Bairachtari, K.A.; Querol, X.; Amato, F.; Alastuey, A.; Karanasiou, A.A.; et al. AIRUSE-LIFE +: Estimation of natural source contributions to urban ambient air PM10 and PM2.5 concentrations in Southern Europe—implications to compliance with limit values. *Atmos. Chem. Phys.* **2017**, *17*, 3673–3685. [CrossRef]
63. Eleftheriadis, K.; Ochsenkuhn, K.M.; Lymperopoulou, T.; Karanasiou, A.; Razos, P.; Ochsenkuhn-Petropoulou, M. Influence of local and regional sources on the observed spatial and temporal variability of size resolved atmospheric aerosol mass concentrations and water-solublespecies in the Athens metropolitan area. *Atmos. Environ.* **2014**, *97*, 252–261. [CrossRef]
64. Kim, B.M.; Park, J.-S.; Kim, S.-W.; Kim, H.J. Source apportionment of PM10 mass and particulate carbon in the Kathmandu Valley, Nepal. *Atmos. Environ.* **2015**, *123*, 190–199. [CrossRef]
65. Stein, A.F.; Draxler, R.R.; Rolph, G.D.; Stunder, B.J.B.; Cohen, M.D.; Ngan, F. NOAA's HYSPLIT atmospheric transport and dispersion modeling system. *Bull. Amer. Meteor. Soc.* **2015**, *96*, 2059–2077. [CrossRef]

© 2020 by the authors. Licensee MDPI, Basel, Switzerland. This article is an open access article distributed under the terms and conditions of the Creative Commons Attribution (CC BY) license (http://creativecommons.org/licenses/by/4.0/).

Article

Combined Eulerian-Lagrangian Hybrid Modelling System for PM2.5 and Elemental Carbon Source Apportionment at the Urban Scale in Milan

Giovanni Lonati [1,*], Nicola Pepe [1,2], Guido Pirovano [2], Alessandra Balzarini [2], Anna Toppetti [2] and Giuseppe Maurizio Riva [2]

[1] Department of Civil and Environmental Engineering, Politecnico di Milano, 20133 Milano, Italy; n.pepe@aria-net.it
[2] Ricerca sul Sistema Energetico—RSE Spa, via Rubattino 54, 20134 Milano, Italy; guido.pirovano@rse-web.it (G.P.); alessandra.balzarini@rse-web.it (A.B.); anna.toppetti@rse-web.it (A.T.); maurizio.riva@rse-web.it (G.M.R.)
* Correspondence: giovanni.lonati@polimi.it

Received: 1 September 2020; Accepted: 7 October 2020; Published: 10 October 2020

Abstract: Air quality modeling at the very local scale within an urban area is performed through a hybrid modeling system (HMS) that combines the CAMx Eulerian model the with AUSTAL2000 Lagrangian model. The enhancements obtained by means of the HMS in the reconstruction of the spatial distribution of fine particles (PM2.5) and elemental carbon (EC) concentration are presented for the case-study of Milan city center in Northern Italy. Modeling results are reported for three receptors (a green area, a residential and shopping area, and a congested crossroad on the inner ring road of the city center) selected in order to represent urban sites characterized by both different features in terms of the surrounding built environment and by different exposure to local emission sources. The peculiarity of the three receptors is further highlighted by source apportionment analysis, developed not only with respect to the kind of emission sources but also to the geographical location of the sources within the whole Northern Italy computational domain. Results show that the outcome of the Eulerian model at the local scale is only representative of a background level, similar to the Lagrangian model's outcome for the green area receptor, but fails to reproduce concentration gradients and hot-spots, driven by local sources' emissions.

Keywords: air quality; hybrid modeling; source apportionment; source regions; CAMx; AUSTAL2000; Milan; Po valley

1. Introduction

Air pollution in urban areas is a significant public health issue because of chronic diseases, principally of respiratory and cardiovascular nature [1], and because of premature deaths resulting from population exposure to atmospheric pollutants [2–4]. According to Lelieveld et al. [5], the loss of life expectancy due to air pollution is higher than those caused by tobacco smoking and AIDS. In 2015, air pollution-related diseases were estimated to be responsible for about 6.4 million premature deaths worldwide, with 4.2 million due to ambient air pollution. Indeed, during that year, about 90% of the population, mostly people living in cities, was exposed to particulate matter in concentrations exceeding the WHO air quality guidelines [2]. Air quality monitoring networks provide data in order to check the compliance with air quality limits. However, monitoring sites are usually limited in number and, even though located at sites supposedly representative of different urban microenvironments, their data may not properly assess the actual air quality over the whole urban area [6].

Air quality models are useful tools to assess pollutant concentration levels and emission sources' contributions in urban areas [7–9]. Situations where models can be applied for air quality assessment instead of, or in combination with fixed measurements have been defined by the European Union Directive 2008/50 on air quality [10]. Moreover, most of models can support epidemiological studies providing concentration data with high spatial and temporal resolution for exposure assessment of selected individuals, allowing both predictions on future exposure and reconstruction of historical exposure [11].

However, air quality modeling in urban areas is quite challenging for a number of reasons:

- The concentration levels of atmospheric pollutants are the result of three additional contributions: a regional background (i.e., the baseline level in the surroundings/region of the urban area); an urban background, representing the increment of concentration due to emission of the urban area itself; and a very local contribution of emission sources at pollution hotspots within the urban area [12].
- The pollutants of interest may be not only of primary origin (i.e., directly emitted by emission sources) but also the result of secondary formation processes, like those affecting O_3, NO_2, and fine particulate matter.
- The spatial distribution of the emission sources in urban areas, essentially road traffic and space heating, depends on the urban structure of the built environment, that, in turn, can affect the dispersion of the locally emitted polluted, with local modifications of the wind conditions induced by buildings and with urban canyon structures favoring the build-up of pollutants at emission hot-spots [13].
- Last but not least, model reliability strongly depends on both emission data accuracy, spatial resolution and temporal modulation [14–17], and on the proper reconstruction of the meteorological conditions driving the motion of air masses and pollutant transport and diffusion [18,19].

Eulerian chemical and transport models (CTMs) are commonly applied for air quality modeling at the regional scale, reasonably well estimating regional and urban background concentration levels for both primary and secondary pollutants. Conversely, they are not able to properly reproduce the local concentration gradients in urban areas due to the relatively large step of the computational grid. In order to overcome this limitation, various modeling frameworks that integrate CTMs outputs with other models have been proposed for better reproducing within-city exposure variability. Geostatistical approaches combining land use regression (LUR) and chemical transport modeling have been proposed for long-term concentrations of NO_2, O_3 and particulate matter [20–24]. Hybrid deterministic models combining a Lagrangian dispersion model (LDM) at the local scale and a CTM at the regional scale have been proposed for NO_X and PM10 [25–27]. Indeed, LDMs can be profitably used for the assessment of local sources' impact at high spatial resolution over a small domain because of the fine computational grid that allows a more realistic spatialization of urban emissions, depending on the road network layout and the built environment structure, also accounting for wind field modifications induced by buildings [28]. As drawbacks, LDMs do not include comprehensive gas and aerosol chemistry modules for secondary pollutants assessment and cannot be used for regional scale modeling due to their computational burden.

This work is focused on air quality modeling at the very local scale within an urban area through a hybrid modeling system that combines the CAMx (Comprehensive Air quality Model with Extensions) Eulerian model with the AUSTAL2000 Lagrangian model. In particular, the work highlights the enhancements in the reconstruction of the spatial distribution of the concentration of fine particles (PM2.5) and elemental carbon in the city center of Milan, obtained by means of the hybrid modeling system in comparison with the CAMx Eulerian model. The city of Milan, located in the Po valley in Northern Italy, is a well-known hotspot for particulate matter (PM) pollution. Modelling results are presented for three receptors selected in order to represent urban sites characterized by both different features in terms of the surrounding built environment and by different exposure to the local emission sources. The peculiarity of the three receptors is further highlighted by the source's apportionment analysis developed not only with respect to the kind of emission sources but also to the geographical locations of the sources within the whole Po valley computational domain.

2. Methods

2.1. HMS Modeling Chain Setup

The hybrid modeling system (HMS) used for air quality simulations and source apportionment analysis is composed by three main components (Figure 1): the Weather Research and Forecasting meteorological model (WRF v3.4.1; [29]), the Comprehensive Air quality Model with Extensions (CAMx v6.30; [30]), as Eulerian chemical and transport model, and AUSTAL2000 (AUSTAL2000 v2.6.9; [31]), as Lagrangian local-scale model. The HMS setup also includes the Sparse Matrix Operator for Kernel Emissions model (SMOKE v3.5; [32]) for emission inventory data processing. CAMx was coupled with AUSTAL2000 in order to better estimate concentration levels and properly assess sources' contribution over the urban local domain, thanks to the strong refinement in the spatialization of local emissions according to the road network layout (for traffic emissions) and of the built environment structure (for commercial and residential combustion emissions).

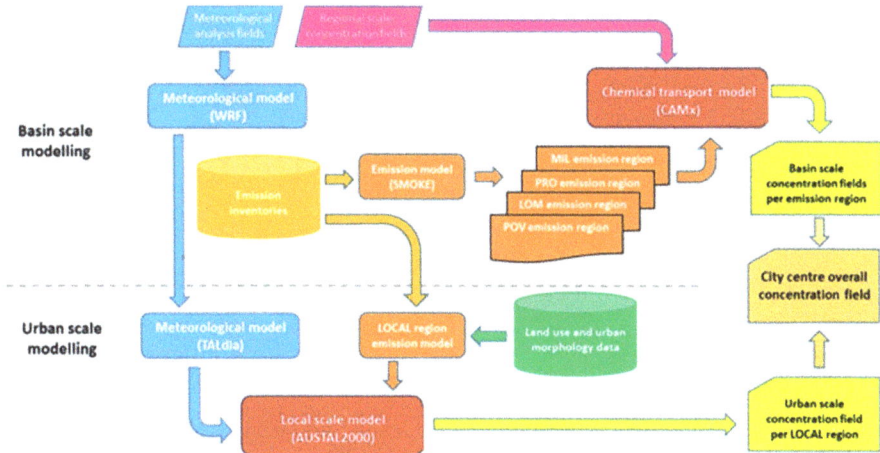

Figure 1. Hybrid modeling system setup.

Input meteorological fields for 2010 were reconstructed by the WRF model, driven by ECMWF (European Centre for Medium-Range Weather Forecasts) analysis fields, "Operational Atmospheric Model Data Set class = od − ANA", at all hours available (00:00, 06:00, 12:00, 18:00 GMT), and set up with 30 vertical layers spanning from 25 m up to 15 km above the ground. Initial and boundary conditions from ECMWF analysis fields, at 0.5° × 0.5° grid size, included 3D (wind speed components, temperature, relative humidity) and 2D surface parameters (sea level pressure and temperature), 2D static parameter for land-sea mask, and 3D soil parameters (temperature and water content), integrated on 4 ground layers (0–7 cm, 7–28 cm, 28 cm–1 m, 1–2.55 m). Horizontally, four nested grids covering Europe, Italy, Po valley, and Milan metropolitan area have been used; grid resolutions are 45 km, 15 km, 5 km, and 1.7 km, respectively. WRF model evaluation was discussed in Pepe et al. [26], with respect to both the Po valley and Milan metropolitan area; the time series of observed and computed values of mixing ratio, temperature and wind speed over Milan metropolitan area are reported in Supplementary Material Figure S1. The wind rose for calendar year 2010 at Milano Linate airport is presented in Figure S2.

CAMx simulations covered the two innermost domains of WRF (Po valley and Milan metropolitan area). CAMx was run at the same spatial resolution of WRF, but with a slight reduction of the domains in order to remove boundary effects (Supplementary Material Table S1). AUSTAL2000 simulation covered one single cell (1.7 × 1.7 km^2) of CAMx Milan metropolitan area domain with 20 m grid resolution. This cell covers part of Milan city center and is characterized by a heterogeneous urban pattern with road arches separating the densely built up commercial and residential areas (Figure 2).

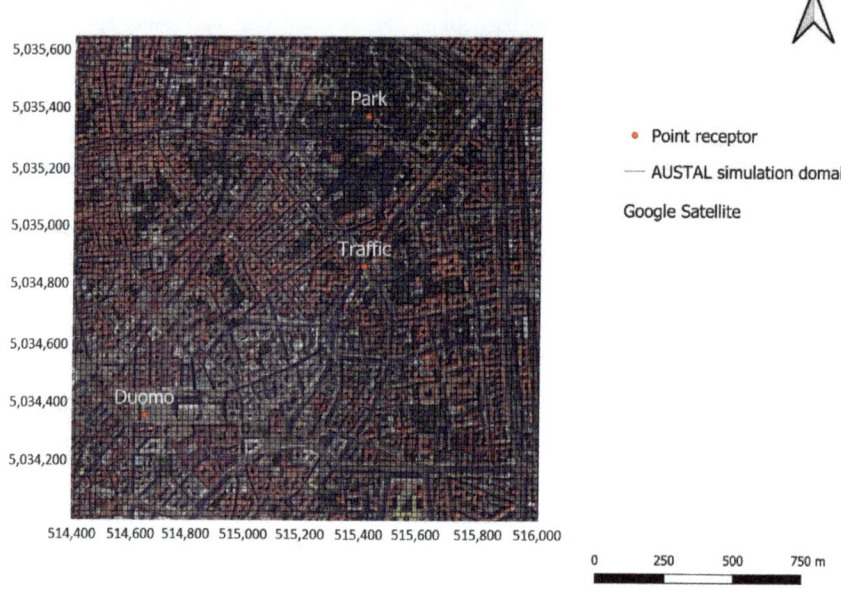

Figure 2. AUSTAL2000 computational domain over Milan city center and selected receptors location.

Input meteorological fields for AUSTAL2000 were calculated by the diagnostic wind field model TALdia based on WRF model output and on the urban canopy features at 20 × 20 m^2 resolution, according to the European land use atlas data [33].

Emissions were derived from inventory data at three different spatial resolution levels: European Monitoring and Evaluation Programme data (EMEP) [34], available over a regular grid of 50 × 50 km^2; ISPRA Italian national inventory data [35] which provides a disaggregation for province; regional inventories data based on INEMAR (INventario EMissioni ARia) methodology [36] for the administrative regions of Lombardy, Veneto and Piedmont, which provide detailed emissions data at single municipality level. Data from each emission inventory were processed using the SMOKE model in order to obtain the hourly time pattern of the emissions. Further details on WRF, CAMx, AUSTAL2000 setup, on meteorological and emissions input data, and on the chemical schemes adopted in this work are reported in Pepe et al. [26], together with the model validation phase for the 2010 calendar year through the comparison between model results and measurements at air quality stations.

2.2. Emission Regions and Emission Categories

HMS source apportionment analyses concurrently considered different emission categories (e.g., road transport, disaggregated by vehicles type, commercial and residential combustion, disaggregated by fuel used) and emission regions i.e., user-defined areas of particular interest for administrative reasons or for the location of specific emission sources. Thus, thanks to the Particulate matter Source Apportionment Technology, (PSAT) algorithm [37], embedded in CAMx, source apportionment analyses could detect contributions from emission categories in combination with geographic information on source locations. The key feature introduced by HMS is the higher detail of source apportionment analysis at urban receptors, where the contributions due to local sources (i.e., located in proximity of the receptors), not correctly disaggregated by CAMx with consequent inaccuracies for source apportionment analyses, are properly accounted for by AUSTAL2000, conversely. In this work, 11 emission categories (Table S2) and 5 emission regions over the Po Valley are considered for the source apportionment (Figure S1). Together with a local urban region (LOCAL), covering the CAMx cell in Milan city center corresponding to the AUSTAL2000 domain, four nested emission regions have been defined: the MIL region covers the municipality of Milan, the PRO region the metropolitan area of Milan, the LOM region the Lombardy region, and the POV region the Po valley. Each emission region accounts for the sources located within its boundaries except those located in the nested inner domains. Additionally, a sixth region that considers sources located within the computational domain but outside of Italy and long-range transport was included in source apportionment analyses. Further details about emission categories and region maps are illustrated in Pepe et al. [25].

2.3. HMS Source Apportionment Output

The HMS approach has been applied in order to assess fine particles (PM2.5) and elemental carbon (EC) ambient concentration levels and related source contributions at three receptors in Milan city center. PM2.5 is one of the regulated pollutants of major concern, due to its health relevance and to its rather high average and episodic concentration levels, especially during the winter months. EC is a primary component of PM particularly critical for human health [38]. The three receptors, all located within the same cell of CAMx's innermost domain (Figure 2), which corresponds to AUSTAL2000 computational domain, were selected in order to represent sites with a different exposure to the local emission sources. Namely, they correspond to a green area (PARK), to a residential and shopping area near the main cathedral square (DUOMO), and to a congested crossroad on the inner ring road of the city center (TRAFFIC). Such a selection was also intended to highlight the capability of AUSTAL2000 for a more reliable source apportionment at the local scale in comparison with CAMx.

Double counting of the emissions from the LOCAL region was avoided thanks to the PSAT source apportionment that allowed to track separately the contributions of the emission regions to the total concentration computed by CAMx. HMS concentration values and source apportionment results at the selected receptors are the combination of CAMx results for the emissions from the most part of the Po valley computational domain (i.e., from POV, LOM, PRO, and MIL region) and of AUSTAL2000 results for the LOCAL region.

3. Results and Discussion

All model results refer to the three abovementioned urban receptors and consider both PM2.5 and EC, focusing on the following issues: (i) AUSTAL2000 output, that is concentration levels and related contribution for the emission sources in the LOCAL region; (ii) comparison between AUSTAL2000 and CAMx output for the concentration levels and for the contributions of the emission sources from the LOCAL region; (iii) HMS concentration levels and sources' contributions for the emissions from the whole Po valley computational domain, also accounting for the geographical locations of the sources (i.e., emission regions).

3.1. AUSTAL2000 Output

This section provides an overview of AUSTAL2000 source apportionment results for PM2.5 and elemental carbon (EC) emitted by local sources of the innermost computational domain (LOCAL region). AUSTAL2000 simulations for PM2.5 consider only its primary fraction (PPM2.5) because the model does not handle aerosol chemistry; that at the very small spatial scale the secondary PM generated by the local emissions of precursors is substantially negligible [39]. According to emission inventory data, commercial and residential combustion (C&R combustion–SNAP category 02) and road traffic (SNAP category 07) are responsible for more than 90% of the emissions within the domain; thus, for the LOCAL region this work is focused on the contributions to ambient levels from these two source categories. Estimated contributions to ambient PM2.5 and EC levels at the three receptors are presented in the panels of Figure 3 as monthly average values for calendar year 2010, with a breakdown based on both source categories and sub-categories (i.e., kind of fuel for C&R combustion, vehicle class for road traffic). Relative contributions, assessed both on annual basis and for the cold and warm season separately, for a deeper insight on seasonality of sources' activity, are summarized in Table 1 for PM2.5 and in Table 2 for EC.

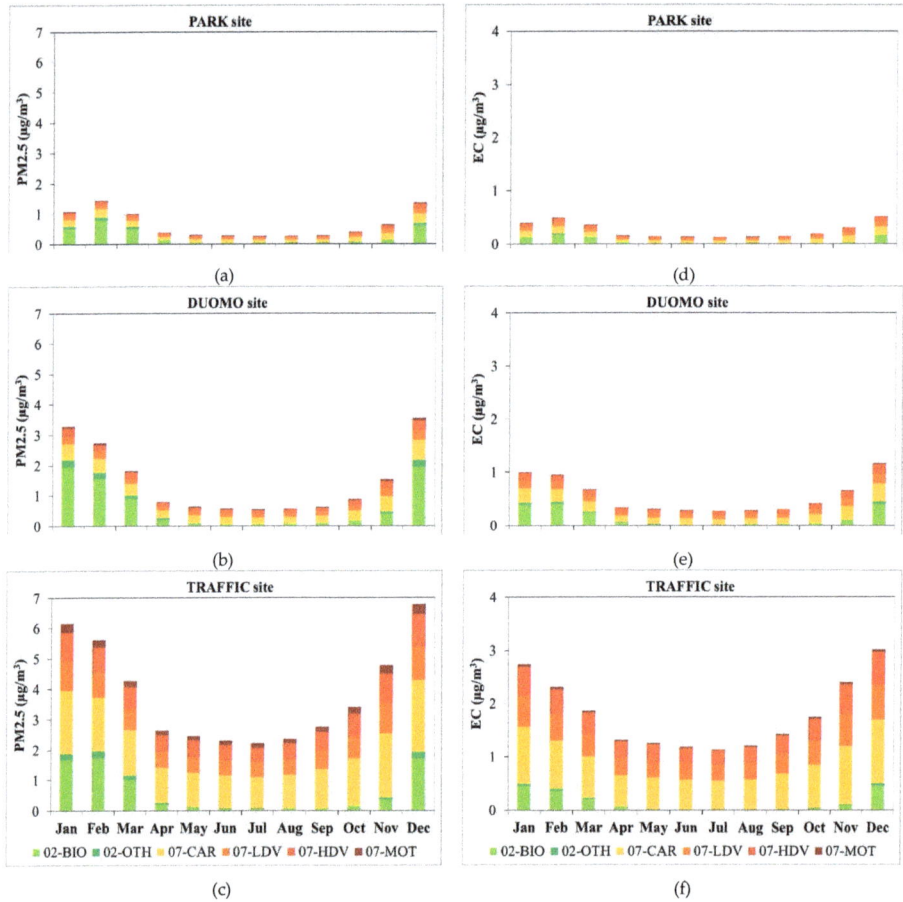

Figure 3. Monthly contributions of C&R combustion and road traffic emissions from the LOCAL area to primary PM2.5 (left column: (**a**) PARK site; (**b**) DUOMO site; (**c**) TRAFFIC site) and EC (right column: (**d**) PARK site; (**e**) DUOMO site; (**f**) TRAFFIC site) ambient levels at the three urban receptors for 2010.

Table 1. Total annual and seasonal average contributions to PM2.5 ambient levels and percentage contributions of emission sources in the LOCAL area at the three urban receptors. (02-BIO: Residential and commercial heating-Biomass burning; 02-OTH: Residential and commercial heating-Fossil fuels burning; 07-CAR: Road traffic-Passenger cars; 07-LDV: Road traffic-Light Duty Vehicles; 07-HDV: Road traffic-Heavy Duty Vehicles; 07-MOT: Road traffic-Mopeds and motorcycles).

Receptor	Period	PM2.5 (µg/m^3)	Emission Sub-Categories Contribution (%)					
			02-BIO	02-OTH	07-CAR	07-LDV	07-HDV	07-MOT
PARK	Annual	0.65	38.6	4.4	27.8	13.0	12.5	3.8
	Cold	1.17	52.3	5.9	20.4	9.5	9.2	2.8
	Warm	0.28	14.5	1.7	40.8	19.0	18.4	5.6
DUOMO	Annual	1.46	42.9	4.9	25.5	11.8	11.4	3.5
	Cold	2.60	56.9	6.4	17.9	8.3	8.0	2.4
	Warm	0.58	10.7	1.2	42.9	20.0	19.3	5.8
TRAFFIC	Annual	3.80	15.9	1.8	40.2	18.6	18.0	5.4
	Cold	5.32	27.9	3.2	33.6	15.6	15.1	4.6
	Warm	2.44	2.7	0.3	47.4	21.9	21.3	6.4

Table 2. Total annual and seasonal average contributions to EC ambient levels and percentage contributions of emission sources in the LOCAL area at the three urban receptors.

Receptor	Period	EC (µg/m^3)	Emission Sub-Categories Contribution (%)					
			02-BIO	02-OTH	07-CAR	07-LDV	07-HDV	07-MOT
PARK	Annual	0.26	22.9	1.7	36.0	19.7	18.0	1.7
	Cold	0.41	35.0	2.4	29.9	16.3	14.9	1.4
	Warm	0.14	7.2	0.7	44.0	24.1	22.0	2.1
DUOMO	Annual	0.55	25.8	1.9	34.5	18.9	17.3	1.7
	Cold	0.88	39.9	2.7	27.4	15.0	13.6	1.3
	Warm	0.28	5.5	0.4	44.9	24.6	22.5	2.2
TRAFFIC	Annual	1.80	8.2	0.6	43.6	23.8	21.8	2.1
	Cold	2.30	15.1	1.1	40.1	21.8	20.0	1.9
	Warm	1.26	1.2	0.1	47.1	25.8	23.6	2.3

3.1.1. PM2.5

PM2.5 results account only for primary PM2.5 because the local scale model AUSTAL2000 does not include a chemical module, thus neglecting chemical transformation process of gaseous precursors that lead to secondary particle formation. The map of the spatial distributions of PM2.5 annual mean concentration over AUSTAL2000 simulation domain (Figure 4a) clearly displays the model ability to account for the structure of the built environment, namely with respect to the main road network. Because of the different exposure to the LOCAL region sources, quite different concentrations levels are estimated for the selected receptors, as expected (0.65 µg/m^3, 1.46 µg/m^3, and 3.80 µg/m^3 at PARK, DUOMO, and TRAFFIC receptor, respectively).

At PARK receptor the overall contribution of the local sources displays a clear seasonal pattern (Figure 3a), with higher contribution in the cold season (>1 µg/m^3 from December to March) and lower contributions in the warm season (<0.5 µg/m^3 from April to October). Peaks up to about 1.5 µg/m^3 are estimated in February and December and the lowest concentrations (about 0.25 µg/m^3) from June to August. The pattern is mainly driven by the seasonality that affects sources' activity, with emissions from C&R combustion in addition to traffic emissions in the cold season. C&R combustion, namely biomass burning (02-BIO), is the first contributor among local sources with a contribution ranging between 0.5 µg/m³ and 0.9 µg/m³; conversely, it is almost negligible in the warm season when its contribution is about 0.04 µg/m³. Road traffic contribution is rather constant throughout the year (ranging between 0.25 µg/m³ and 0.5 µg/m³). In the cold season (January–March), 58.2% of PM2.5 from local sources comes from C&R combustion (category 02), mostly from biomass burning (52.3%); in the warm season (July–September), its contribution drops down to 16.2%, still almost entirely due to biomass combustion (14.5%). Road traffic is responsible for 41.9% of PM2.5 mass in the cold season and for 83.8% in the warm season. However, notwithstanding the higher values computed during winter for C&R combustion, on annual basis road traffic (57.1%) prevails on domestic heating (42.9%). In details, on annual basis passenger cars contribute for almost half of the road traffic share (27.8%), light- and heavy-duty vehicles produce almost the same contribution (13% each), while mopeds and motorcycles account for about 4%.

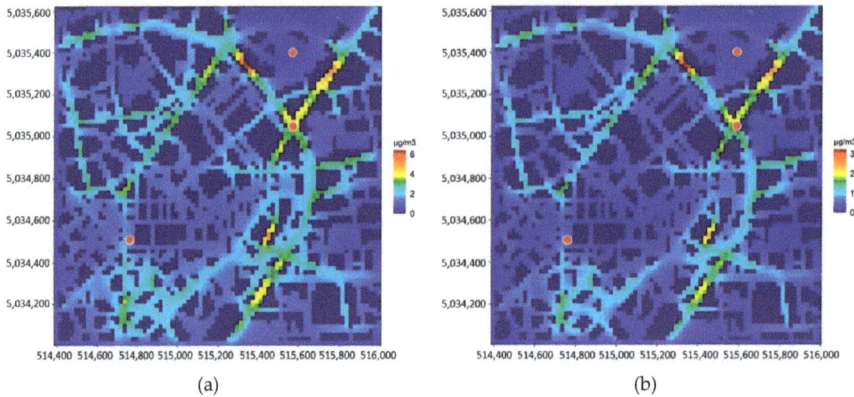

Figure 4. Spatial distribution of PM2.5 (**a**) and EC (**b**) annual mean concentrations generated by LOCAL source emissions over AUSTAL2000 computational domain.

At DUOMO receptor the overall contribution of the local sources follows the same seasonal pattern observed at PARK site, but concentration levels are twice as high in each month (Figure 3b). During the winter months the total contribution generated by vehicles and heating is always in excess of 1.5 µg/m³, but even higher than 3 µg/m³ in January and up to 3.5 µg/m³ in December. From April to October PM2.5 levels are always below 0.9 µg/m³, but they are around 0.5 µg/m³ from June to August. In spite of the limited distance between PARK and DUOMO receptor (less than 2 km as crow fly distance) the model is able to reproduce the concentration gradient due to the activity of very local sources active in the neighborhoods of the residential DUOMO site. As for the PARK receptor, the biomass burning contribution prevails over road traffic during the winter period while the latter takes over during summer. Contribution from sub-category 02-BIO is in the 1–2 µg/m³ range in the cold season but falls down to about 0.05 µg/m³ in the warm season; road traffic contribution ranges between summer values around 0.5 µg/m³ and winter values up to 1.3 µg/m³. Relative contributions show seasonal figures similar to those of PARK receptor: in the cold season, category 02 accounts for 63.3%, with a 56.9% from biomass burning, in the warm season for 11.9%, again almost entirely due to biomass combustion (10.7%); road traffic is responsible for 36.7% of PM2.5 mass in the cold season and for 88% in the warm season, a few percentage points more than at the PARK receptor, coherently with the higher exposure to traffic emission of the DUOMO receptor. Thus, on annual basis the two sources categories provide almost equal contributions to ambient PM2.5: road traffic contributes for 52.2%, with vehicle classes' share similar to those of the PARK site (25.5% from passenger cars, about 11.5% from light- and heavy-duty vehicles, and 3.5% from mopeds and motorcycles) and C&R combustion generates the remainder 47.8%.

At TRAFFIC receptor the monthly time series of estimated contributions shows the same seasonal pattern as at the two previous sites, but with values 4 to 8 times higher than at the PARK receptor and roughly 2 to 4 times higher than at DUOMO receptor (Figure 3c). During the winter months the local sources is always in excess of 4 µg/m^3, but even higher than 6 µg/m^3 in January and December; from April to September, PM2.5 levels are always in the 2–3 µg/m^3 range. The highest contribution (6.8 µg/m^3 in December) is about 5× and 2× the contribution estimated at PARK and DUOMO receptors in the same month; the lowest (2.2 µg/m^3 in July) is 8× and 4× the contributions estimated at the latter sites. Differently from PARK and DUOMO, road traffic is the main contributor not only in the warm season but also in the cold season. Estimated contribution from traffic are in the 4–5 µg/m^3 range in wintertime and about 2.5 µg/m^3 in summertime; C&R combustion contribution is in the same order as at DUOMO (2 µg/m^3) in wintertime and becomes practically insignificant in summertime (<0.06 µg/m^3) as for the other two receptors. Actually, in the warm season the contribution of C&R combustion to primary PM2.5 is almost negligible (3%), the remainder 97% coming from road traffic. In the cold season, the overall contribution of category 02 rises up to 31.1% (27.9% from biomass burning) but still largely ancillary with respect to road traffic, and even to the contribution of passenger cars alone (33.6%). On an annual, basis traffic is responsible for 82.3% of annual mean concentration with 40% due to passenger cars, 18.5% from light- and heavy-duty vehicles, and 5% from mopeds and motorcycles; the residual 17.7% is described by C&R combustion, once again almost entirely due to biomass burning (15.9%).

It is worth noting that the contribution of C&R combustion to primary PM2.5 due to LOCAL sources is probably affected by a partial overestimation because of both the spatial disaggregation and temporal modulation procedures. Indeed, due to the limitation on proxy data for emission spatialization, the emission density was kept constant at municipality level, meaning that it was not possible discriminate differences among Milan city center (i.e., LOCAL region) and the city outskirts (i.e., MIL region). Thus, a possible overestimation of the total emission of biomass burning in the LOCAL cell could have taken place. Furthermore, due to a limitation of the data availability C&R combustion emissions were modulated using a unique profile, more representative of domestic heating than of commercial activities. As a consequence, commercial activity emissions, such as wood oven pizzerias, that are the prevailing source of primary PM for non-residential biomass burning in SNAP category 02, are probably overestimated during the winter season.

3.1.2. EC

Results for EC confirm that the model is able to reproduce the seasonality that affects both sources' activity and atmospheric dispersion, as well the spatial variability of sources' contributions within the urban area due to the different exposure to emission sources of the receptor sites. Seasonality effects are pointed out by the same U-shaped pattern as primary PM2.5 computed at all sites for the time series of monthly EC contributions and by the higher EC/PM2.5 ratios in the warm season, when road traffic is practically the only contributor to PM2.5 levels. Site-dependent exposure to emission sources, unevenly distributed in the urban area because of the features of the built environment and of the road network, is highlighted by the different concentration levels and related gradients at the very local scale (Figure 4b). Indeed, the EC annual mean concentrations estimated at PARK, DUOMO, and TRAFFIC (0.26 µg/m^3, 0.55 µg/m^3, and 1.80 µg/m^3, respectively) reflect their different exposure to the traffic emission of the LOCAL region sources.

At PARK receptor EC levels are in the 0.13–0.14 µg/m^3 and in the 0.4–0.5 µg/m^3 range during the warm and cold season, respectively (Figure 3d). As for primary PM2.5, road traffic is the main contributor (>90%) in the warm season but, differently, it is also the main contributor in the cold season (62.5%); on annual basis, local traffic is responsible for 75.5% of ambient EC, the remainder 24.5% almost entirely due to biomass burning in C&R combustion processes (22.9%) (Table 2). The breakdown of road traffic contribution by vehicle class shows figures similar to those of primary PM2.5 but with a slightly higher share (few % points) for light- and heavy-duty vehicles, all powered by diesel engines.

Similar considerations hold for the DUOMO receptor despite the fact that, as for PM2.5, monthly concentration levels are 2×–3× higher than at PARK (Figure 3e), ranging between 0.27 and 0.42 µg/m^3 and 0.7 and 1.2 µg/m^3 during the warm and cold season, respectively. Like at the PARK receptor, road traffic is the first contributor on both seasons, responsible for 57.4% in wintertime and 94.1% in summertime, with the contribution of passenger cars almost equal to Light Duty Vehicles (LDV) and Heavy Duty Vehicles (HDV) contributions altogether.

At the TRAFFIC receptor, monthly EC levels are 4 to 10 times higher than at the PARK receptor and 2.4 to 4.6 times higher at DUOMO (Figure 3f), thus pointing out spatial gradients even larger than for PM2.5, as a consequence of the greater exposure to primary emissions from traffic. Estimated EC values are in the 1.2–3 µg/m^3 range, averaging about 2.3 µg/m^3 and 1.3 µg/m^3 in wintertime and summertime, respectively, and leading to an annual average of 1.8 µg/m^3. Road traffic is by far the main contributor to EC levels (cold season: 83.8%; warm season: 98.7%; annual basis 91.2%) with single SNAP 07 sub-categories contributions prevailing on biomass burning (02-BIO) even in the cold season (Table 2).

3.2. Comparison between AUSTAL2000 and CAMx Output

Results for ambient PM2.5 and EC levels estimated by AUSTAL2000 have been compared to the corresponding estimates provided by CAMx for the three receptors of the urban area in order to point out the accuracy increase that can arise from coupling the local-scale Lagrangian approach with the mesoscale Eulerian approach within the hybrid modeling system. Actually, CAMx (Eulerian approach) is able to reproduce the large part of chemical processes in the atmosphere, including the secondary formation of aerosols, but has a limited spatial resolution; conversely, AUSTAL2000 (Lagrangian approach) is expected to reconstruct more faithfully the spatial resolution of concentrations, though neglecting chemical transformations.

Local sources' contributions and source apportionment results at the three urban receptors from CAMx and AUSTAL2000 runs are compared in Figure 5 for primary PM2.5 and in Figure 6 for EC. Comparisons, presented for both annual and seasonal average concentrations, point out the relatively "flat" output of CAMx with very limited spatial variability of PPM2.5 and EC concentration levels, notwithstanding the radical change of features of the urban receptors. Conversely, AUSTAL2000 results highlight the difference between the receptor strongly affected by local sources (TRAFFIC) and the receptors less exposed to traffic emissions (DUOMO) and in an urban background location (PARK). PPM2.5 concentrations estimated by CAMx vary in rather narrow ranges (0.7–1.1 µg/m^3 in wintertime, 0.3–0.4 µg/m^3 in summertime) whereas AUSTAL2000 results display variations in the order of a 2× factor between DUOMO and PARK and, in turn, between TRAFFIC and DUOMO. Similarly, for EC, a well-known tracer of traffic emissions, CAMx predictions do not show significant spatial gradients (0.40–0.50 µg/m^3 in wintertime, 0.17–0.26 µg/m^3 in summertime) whilst AUSTAL2000 results indicate much wider ranges (0.41–2.30 µg/m^3 in wintertime, 0.14–1.26 µg/m^3 in summertime).

CAMx limits in properly reproducing the impact of local sources are also confirmed by source apportionment results. Regardless for the period considered, CAMx provides very similar percentage contribution for source categories and sub-categories at the three receptors. Namely, for PPM2.5 traffic shares are 45.0–47.3%, 28.6–30.1%, and 72.6–73.7% respectively on annual, cold season, and warm season basis; corresponding figures for AUSTAL2000 are 52.2–82.3%, 36.7–68.9%, and 83.8–97%. Additionally, such a limited capability of reproducing the exposure to primary emission from road traffic is further highlighted by EC source apportionment, with CAMx traffic shares insensitive to receptors' features (66%, 48%, and 86% on annual, cold season, and warm season basis) whereas AUSTAL2000 provides receptor-dependent shares (72.3–91.2%, 57.4–83.8%, and 92.2–98.7%).

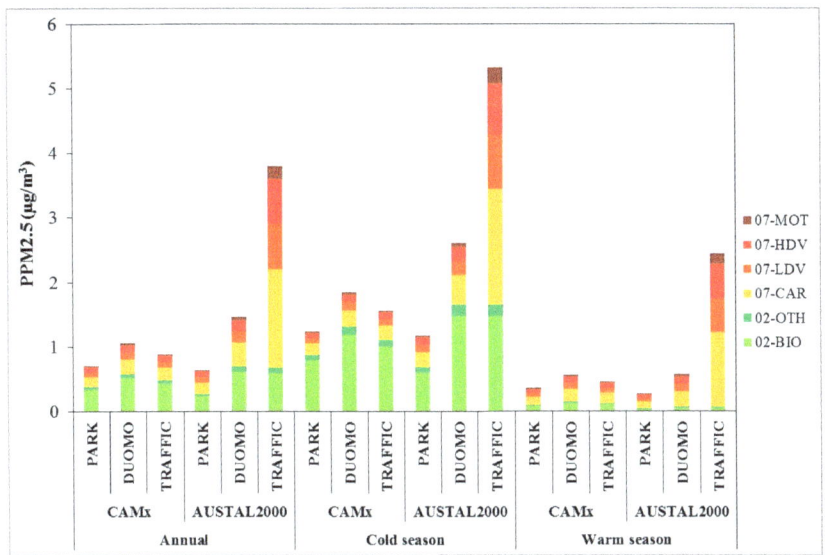

Figure 5. Primary PM2.5 levels and source apportionment comparison between CAMx and AUSTAL2000 at the urban receptors (PARK, DUOMO, TRAFFIC) for LOCAL region emissions.

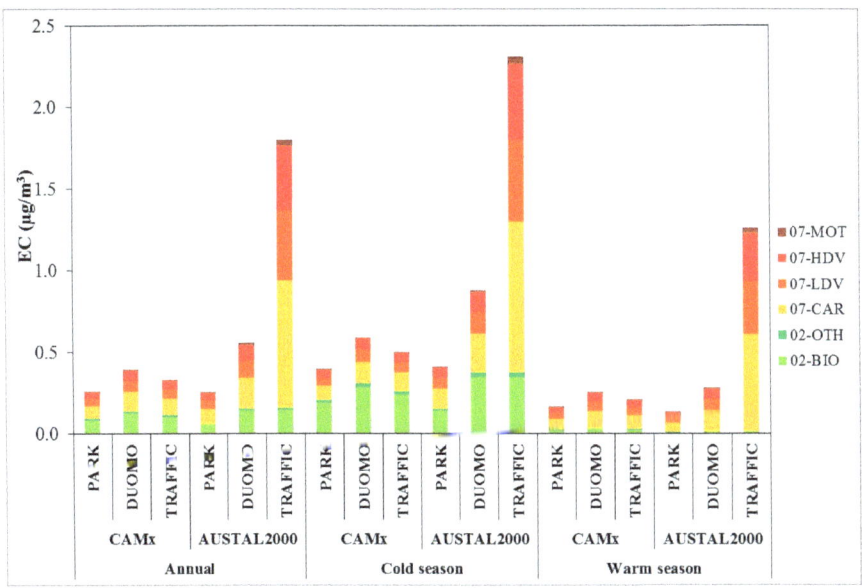

Figure 6. EC levels and source apportionment comparison between CAMx and AUSTAL2000 at the urban receptors (PARK, DUOMO, TRAFFIC) for LOCAL region emissions.

In spite of the abovementioned limits of CAMx and notwithstanding the fact that two models are based on different approaches, their results are in rather good agreement at PARK receptor for both PPM2.5 (0.65 μg/m^3 vs. 0.76 μg/m^3 for the annual average and 1.17 μg/m^3 vs. 1.32 μg/m^3 and 0.28 μg/m^3 vs. 0.40 μg/m^3 for seasonal values) and especially for EC (both models predict 0.26 μg/m^3 for EC annual average and 0.41 μg/m^3 vs. 0.39 μg/m^3, 0.14 μg/m^3 vs. 0.17 μg/m^3 for the seasonal values). Deviations between model outputs become larger and larger as the density of emission sources nearby the receptor increases (i.e., passing from PARK to DUOMO and to TRAFFIC receptor). Thus, CAMx results can be regarded as concentrations levels representative of receptors not directly exposed to local emissions but poorly representative of both residential areas and, most of all, of traffic exposed receptors, where the impact of local sources is missed.

These outcomes are a direct consequence of the different emission spatialization degree within the LOCAL region adopted by the two models: AUSTAL2000 locates road traffic and domestic heating emissions in correspondence with the arches of the urban street network and of built areas; conversely, CAMx evenly distributes the emissions all over the region. Additionally, AUSTAL2000 really calculates concentrations at the receptors points, whereas CAMx determines the concentrations as a distance-based weighed average of the concentrations computed for the four grid cells nearest to the receptor. The drawback of AUSTAL2000 results is that they neglect the secondary PM generated by precursors emissions from the sources within the LOCAL region, that conversely CAMx is able to separately predict and apportion. However, as shown in Figure 7, the contribution of secondary PM to the total PM2.5 is very limited, roughly in the order of 6–7%. Thus, AUSTAL2000 enhanced capability of better reproducing the variability of the concentration levels at the local urban scale, according to receptors' exposure to the emission sources, almost totally offsets the limitation of the neglected secondary PM formation processes.

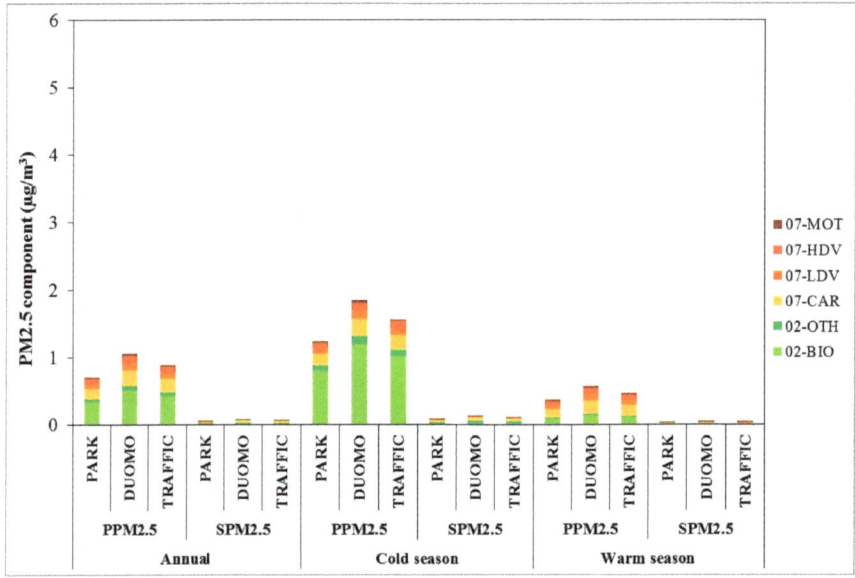

Figure 7. CAMx primary (PPM2.5) and secondary (SPM2.5) PM2.5 levels and source apportionment comparison at the urban receptors (PARK, DUOMO, TRAFFIC) for LOCAL region emissions.

3.3. HMS Output

Previous results demonstrate that AUSTAL2000 has a greater ability in reproducing the variability of concentration levels and source contributions at the local scale, thus supporting the setup of a Hybrid Modeling System (HMS) that combines the key features of CAMx and AUSTAL2000. At the receptors of the local scale urban domain the concentration predicted by the HMS is given by the sum of two terms: the former representing the contribution of sources located outside the urban domain, which accounts for PM secondary formation processes too, the latter representing the contribution of local sources, which neglects chemical processes but accounts for the receptors' exposure to the emission sources. In practice, the spatially-variable and site-dependent contribution of local sources computed by AUSTAL2000 is superimposed to a common base produced by sources located elsewhere (CAMx Background). For the sake of accuracy, the secondary PM2.5 generated by the emissions from the LOCAL region estimated by CAMx was also accounted for within the LOCAL region contribution, in spite of its marginal contribution

3.3.1. PM2.5

HMS results for PM2.5 source apportionment are shown in Figure 8 that allows appreciating both source categories and emission regions contributions concurrently. Summary Tables for the estimated contributions are reported in Supplementary materials. As pointed out, the difference among the estimated concentration levels is driven by the contribution of the local sources, because the CAMx Background is the same at the three receptors, all located in the same cell of CAMx computational domain.

On annual basis (Figure 8a), the sources located outside the local domain (CAMx Background) are responsible for about 17 µg/m^3, that is 96%, 92%, and 82% of the total estimated PM2.5 (17.6 µg/m^3 at PARK, 18.4 µg/m^3 at DUOMO, 20.6 µg/m^3 at TRAFFIC); LOCAL contributions are in the 0.7–3.7 µg/m^3 range. Among emission regions, MIL region is the first contributor (5.2 µg/m^3, 25.5–29.9%), followed by LOM region (4.0 µg/m^3, 19.5–22.9%) and by POV and PRO (both around 1.9 µg/m^3, 9.0–10.7%); long range transport contributes with a 3rd-highest 3.9 µg/m^3 (19.0–22.2%). In terms of emission categories, road traffic (5.2–7.8 µg/m^3 range, 31.6–38.0%) slightly prevails on C&R combustion (4.8–5.2 µg/m^3 range, 25.2–28.4%), both by far greater than the long-range transport and the other anthropogenic sources (OTHAS, 2.9 µg/m^3, 14.3–16.8%). Figures for road traffic at the TRAFFIC receptor are higher than at the two other receptors both as concentration (7.8 µg/m^3) and as percent share (39.0%) because of the additional contribution (3 µg/m^3, 14.5% of total PM2.5 at TRAFFIC) from local traffic estimated by AUSTAL2000, sensitively greater than at DUOMO and, most of all, at PARK (0.38 µg/m^3, 2.2% of total PM2.5 at PARK) as shown in Figure 7. Road traffic contribution mainly derives for ~40% from passenger cars emissions (07-CAR) and for ~30% from light- and heavy-duty vehicles; motorcycles are responsible for 2% only. C&R combustion almost entirely derives from biomass burning (02-BIO). When both emission regions and categories are considered, the features of the receptors come clearly to the light: at PARK and DUOMO road traffic (2.3 µg/m^3) and biomass burning (2.1 µg/m^3) from the MIL region are the first two contributors; conversely, at TRAFFIC road traffic from the LOCAL region (3.0 µg/m^3) takes over. However, all these contributions are smaller than the one coming from long range transport (3.9 µg/m^3), that accounts for about 20% of the total PM2.5.

Seasonal results (Figure 8b,c) point out the different activity of biomass burning, not only in the local region but also in all the other emission regions, sensitively affecting the total PM2.5 levels. Actually, in the cold season CAMx Background (24.4 µg/m^3) is about 2.5 times as high as in the warm season (10.2 µg/m^3) and LOCAL region contributions are from 2 to 4 times higher (cold season range 1.3–5.4 µg/m^3 vs. warm season range 0.3–2.5 µg/m^3). MIL region (8.7 µg/m^3) and LOM region (6.0 µg/m^3) are the two first contributors to PM2.5 levels: LOCAL region contribution is 3rd-highest at TRAFFIC (5.4 µg/m^3) and at DUOMO (2.7 µg/m^3), but is the lowest at PARK (1.3 µg/m^3) where is doubled by both POV and PRO contributions (~2.6 µg/m^3). Contribution from biomass burning is about 10 times higher than in the warm season (~10 µg/m^3 vs. ~1 µg/m^3), making this source the first contributor to PM2.5 at all the receptors, even at TRAFFIC site where it takes over road traffic (10.6 µg/m^3 vs. 9.1 µg/m^3). Overall, biomass burning accounts for 35.7–39.3% of total PM2.5, mostly deriving from the MIL and LOM region (45–49% and 25.7–28% of 02-BIO contribution, respectively). Biomass burning from the MIL (4.8 µg/m^3) and LOM region (2.7 µg/m^3) are the two top contributors in the joint breakdown by emission regions and categories, followed by OTHAS contribution from LOM region (2.9 µg/m^3) at PARK and DUOMO and by passenger cars contribution (07-CAR, 1.8 µg/m^3) from the LOCAL region at TRAFFIC receptor.

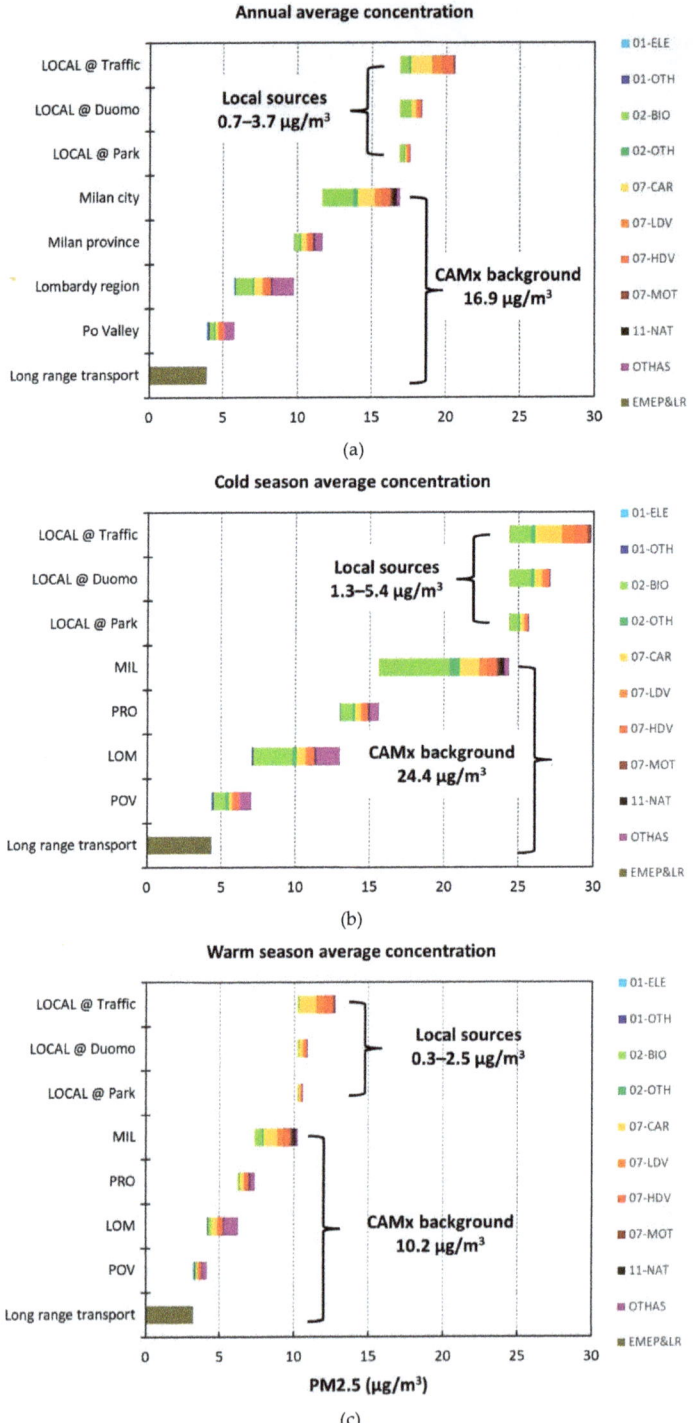

Figure 8. PM2.5 source apportionment by HMS for PARK, DUOMO, and TRAFFIC receptors for annual average (**a**), cold season average (**b**), warm season average concentration (**c**).

In the warm season the contributions from the 4 outermost regions prevails on that from the LOCAL region, still with MIL and LOM regions as top contributors (2.8 µg/m^3 and 2.1 µg/m^3 respectively); conversely, at TRAFFIC receptor road traffic emissions from the LOCAL region become the 2nd-most contributor (2.4 µg/m^3). Almost missing the C&R combustion, road traffic is the first source of PM2.5, accounting for about 36% at PARK and DUOMO, but up to 46% (5.9 µg/m^3) at TRAFFIC, due to the relevant contribution from the LOCAL region, followed by long range transport (25–30%) and OTHAS (15–17%). The primary role of road traffic is also confirmed by joint regions-categories breakdown: traffic emissions from the MIL region are responsible for a top 1.8 µg/m^3 at PARK and DUOMO (17% of total PM2.5), whereas those from the LOCAL region are the 1st-contributor at TRAFFIC (2.5 µg/m^3 of total PM2.5). Overall, traffic emissions from the entire Milan municipality (i.e., LOCAL + MIL region) determine 2.2 µg/m^3 (PARK), 2.5 µg/m^3 (DUOMO), and 4.3 µg/m^3 (TRAFFIC), that is, respectively, 20%, 23%, and 34% of the total estimated PM2.5 at the three receptors.

3.3.2. EC

Estimated EC levels and related source and region apportionment are presented in the panels of Figure 9. As annual average, EC concentration is in the 3.3–4.7 µg/m^3 range, strongly receptor-dependent because of the strong variation in the LOCAL region contributions (0.3–1.7 µg/m^3) in addition to the CAMx Background of 3.0 µg/m^3 (Figure 9a). Such receptor-dependency is due to the great difference among local road traffic contributions, ranging between a minimum of 0.2 µg/m^3 at PARK up to a maximum of 1.6 µg/m^3 at TRAFFIC receptor. Overall, road traffic emissions are responsible for about 57% of EC at PARK and DUOMO (~1.9 µg/m^3) and of 68% (3.2 µg/m^3) at TRAFFIC, but at the two former receptors the traffic contribution originates from the MIL region traffic (30–32% of total EC) whereas at the latter receptor LOCAL region (33%) prevails on MIL region contribution (23%). The contribution of C&R combustion (~1.1 µg/m^3) is in the order of 31% at PARK and DUOMO and down to 23% at TRAFFIC, with the MIL region responsible for about half of these contributions.

The comparison between seasonal results (Figure 9b,c) highlights the wintertime contribution of biomass burning: at PARK and DUOMO biomass combustion (~2.5 µg/m^3, 48%) slightly prevails on road traffic provide almost the same contribution (2.2 µg/m^3, 43%); at TRAFFIC, the latter source prevails (3.8 µg/m^3 vs. 2.6 µg/m^3, that is 56% vs. 38% of total EC). Conversely, in summertime biomass combustion displays a much lower contribution (0.25 µg/m^3), almost entirely deriving from sources outside the LOCAL region, and at all receptors road traffic is, by far, the first contributor overall (1.5–2.6 µg/m^3, that is 70–80% of EC). As far as traffic contribution is concerned, on both in the cold and warm season the LOCAL region contribution prevails on the MIL region one at the TRAFFIC receptor (1.9 µg/m^3 vs. 1.2 µg/m^3 and 1.2 µg/m^3 vs. 0.9 µg/m^3, respectively) contrary to the two other receptors, and especially at PARK, where LOCAL region (0.26 µg/m^3 and 0.12 µg/m^3) is by far exceeded by the MIL region contribution.

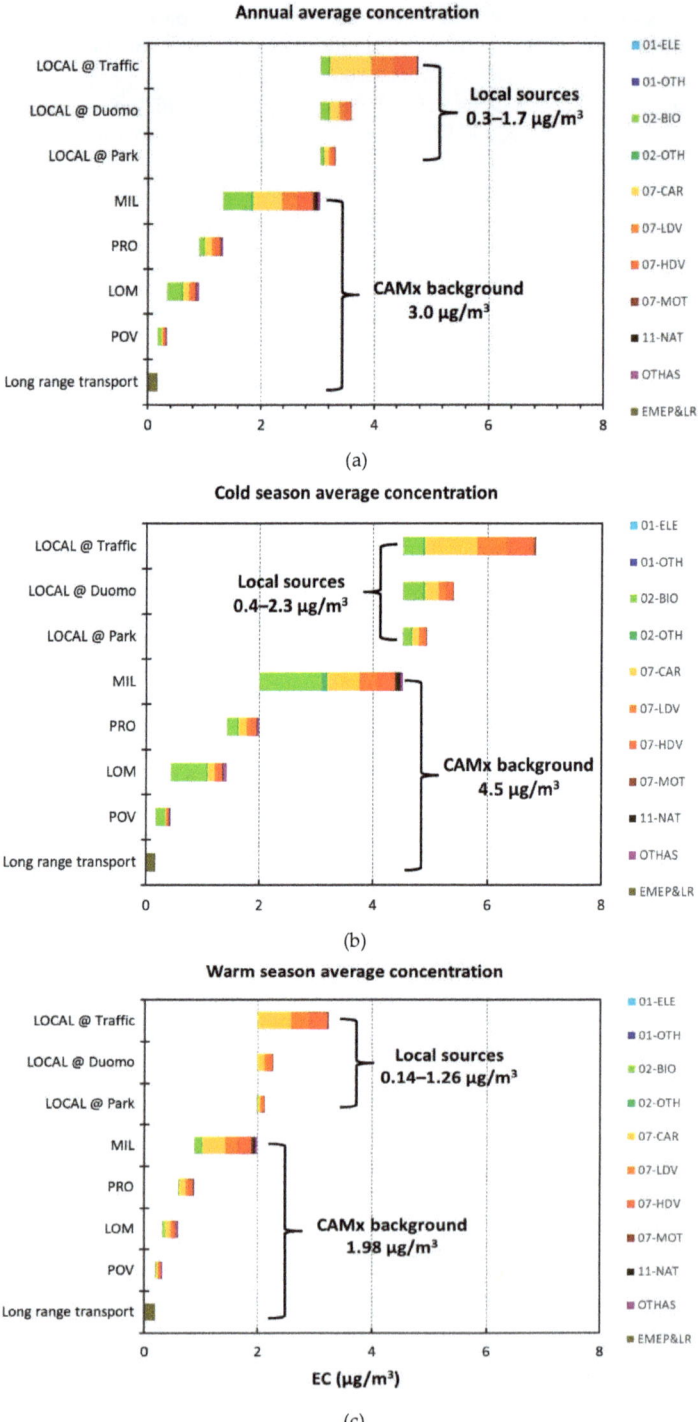

Figure 9. EC source apportionment by HMS for PARK, DUOMO and TRAFFIC receptors for annual average (**a**), cold season average (**b**), warm season average concentration (**c**).

4. Conclusions

A hybrid modeling system (HMS), which couples the Eulerian chemical and transport model CAMx with the local-scale Lagrangian model AUSTAL2000, was set up for air quality assessment and source apportionment in urban areas. In the HMS approach, estimated concentration levels come from the superposition of the background contribution of emission sources located outside the local-scale urban domain, assessed by the Eulerian component, and of the contribution of sources within this latter domain, assessed by the Lagrangian component. In comparison with the Eulerian model, the Lagrangian component relies on a better spatial distribution of emission sources within its computational domain, unevenly distributed because of the road network layout (for traffic emissions) and of the built environment structure (for space heating). Such refinement in the spatialization of local emissions allows the HMS to reproduce more realistically the variability of concentration levels among urban receptors (within the local-scale domain), according to their actual exposure to local sources. In this work, the enhanced performance of the HMS has been assessed for three selected receptors, located in a relatively small area of Milan city center and representative of different exposure to emission sources (PARK receptor: urban background; TRAFFIC receptor: heavy urban traffic; DUOMO receptor: urban traffic, commercial and residential emissions). However, this promising result is someway limited by the few receptors considered and the limited extension of the modelled area by the Lagrangian component. Therefore, these first outcomes need to be further confirmed by future work where, extending the computational domain of the Lagrangian component, a larger number of receptors (e.g., residential areas outside city center, city outskirts, main access roads to the city) can be considered, in order to account for the complex spatial variability of air pollution within the whole urban area. Future work could also include the development of alternative local scale modeling layers, based either on Lagrangian modeling approach, such as puff models, as well as on geostatistical-based methods, both allowing the extensions of the analysis to larger portions of the urban area.

For the presented case-study, the HMS results in almost a 6× factor between PM2.5 annual average at the traffic exposed receptor and at the background receptor in a green area, whereas the Eulerian model, missing the impact of local traffic at the former receptor, predicts only a 26% difference. In practice, the outcome of the Eulerian model at the local scale is only representative of a background level, similar to HMS outcome for the green area receptor, but fails to reproduce concentration gradients and hot-spots, driven by local sources' emissions.

The HMS shows the potential for source apportionment evaluations based on both emission source categories, emission regions (i.e., user-defined areas where sources ae located), and on the combination of categories with regions. This feature enables to assess the overall contribution to ambient concentration levels of the different sources, but also to split the contribution based on geographical location of the sources, that may significantly vary in relation to the receptors' exposure to local sources. In our case-study, at the two receptors less exposed to vehicular emissions, domestic heating and road traffic almost equally contribute to PM2.5 annual average concentration (30%), Milan's urban area is the top contributor among emission regions (29.6%), and road traffic from Milan's urban area is the top contributor (13%) when both emission categories and regions are considered. At the traffic-exposed receptor, road traffic is the top contributor (38%), the Milan region is still the main origin for PM2.5, but with a lower share (25%), and local road traffic is the first contributor (14.5%) in the joint emission categories and regions analysis. The different features of the receptors, namely their exposure to local traffic emissions, are confirmed by PM2.5 source apportionment results on seasonal basis and for elemental carbon too. Thus, the enhanced reproduction of the impact of local sources ensured by the Lagrangian component of the HMS allows to better assess and properly put in context sources' contributions. Additionally, the joint apportionment by categories and regions provides a piece of information that receptor models cannot provide, as their outcome is limited to sources' identification and overall contribution estimation.

This peculiar output of the HMS is particularly useful for scenario evaluations in the framework of air quality remediation and management plans, because it allows the prediction of the outcome of interventions and policies at different spatial scales. Implementation of a small low-emission-zone can affect traffic emissions at the very local scale, whereas the adoption of a congestion/pollution charging scheme, or the ban for diesel vehicles prospected for some European large cities in the near future, has a wider impact, affecting emissions at the urban scale. From the space heating standpoint, implementation/enlargement of district heating at the municipality level may affect local/urban scale emissions, whereas policies on fuels, like limitations or bans in biomass use for domestic heating, usually act on larger administrative areas, therefore potentially affecting emissions even up to the regional scale.

Supplementary Materials: The following are available online at http://www.mdpi.com/2073-4433/11/10/1078/s1, Table S1. Lambert Conformal coordinates for Po valley and Milan metropolitan area nested domains in WRF and CAMx model. Table S2. Emission categories defined considered for source apportionment. Table S3. Annual and seasonal average concentrations of PM2.5 and relative contributions by source regions. Table S4. Contributions (μg/m^3) by source regions to the annual PM2.5 concentrations. Figure S1. Time series of the box and whisker plots for the distribution of the observed (black/grey) and computed (red/orange) daily mean values of mixing ratio, temperature and wind speed over Milan metropolitan area for 2010. Bars show the interquartile range (IR), lines the median values, dashed vertical bars (25th − 1.5 · IR) and the (75th + 1.5 · IR) value. Values for the 25th, 50th, 75th, and 95th quantiles of the whole yearly time series are reported too. (Pepe et al., 2016). Figure S2. Wind rose for calendar year 2010 at Milano Linate airport (5 km crow-fly distance East from Milano city center). Figure S3. Emission regions within Po Valley (top panel, 5 km grid step) and Milan metropolitan area (bottom panel, 1.7 km grid step) computational domains. Figure S4. Google street views of the surroundings of DUOMO receptor. Figure S5. Google street views of the surroundings of PARK receptor. Figure S6. Google street views of the surroundings of TRAFFIC receptor.

Author Contributions: Conceptualization and methodology: G.P.; Software and Modelling simulations: N.P., A.B., A.T., G.M.R.; Formal Analysis: G.L.; Data Curation: N.P. and G.L.; Writing—Original Draft Preparation: G.L.; Writing—Review & Editing, G.L. and G.P. All authors have read and agreed to the published version of the manuscript.

Funding: This research received no external funding.

Acknowledgments: This work has been financed by the Research Fund for the Italian Electrical System under the Contract Agreement between RSE S.p.A. and the Ministry of Economic Development—General Directorate for Nuclear Energy, Renewable Energy and Energy Efficiency in compliance with the Decree of March 8, 2006.

Conflicts of Interest: The authors declare no conflict of interest.

References

1. Forouzanfar, M.H.; Afshin, A.; Alexander, L.T.; Anderson, H.R.; Bhutta, Z.A.; Biryukov, S.; Brauer, M.; Burnett, R.; Cercy, K.; Charlson, F.J.; et al. Global, regional, and national comparative risk assessment of 79 behavioural, environmental and occupational, and metabolic risks or clusters of risks, 1990–2015: A systematic analysis for the Global Burden of Disease Study 2015. *Lancet* **2016**, *388*, 1659–1724. [CrossRef]
2. World Health Organization (WHO). *Ambient Air Pollution: A Global Assessment of Exposure and Burden of Disease*; WHO: Geneva, Switzerland, 2016.
3. Kelly, F.J.; Fussel, J.C. Air pollution and public health: Emerging hazards and improved understanding of risk. *Environ. Geochem. Health* **2015**, *37*, 631–649. [CrossRef] [PubMed]
4. World Health Organization (WHO). *Regional Office for Europe, 2013. Review of Evidence on Health Aspects of Air Pollution—REVIHAAP Project, Technical Report*; WHO: Geneva, Switzerland, 2013.
5. Lelieveld, J.; Pozzer, A.; Pöschl, U.; Fnais, M.; Haines, A.; Münzel, T. Loss of life expectancy from air pollution compared to other risk factors: A worldwide perspective. *Cardiovasc. Res.* **2020**, *116*, 1910–1917. [CrossRef] [PubMed]
6. Duyzer, J.; van der Hout, D.; Zandveld, P.; van Ratigen, S. Representativeness of air quality monitoring networks. *Atmos. Environ.* **2015**, *104*, 88–101. [CrossRef]
7. Ciarelli, G.; Aksoyoglu, S.; El Haddad, I.; Bruns, E.A.; Crippa, M.; Poulain, L.; Äijäkä, M.; Carbone, S.; Freney, E.; O'Dowd, C.; et al. Modelling winter organic aerosol at the European scale with CAMx: Evaluation and source apportionment with a VBS parameterization based on novel wood burning smog chamber experiments. *Atmos. Chem. Phys.* **2017**, *17*, 7653–7669. [CrossRef]

8. Squizzato, S.; Cazzaro, M.; Innocente, E.; Visin, F.; Hopke, P.; Rampazzo, G. Urban air quality in a mid-size city – PM2.5 composition, sources and identification of impact areas: From local to long range contributions. *Atmos. Res.* **2017**, *186*, 51–62. [CrossRef]
9. Pirovano, G.; Colombi, C.; Balzarini, A.; Riva, G.M.; Gianelle, V.; Lonati, G. PM2.5 source apportionment in Lombardy (Italy): Comparison of receptor and chemistry transport modelling results. *Atmos. Environ.* **2015**, *106*, 56–70. [CrossRef]
10. European Union. Directive 2008/50/EC of the European Parliament and of the Council of 21 May 2008 on ambient air quality and cleaner air for Europe. *Off. J. Eur. Union* **2008**, *152*, 1–44.
11. Zou, B.; Wilson, J.G.; Zhan, F.B.; Zeng, Y. Air pollution exposure assessment methods utilized in epidemiological studies. *J. Environ. Monitor.* **2009**, *11*, 475–490. [CrossRef]
12. Lenschow, P.; Abraham, H.-J.; Kutzner, K.; Lutz, M.; Preuß, J.-D.; Reichenbächer, W. Some ideas about the sources of PM10. *Atmos. Environ.* **2001**, *35*, S23–S33. [CrossRef]
13. Vardoulakis, S.; Fisher, B.E.; Pericleous, K.; Gonzalez-Flesca, N. Modelling air quality in street canyons: A review. *Atmos. Environ.* **2003**, *37*, 155–182. [CrossRef]
14. Guevara, M.; Jorba, O.; Tena, C.; van der Gon, H.D.; Kuenen, J.; Elguindi-Solmon, N.; Darras, S.; Granier, C.; García-Pando, C.P. CAMS-TEMPO: Global and European emission temporal profile maps for atmospheric chemistry modelling. *Earth Syst. Sci. Data Discuss.* **2020**. [CrossRef]
15. López-Aparicio, S.; Guevara, M.; Thunis, P.; Cuvelier, K.; Tarrasón, L. Assessment of discrepancies between bottom-up and regional emission inventories in Norwegian urban areas. *Atmos. Environ.* **2017**, *154*, 285–296. [CrossRef]
16. van der Gon, H.A.; Bergström, R.; Fountoukis, C.; Johansson, C.; Pandis, S.N.; Simpson, D.; Visschedijk, A.J.H. Particulate emissions from residential wood combustion in Europe - revised estimates and an evaluation. *Atmos. Chem. Phys.* **2015**, *15*, 6503–6519. [CrossRef]
17. Guevara, M.; Pay, M.T.; Martínez, F.; Soret, A.; Denier van der Gon, H.A.C.; Baldasano, J.M. Inter-comparison between HERMESv2.0 and TNO-MACC-II emission data using the CALIOPE air quality system (Spain). *Atmos. Environ.* **2014**, *98*, 134–145. [CrossRef]
18. Bessagnet, B.; Pirovano, G.; Mircea, M.; Cuvelier, C.; Aulinger, A.; Calori, G.; Ciarelli, G.; Manders, A.; Stern, R.; Tsyro, S.; et al. Presentation of the EURODELTA III intercomparison exercise – evaluation of the chemistry transport models' performance on criteria pollutants and joint analysis with meteorology. *Atmos. Chem. Phys.* **2016**, *16*, 12667–12701. [CrossRef]
19. Pernigotti, D.; Thunis, P.; Cuvelier, C.; Georgieva, E.; Gsella, A.; De Meij, A.; Pirovano, G.; Balzarini, A.; Riva, G.M.; Carnevale, C.; et al. POMI: A model inter-comparison exercise over the Po valley. *Air Qual. Atmos. Health* **2013**, *6*, 701–715. [CrossRef]
20. Cowie, C.T.; Garden, F.; Jegasothy, E.; Knibbs, L.D.; Hanigan, I.; Morley, D.; Hansell, A.; Hoek, G.; Marks, G.B. Comparison of model estimates from an intra-city land use regression model with a national satellite-LUR and a regional Bayesian Maximum Entropy model, in estimating NO2 for a birth cohort in Sydney, Australia. *Environ. Res.* **2019**, *174*, 24–34. [CrossRef]
21. He, B.; Heal, M.R.; Reis, S. Land-Use Regression Modelling of Intra-Urban Air Pollution Variation in China: Current Status and Future Needs. *Atmosphere* **2018**, *9*, 134. [CrossRef]
22. Hennig, F.; Sugiri, D.; Tzivian, L.; Fuks KMoebus, S.; Jöckel, K.-H.; Vienneau, D.; Kuhlbusch, T.; de Hoogh, K.; Memmesheimer, M.; Jakobs, H.; et al. Comparison of Land-Use Regression modeling with dispersion and chemistry transport modeling to assign air pollution concentrations within the Ruhr area. *Atmosphere* **2016**, *7*, 48. [CrossRef]
23. Wang, M.; Sampson, P.D.; Hu, J.; Kleeman, M.; Keller, J.P.; Olives, C.; Szpiro, A.A.; Vedal, S.; Kaufman, J.D. Combining Land-Use regression and chemical transport modeling in a spatiotemporal geostatistical model for ozone and PM2.5. *Environ. Sci. Technol.* **2016**, *50*, 5111–5118. [CrossRef] [PubMed]
24. Akita, Y.; Baldasano, J.M.; Beelen, R.; Cirach, M.; de Hoogh, K.; Hoek, G.; Nieuwenhuijsen, M.; Serre, M.L.; de Nazelle, A. Large scale air pollution estimation method combining land use regression and chemical transport modeling in a geostatistical framework. *Environ. Sci. Technol.* **2014**, *48*, 4452–4459. [CrossRef] [PubMed]
25. Pepe, N.; Pirovano, G.; Balzarini, A.; Toppetti, A.; Riva, G.M.; Amato, F.; Lonati, G. Enhanced CAMx source apportionment analysis at an urban receptor in Milan based on source categories and emission regions. *Atmos. Environ.* **2019**, *X2*, 100020. [CrossRef]

26. Pepe, N.; Pirovano, G.; Lonati, G.; Balzarini, A.; Toppetti, A.; Riva, G.M.; Bedogni, M. Development and application of a high resolution hybrid modelling system for the evaluation of urban air quality. *Atmos. Environ.* **2016**, *141*, 297–311. [CrossRef]
27. Berchet, A.; Zink, K.; Arfire, A.; Marjovi, A.; Martinoli, A.; Emmenegger, L.; Brunner, D. High-resolution air pollution modeling for urban environments in support of dense multi-platform networks. *Geophys. Res. Abstr.* **2015**, *17*, EGU2015-11161.
28. Ghermandi, G.; Fabbi, S.; Bigi, A.; Veratti, G.; Despini, F.; Teggi, S.; Barbieri, C.; Torreggiani, L. Impact assessment of vehicular exhaust emissions by microscale simulation using automatic traffic flow measurements. *Atmos. Pollut. Res.* **2019**. [CrossRef]
29. Skamarock, W.C.; Klemp, J.B.; Dudhia, J.; Gill, D.O.; Barker, D.M.; Duda, M.G.; Huang, X.-Y.; Wang, W.; Powers, J.G. *A Description of the Advanced Research WRF Version 3, NCAR Technical Note NCAR/TN-475+STR*; University Corporation for Atmospheric Research: Boulder, CO, USA, 2008.
30. ENVIRON. *CAMx (Comprehensive Air Quality Model with extensions) User's Guide Version 6.3*; Environ International Corporation: Novato, CA, USA, 2016.
31. Janicke Consulting. *AUSTAL2000 Program Documentation of Version 2.6 2014-02-24*; Federal Environmental Agency (UBA): Dessau-Roßlau, Germany, 2014.
32. University of North Carolina at Chapel Hill (UNC). SMOKE v3.5 User's Manual. 2013. Available online: https://www.cmascenter.org/smoke/documentation/3.5/manual_smokev35.pdf (accessed on 9 October 2020).
33. European Land Use Atlas Data. Available online: http://www.eea.europa.eu/data-and-maps/data/urban-atlas (accessed on 9 October 2020).
34. European Monitoring and Evaluation Programme data (EMEP). Available online: https://www.ceip.at/webdab-emission-database (accessed on 9 October 2020).
35. Italian National Inventory Data. Available online: http://www.sinanet.isprambiente.it/it/sia-ispra/inventaria/disaggregazione-dellinventario-nazionale-2015/view (accessed on 9 October 2020).
36. ARPA Lombardia. INEMAR, Inventario Emissioni in Atmosfera: Emissioni in Regione Lombardia Nell'anno 2012—Revisione Pubblica. *ARPA Lombardia Settore Monitoraggi Ambientali*. Available online: https://www.inemar.eu/xwiki/bin/view/InemarDatiWeb/Aggiornamenti+dell%27inventario+2012 (accessed on 9 October 2020).
37. Yarwood, G.; Morris, R.E.; Wilson, G.M. Particulate matter source apportionment technology (PSAT) in the CAMx photochemical grid model. In Proceedings of the 27th NATO/CCMS International Technical Meeting on Air Pollution Modeling and Application, Banff, AB, Canada, 24–29 October 2004.
38. Janssen, N.A.H.; Gerlofs-Nijland, M.E.; Lanki, T.; Salonen, R.O.; Cassee, F.; Hoek, G.; Fischer, P.; Brunekreef, B.; Krzyzanowski, M. *Health Effects of Black Carbon*; WHO: Copenhagen, Denmark, 2012.
39. Amann, M.; Bertok, I.; Borken-Kleefeld, J.; Cofala, J.; Heyes, C.; Höglund-Isaksson, L.; Klimont, Z.; Nguyen, B.; Posch, M.; Rafaj, P.; et al. Cost-effective control of air quality and greenhouse gases in Europe: Modeling and policy applications. *Environ. Model. Softw.* **2011**, *26*, 1489–1501. [CrossRef]

© 2020 by the authors. Licensee MDPI, Basel, Switzerland. This article is an open access article distributed under the terms and conditions of the Creative Commons Attribution (CC BY) license (http://creativecommons.org/licenses/by/4.0/).

Article

Seasonal Variation in the Chemical Composition and Oxidative Potential of PM$_{2.5}$

Alex Vinson [1], Allie Sidwell [2], Oscar Black [1] and Courtney Roper [1,*]

[1] Department of BioMolecular Sciences, University of Mississippi, Oxford, MS 38677, USA; atvinson@go.olemiss.edu (A.V.); obblack@olemiss.edu (O.B.)
[2] Department of Biology, University of Mississippi, Oxford, MS 38677, USA; amsidwel@go.olemiss.edu
* Correspondence: clroper@olemiss.edu; Tel.: +1-662-915-1273

Received: 11 September 2020; Accepted: 9 October 2020; Published: 13 October 2020

Abstract: Exposure to fine particulate matter (PM$_{2.5}$) has well-established systemic human health effects due in part to the chemical components associated with these exposures. Oxidative stress is a hypothesized mechanism for the health effects associated with PM$_{2.5}$ exposures. The oxidative potential of PM$_{2.5}$ has recently been suggested as a metric that is more indicative of human health effects than the routinely measured PM$_{2.5}$ concentration. The purpose of this study was to analyze and compare the oxidative potential and elemental composition of PM$_{2.5}$ collected at two locations during different seasons. PM$_{2.5}$ was collected onto PTFE-coated filters ($n = 16$) along two highways in central Oregon, USA in the Winter (January) and Summer (July/August). PM$_{2.5}$ was extracted from each filter via sonication in methanol. An aliquot of the extraction solution was used to measure oxidative potential using the dithiothreitol (DTT) assay. An additional aliquot underwent analysis via inductively coupled plasma—mass spectrometry (ICP-MS) to quantify elements ($n = 20$). Differences in PM$_{2.5}$ elemental composition were observed between locations and seasons as well as between days in the same season. Overall, concentrations were highest in the winter samples but the contribution to total PM$_{2.5}$ mass was higher for elements in the summer. Notably, the oxidative potential (nM DTT consumed/µg PM$_{2.5}$/min) differed between seasons with summer samples having nearly a two-fold increase when compared to the winter. Significant negative correlations that were observed between DTT consumption and several elements as well as with PM$_{2.5}$ mass but these findings were dependent on if the data was normalized by PM$_{2.5}$ mass. This research adds to the growing evidence and justification for investigating the oxidative potential and composition of PM$_{2.5}$ while also highlighting the seasonal variability of these factors.

Keywords: fine particulate matter; oxidative potential; particulate matter composition; filter extraction; seasonal differences

1. Introduction

Fine particulate matter (PM$_{2.5}$) is a component of air pollution with well-established systemic health effects. There is strong evidence linking increased concentrations of PM$_{2.5}$ to cardiovascular and pulmonary comorbidities, especially amongst the elderly and those with compromised pulmonary function [1,2]. Many regulatory agencies throughout the world set standards and guidelines based on PM$_{2.5}$ mass concentrations [3–5]. While the adverse health effects related to concentration have been well-studied, there is comparatively less known about the underlying mechanisms of the health effects, as well as connections of these effects to PM$_{2.5}$ composition.

The relationship between the concentration of PM$_{2.5}$ and its adverse health effects has been investigated [6–8], but it is also important to note that particulate matter is a heterogenous mixture with temporally and spatially varied compositions. Factors that impact PM$_{2.5}$ composition include weather patterns and emission sources [9]. In coastal regions, for example, mass fractions of crustal material, trace elements, and organic matter are higher during the Fall and Winter months while sea salt particles were more prevalent in the Spring and Summer [9]. Throughout a single day, shifts in $PM_{2.5}$ composition can be observed at a location. This was demonstrated in the Southeast United States where higher concentrations of carbonaceous materials were observed in the morning while higher concentrations of sulfates were present in the afternoon [10]. In addition to temporal factors, location can also impact compositional differences in $PM_{2.5}$ which are strongly influenced by emission sources. Both the concentration and composition of $PM_{2.5}$ in rural and urban spaces varies widely due to differing types and levels of human activity (e.g., farming, traffic, etc.) and meteorological factors [11].

Due to the established variability in $PM_{2.5}$ composition, research has begun to investigate the health impacts of these differences. Epidemiology studies have shown a positive correlation between elemental concentrations (primarily C, Ni, and V) and hospital admissions for cardiovascular and respiratory issues [12,13]. Results from several studies suggest that increased toxicity is associated with $PM_{2.5}$ emitted from traffic, which contains higher concentrations of carbon and specific metals than standard urban background [14,15]. It has also been noted that compositional differences in inorganic (Al, Ca, Fe, K, Mg, and Pb) and organic (polycyclic aromatic hydrocarbons) content may lead to varying mechanisms of cell death (necrosis, apoptosis, or autophagy) [16]. Establishing not only the compositional differences and resultant toxicity outcomes but also identifying the underlying mechanisms is critical for providing targeted regulations to protect human health.

Oxidative stress occurs when the accumulation of reactive oxygen species (ROS) overwhelms the body's mechanisms for neutralizing them. This state can contribute to a number of adverse health effects including cardiovascular, neurological, respiratory, reproductive and kidney diseases, as well as cancer [17]. $PM_{2.5}$ is known to cause oxidative stress via the generation of ROS such as hydrogen peroxide (H_2O_2) or hydroxyl radicals ($^\bullet OH$) [18]. The tendency of a chemical species to oxidize a target molecule is known as its oxidative potential [19]. Analysis of oxidative potential has been conducted for a number of $PM_{2.5}$ studies using various assays including the dithiothreitol (DTT) assay [20,21]. Understanding the role of $PM_{2.5}$ composition and source contributions to oxidative potential is an area of growing interest particularly due to the recent hypothesis that oxidative potential and composition are more health relevant metrics of $PM_{2.5}$ than mass [22]. Previous research has observed differences in oxidative potential based on source contributions with increased oxidative potential for $PM_{2.5}$ from traffic and underground railway stations as compared to rural and lower traffic locations [22]. Seasonal variations have also been explored with several studies identifying increased oxidative potential in the summer compared to the winter [23–26]. Other studies have investigated the connection between oxidative potential and fine particulate matter composition and identified significant positive correlations between oxidative potential measured by DTT assays and the concentration of transition metals present [27]. While there is growing research in this field there is a need to further establish the trends of oxidative potential based on $PM_{2.5}$ source contributions, seasonal differences, and chemical composition particularly in understudied areas of the world like the state of Oregon in the United States.

In this study we explore the connection between $PM_{2.5}$ composition and oxidative potential at two sampling locations across seasons. $PM_{2.5}$ filter-based samples were extracted and aliquots of each sample were used for chemical composition analysis and oxidative potential analysis. We hypothesized that differences would be observed in chemical composition and oxidative potential between the sampling locations and seasons. The study also was designed to explore correlations between oxidative potential and composition of $PM_{2.5}$ and analyze how these factors may change between seasons.

2. Experiments

2.1. Reagents

Solvents and reagents included methanol, potassium phosphate monobasic (KH_2PO_4), and 1,4-Dithiothreitol (DTT) (Thermo Fisher Scientific, Waltham, MA, USA), as well as 5,5′-Dithiobis (2′-nitrobenzoic acid) (DTNB) (Sigma Aldrich, St. Louis, MO, USA).

2.2. $PM_{2.5}$ Samples

Sampling sites were established monitoring locations of the Lane Regional Air Protection Agency that had similar potential traffic sources as both were adjacent to state highways in central Oregon, USA in residential areas (Site A: OR Route 58 near Oakridge, Oregon and Site B: OR Route 99 in Eugene, Oregon). All sampling procedures and gravimetric procedures were according to United States Environmental Protection Agency (EPA) standards with federal reference methods employed [28,29]. $PM_{2.5}$ samples were collected on 47 mm PTFE-coated filters during 2016 in the winter (January) and summer (July and August). Sampling occurred at both Sites A and B on 4 days for each season (winter—1, 2, 3, and 4 January; summer—14, 17, 20 July and 1 August). Samples were collected for 24 h at 16.7 L per minute (LPM) flowrate. A blank 47 mm PTFE-coated filter, without collected $PM_{2.5}$, underwent all analyses to serve as a methods control. Samples underwent gravimetric analysis in a temperature and humidity-controlled chamber to determine $PM_{2.5}$ masses by pre- and post-sampling filter weights and were then stored away from light at −20 °C.

2.3. Extraction

Each filter and control were sonicated (60 Hz, Branson Ultrasonics Corporation, Brookfield, CT) for 60 min in approximately 8 mL of methanol. Each filter was then removed from the tube and rinsed with additional methanol to collect any residual $PM_{2.5}$ remaining on the filter [21]. The resulting $PM_{2.5}$ solution was used for subsequent analyses and represents compounds that were methanol-soluble.

2.4. Analysis of Oxidative Potential

A 96-well plate version of the DTT assay was used for all analyses as previously described [21] with minor modifications. Briefly, each $PM_{2.5}$ sample in methanol and controls ((a) blank filter, (b) vehicle (phosphate buffer), and (c) all reagents but the quenching reagent) underwent the DTT assay. All samples and controls were mixed with 100 µL of phosphate buffer and 5 µL of 0.5 mM DTT solution prior to incubation at 37 °C for 20 min. After incubation, 10 µL of 1 mM DTNB was added to quench the reaction. The plate was then read on a plate reader at 412 nm. DTT consumption was determined based on a DTT calibration curve (0, 0.2, 0.4, 0.6, 0.8, and 1.0 mM) prepared with stock DTT and methanol to adjust for sample volume. All samples, controls, and calibration standards were run in triplicate.

2.5. Chemical Analysis by ICP-MS

Aliquots of each extracted $PM_{2.5}$ and control solution were blown to dryness with N_2, re-suspended in 10 mL of milli-Q water, and sonicated for 5 min. Samples were then analyzed for trace metal content by Inductively Coupled Plasma—Mass Spectrometry (Thermo Fisher Element XR ICP-MS). Quantitation consisted of three runs and three passes per sample to ensure a representative average of element concentrations. Quantitative data in parts per billion (ppb) was collected for Ag, Ba, Ca, Cd, Ce, Co, Cr, Cs, Cu, Fe, Ga, Mn, Ni, P, Pb, Sr, Tl, U, V, and Zn with calibration curves generated using Multielement Calibration Standard Solution 2A (Spex Certiprep, Metuchen, NJ, USA). ICP-MS instrumental parameters are displayed in Supplemental Table S1. Reagent and laboratory blank filter controls were analyzed alongside samples to facilitate background subtraction and to evaluate instrumental drift during analysis. Method accuracy was evaluated by analyzing NIST certified standard reference material 1640a "Trace Elements in Water".

2.6. Statistical Analysis

Statistical analysis for all data was performed with Sigmaplot 14.0 (Systat Software, Inc., San Jose, CA, USA) and Excel 16.0 (Microsoft Corporation, Redmond, WA, USA). All data were reported as a mean ± standard deviation (SD) and corrected with blank filters or other appropriate controls. Data was analyzed using a one- or two-way analysis of variance (ANOVA) or Student's t-test to determine differences between $PM_{2.5}$ samples. Differences with p values ≤ 0.05 were considered significant, unless otherwise noted. Pearson correlation coefficients were calculated and corrected for multiple tests using the Bonferroni correction to determine the p-value that was statistically significant.

3. Results and Discussion

3.1. Chemical Constituents of $PM_{2.5}$

Elements normalized by $PM_{2.5}$ mass were determined for each sampling site and day (Figure 1). These values represent the summed total of all quantifiable elements that were soluble in methanol. Significant differences were observed between the two Sites on the sampling days in both the winter and summer; however, there was not a consistent pattern on which Site had elevated concentrations. In the winter, there was over a 2.8-fold increase in the average summed total element concentration at Site B compared to Site A. This pattern was not observed in the summer. Comparing between seasons, elemental concentrations were nearly 4-fold higher in the summer samples across both Sites. These findings are consistent with the 5 highest total element days all being observed in the summer. Significant differences were also observed between Sites on the same days (indicated by * in Figure 1) and a significant difference across all sampling dates was observed between Sites A and B. Specific factors that may impact the location and seasonal differences are discussed in detail below. Overall, variability in elemental concentrations across seasons, locations, and even days was observed in this research, highlighting the importance of routine monitoring to understand the chemical composition of $PM_{2.5}$.

Individual element concentrations at Sites A and B (Figure 2) demonstrate the variability in concentrations across days, sampling sites, and seasons. The most abundant element across all sites and seasons was Ca, however this element was excluded from the figure as it was 5 times higher than all other individual elemental concentrations. The second highest element by concentration was Fe in both the winter (Figure 2A, average concentration across sites/days of 4.7 ng/m^3) and in the summer (Figure 2B, average concentration across sites/days of 3.5 ng/m^3). One exception to this was the $PM_{2.5}$ collected at site B on 1 January which had a higher Sr concentration (6.6 ng/m^3). Sr concentrations in the winter were variable (0.21–6.6 ng/m^3, 2 samples below quantification limits) and were below detection limits in the summer for all samples. Pb concentrations in the winter were variable (0.02–1.4 ng/m^3, 1 sample below quantification limits) but similar to Sr were much less prevalent in the summer (0.01–0.3 ng/m^3, 4 samples below quantification limits). Pb concentrations are of particular concern due to the established health effects, including cardiovascular-related hospital admissions, following increased exposure to Pb in $PM_{2.5}$ [12]. All individual element concentrations with standard deviations are reported in SI Tables S2–S4.

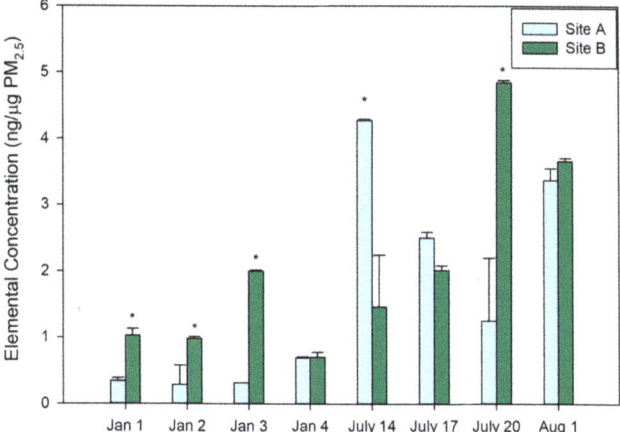

Figure 1. Total elements/PM$_{2.5}$ mass (ng/µg) for Site A and B in winter and summer. Elements (ng) per PM$_{2.5}$ mass (µg) for each sampling day for Sites A and B are reported for January (Jan.) and July. Element concentrations represent the summed totals of all elements that were quantified for a given sample following blank correction. The legend represents Site A and Site B with their corresponding colors. Mean concentrations with standard deviation, represented by error bars, are reported based on triplicate measurements. A two-way ANOVA was used to determine significant differences ($p \leq 0.05$) between the sampling sites and days denoted by *.

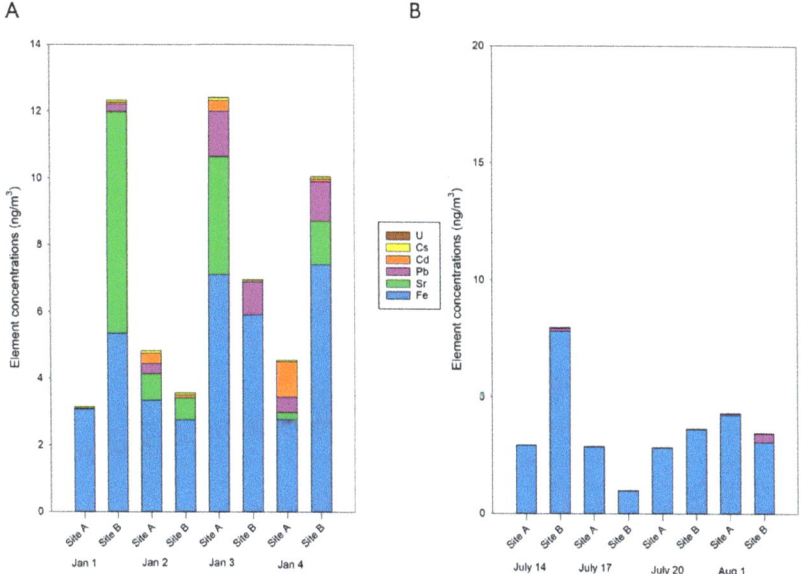

Figure 2. Concentrations (ng) of elements in PM$_{2.5}$ collected at Sites A and B. Bar graphs detailing the elemental composition of PM$_{2.5}$ collected at Sites A and B in the winter (**A**) and summer (**B**). Concentrations are reported for each individual element, represented by differing colors, in ng. Ca was excluded from this figure due to its elevated concentrations in comparison to the other elements quantified. Ca concentrations are presented in Table S2.

Concentrations of all quantified elements in the winter ranged from 11.9–54.3 ng/m^3 and from 7.6–38.0 ng/m^3 in the summer. When excluding the two most abundant elements, Ca and Fe, these ranges were 0.09–5.4 ng/m^3 for the winter and 0.007–0.02 ng/m^3 for the summer. Several elements were present only in the winter (Ag, Cd and Sr), demonstrating the reduced concentration and variety of elements in the summer. These findings are consistent with previous studies that observed higher concentrations and more variety in elements in the winter compared to other seasons [30–32]. One rationale for these findings may be inversion events, which are more common in the winter months particularly in low lying areas like basins [33,34]. Sites A and B are both in the Willamette Basin [35], making them in an inversion-prone area and thus the potential for trapping of air pollutants during these events. Unfortunately, additional data, including ozone concentrations, was not available at these locations, making it difficult to confirm if these events occurred. Additional contributing factors to the increased total elements in the winter at both sites may be source dependent, but a likely influence is the overall increased mass loadings collected in the winter (Table S5). This indicates that while a greater concentration of elements was present in the winter, on a per µg of PM$_{2.5}$ basis, the elemental concentrations were higher in the summer. And thus, in the summer elements contributed more to the total PM$_{2.5}$ mass compared to the winter suggesting that components, outside of the elements measured, are impacting PM$_{2.5}$ mass in the winter.

The elemental profiles were assessed at each of the locations in the winter (Figure 3A) and summer (Figure 3B). In general, across seasons Ca was the largest contributor (>65% in the winter and >75% in the summer) followed by Fe (≥18% in both seasons) for all but one sample (Site B on 2 January). The higher percentage of Ca in the summer may be due in part to the reduced variety and concentrations of detected elements in the summer compared to the winter. Ca and Fe are markers of brake dust [36,37]. Since both sites were located along highways, this likely explains the substantial contributions of these two elements in all samples, independent of season. Additionally, Fe is released during steel production [38] and the proximity of several steel corporations in the sampling area may contribute to the concentration of Fe in the samples. Our findings of high contributions of Ca and Fe in PM$_{2.5}$ samples are in alignment with previous research [39]. There was an anomaly from the observed compositional profile trends at Site B on 20 July when an increased contribution of Fe (46.3% compared to the average of 18.1%) was observed. Subsequently this resulted in a decreased contribution of Ca relative to the other samples. Meteorological factors including wind speed/direction, temperature, and humidity may have also contributed to the observed differences. Average PM$_{2.5}$ mass concentrations for the summer sampling sites were within 0.25 µg/m^3 of the seasonal averages but the representativeness of this data for the entire winter and summer seasons for elemental concentrations is unknown. However, the daily variations observed in this research suggests that assessing daily samples is important as seasonal averages may not fully display daily differences in PM$_{2.5}$ composition.

The variety of quantified elements in the winter was greater than the summer. Several elements including Ag, Cd, Cs, Pb, Sr, and Tl were quantifiable in the winter while the only element outside of Ca and Fe quantified in the summer was Pb on a single day at a single location (Site B on 1 August). A potential rationale for the increased contributions of elements outside of Ca and Fe in the winter may be the use of rock salt for treatment of snow and ice. Meteorological data collected from the region during the sampling periods reported light precipitation before and during the winter sampling dates with minimum temperatures consistently below freezing. Commercial rock salt can potentially contain trace amounts of Ag, Cd, and Sr [40], which all contributed to the compositional profile in the winter but not the summer. When considering the summer compositional profiles, it is important to note that these only include elements that were quantifiable based on instrument limits of quantification. Furthermore, the concentrations of elements not quantified in this research (i.e., Ni, V, Zn) may make a substantial contribution to the summer elemental profile but this is unknown for our study.

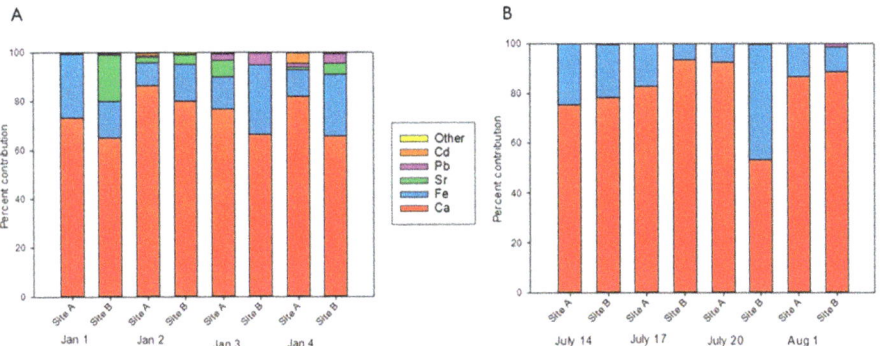

Figure 3. Elemental compositional profiles. Percent contributions of elements for each day and site are represented for winter (**A**) and summer (**B**) in January (Jan) and July. All values are reported out of 100% of the total elemental concentration for individual sampling days/sites. The legend represents elements with their corresponding colors. "Other" represents the combined contribution of Cs, Tl, U, and Ag.

3.2. Oxidative Potential of $PM_{2.5}$

The oxidative potential of each sample was assessed by measuring the DTT consumption based on a standard DTT curve (Figure 4). Significant differences were observed between Sites A and B on both 2 and 3 January. This suggests that spatial differences in the $PM_{2.5}$, even in relatively close proximity with similar source contributions, can result in differences in oxidative potential measurements. This emphasizes the need for further research into a variety of locations and daily, not just seasonal, impacts.

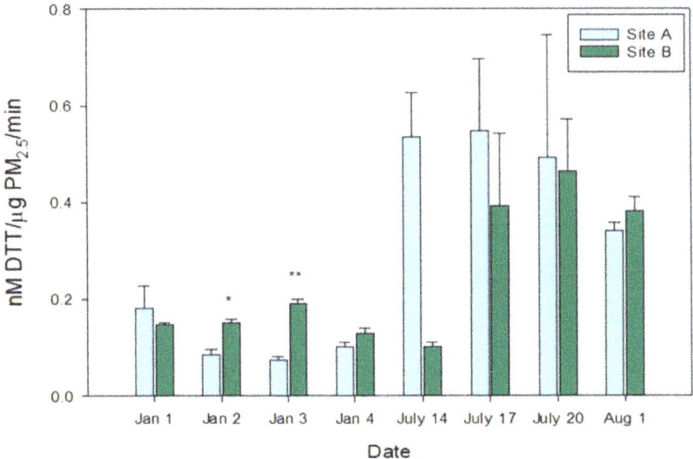

Figure 4. DTT Consumption per $PM_{2.5}$ mass per minute. DTT (dithiothreitol) consumption (nM) per $PM_{2.5}$ mass (µg) per minute of reaction incubation for each sampling day for Sites A and B are reported following blank filter corrections in January (Jan) and July. The legend represents Site A and Site B with their corresponding colors. Mean DTT consumption values with standard deviation, represented by error bars, are reported based on triplicate measurements. Students t-tests were run to determine significant differences ($p \leq 0.05$) between the two sampling sites for each sampling day, denoted by * or ** for $p < 0.001$.

Overall, the summer samples consumed more DTT per µg of methanol-soluble $PM_{2.5}$ at both Sites when compared to the winter samples, except for 14 July at Site B. The differences between seasons were not statistically significant, due largely in part to the increased variability within the triplicate measurements for the summer samples. Significant differences on the same day of sampling were observed between Sites in the winter, suggesting that compositional differences between Sites may be driving DTT consumption. These findings are consistent with those observed in the total element concentrations (ng/µg $PM_{2.5}$, Figure 1) where summer $PM_{2.5}$ had an increased concentration of elements compared to the winter. Beyond the higher per mass contribution of elements in the summer, there were potentially other compounds including DTT active compounds (i.e., Cu, quinones) that may have been present in high concentrations in the summer but not the winter [41,42]. Further exploration into the connections between DTT consumption and chemical constituents of $PM_{2.5}$ are detailed below.

Our results are consistent with previous work in Fresno, CA, that observed a higher average of DTT consumption during the summer [43]. Conversely, a study from China found higher concentrations of $PM_{2.5}$ and DTT consumption during the winter [23]. One potential rationale for these differences may be regional variability as our study findings aligned with another study in the western United States. Additional factors to consider are variation in the meteorology, sampling design, and source contributions. Finally, $PM_{2.5}$ mass may again play a role in the interpretation of this data. For consistency across previous studies DTT consumption is reported as consumption/µg $PM_{2.5}$/min. However, if not considering $PM_{2.5}$ mass, the amount of DTT consumed was not significantly different between Sites (SI Figure S1), indicating that the filter mass loadings may be driving the oxidative potential responses. This is discussed further in the section below.

3.3. Connections between Chemical Components and Oxidative Potential of $PM_{2.5}$

The relationship between DTT consumption, elemental concentrations, and $PM_{2.5}$ mass was assessed (Tables 1 and 2). We found multiple significant Pearson correlation coefficients (R) between the concentrations of elements, the total sum of elements, and $PM_{2.5}$. Several elements had significant positive correlations with the total sum of elements (in ng), including Ca (0.968), Fe (0.557), Sr (0.490), and Tl (0.472); data reported as (element (R)). These correlations are consistent with the increased concentrations of these elements relative to the others quantified. When considering the element concentrations normalized by $PM_{2.5}$ mass (ng/µg, Table 2) similar trends were seen for Ca (0.995) and Fe (0.730). Significant positive correlations were also observed between individual elements including Pb with 5 other elements (Ag, Cs, Fe, Tl, and U). Previous research has observed significant relationships between Ca and Fe, Ca and Sr, and Fe and Sr [44]. Our study found no significant correlations between these elements when considering concentrations in ng. There are a number of reasons for this including the locations of the studies as well as sources and meteorology present.

$PM_{2.5}$ mass and elements showed significant positive correlations for several individual elements (ng) that were only quantified in the winter samples (Cd, Sr, Tl, U) or at concentrations over 2-fold larger in the winter (Cs, Pb). This is consistent with previous work that has observed significant correlations between elements and $PM_{2.5}$ mass [45]. These correlations were not observed when normalizing the elements (ng/µg $PM_{2.5}$) which is consistent with our findings that the summer had elevated normalized total concentrations but a reduced number of quantifiable elements.

Correlations between DTT consumption and $PM_{2.5}$ mass and composition were made with $PM_{2.5}$ mass normalized and non-normalized values for element concentrations (Tables 1 and 2, respectively). For DTT consumption (nM/min) we found significant positive correlations with several elements (Cs, Tl) and $PM_{2.5}$ mass (Table 1). This was consistent with previous research identifying similar positive correlations with DTT consumption [46]. However, when using normalized DTT consumption (nM DTT/µg $PM_{2.5}$/min), we observed that $PM_{2.5}$ mass had a significant negative correlation with DTT consumed. This was not surprising, as the summer samples had lower $PM_{2.5}$ mass but higher DTT values per µg $PM_{2.5}$. Since the inverse was found when not normalizing the data (nM DTT/min or ng), these findings highlight the need to interpret data from various studies that use both normalized [46,47] and non-normalized [48] values for DTT or elements as the results can significantly differ. This is of particular concern if the $PM_{2.5}$ masses between samples have stark differences, as was observed in our study (Table S5), leading to different interpretations of the data. For example, the sum total of elements compared to normalized DTT, if using the non-normalized elemental concentration (ng), shows a slightly negative not statistically significant correlation to DTT. However if using the normalized elemental concentration (ng/µg) there is a significant positive correlation to DTT consumption. Similar findings were observed with Ca, where using non-normalized elemental concentrations had a slightly non-significant negative correlation with DTT consumption but when normalizing the Ca concentration, a significant positive correlation was observed with DTT. This emphasizes the need to consider the effects of reported units in studies looking at chemical composition and oxidative potential.

Table 1. Pearson's Correlations for DTT, elements in ng, and $PM_{2.5}$.

	Sr	Ag	Cd	Cs	Tl	Pb	U	Ca	Fe	Total	$PM_{2.5}$ Mass
DTT/min	0.251	0.0557	0.126	0.569	0.525	0.223	0.188	−0.073	0.246	0.0263	**0.556**
DTT/µg/min	−0.416	−0.218	−0.452	**−0.779**	**−0.761**	−0.517	−0.432	−0.155	−0.197	−0.253	**−0.793**
Sr		0.0714	0.0845	0.562	0.51	0.326	0.225	0.318	0.394	**0.490**	0.474
Ag			−0.0512	0.354	0.0522	0.505	**0.721**	−0.042	0.441	0.0641	0.139
Cd				0.299	**0.548**	0.270	0.132	0.219	−0.065	0.220	0.547
Cs					**0.895**	0.642	0.595	0.284	0.417	0.420	**0.904**
Tl						0.513	0.445	0.388	0.235	0.472	**0.981**
Pb							**0.845**	0.364	**0.675**	0.506	0.589
U								0.248	0.573	0.369	0.495
Ca									0.375	**0.968**	0.356
Fe										0.557	0.293
Total											0.452

Element concentrations in ng and "total" refers to the sum total concentration of elements. $PM_{2.5}$ mass (µg) and DTT consumed is reported for both non-normalized (nM/min) and normalized (nM/µg $PM_{2.5}$/min). The **bolded** numbers signify a significant p-value with Bonferroni multiple comparisons adjustment ($p < 0.001$), p-values for all correlations are reported in Table S6.

Table 2. Pearson's Correlations for DTT, elements in ng/µg $PM_{2.5}$, and $PM_{2.5}$.

	Sr/µg	Ag/µg	Cd/µg	Cs/µg	Tl/µg	Pb/µg	U/µg	Ca/µg	Fe/µg	Total/µg	$PM_{2.5}$ Mass
DTT/min	0.228	0.056	0.104	−0.007	0.611	0.027	−0.102	−0.512	−0.351	−0.508	**0.556**
DTT/µg/min	−0.354	−0.218	−0.415	0.005	**−0.804**	−0.210	0.008	0.603	**0.699**	**0.633**	**−0.793**
Sr/µg		0.090	0.023	0.136	0.482	0.012	−0.083	−0.261	−0.260	−0.245	0.346
Ag/µg			−0.029	0.227	0.128	0.395	**0.493**	−0.174	−0.087	−0.165	0.139
Cd/µg				−0.261	0.466	0.006	−0.097	−0.293	−0.379	−0.315	0.464
Cs/µg					−0.042	**0.708**	−0.212	0.043	0.129	0.067	−0.115
Tl/µg						0.079	0.007	−0.636	**0.710**	−0.662	**0.910**
Pb/µg							0.130	−0.011	0.038	0.006	0.129
U/µg								0.242	−0.007	0.219	−0.018
Ca/µg									0.658	**0.995**	−0.617
Fe/µg										**0.730**	−0.674
Total/µg											−0.643

Element concentrations in ng/µg $PM_{2.5}$ and "total" refers to the sum total concentration of elements. $PM_{2.5}$ mass (µg) and DTT consumed is reported for both non-normalized (nM/min) and normalized (nM/µg $PM_{2.5}$/min). The **bolded** numbers signify a significant p-value with Bonferroni multiple comparisons adjustment ($p < 0.001$), p-values for all correlations are reported in Table S7.

While the total and individual element concentrations were elevated in the winter sampling period for $PM_{2.5}$ this was not reflected in the $PM_{2.5}$ mass normalized element concentrations or the oxidative potential assessment. Several elements and total $PM_{2.5}$ mass concentration had significant negative correlations with normalized DTT consumption, suggesting that components of $PM_{2.5}$, other than elements, play a role in DTT consumption.

One limitation of our work is the absence of additional constituents of $PM_{2.5}$. Elements, particularly transition metals, previously shown to be associated with oxidative potential, including Al, Cu, Mn, Ni, V and Zn [15,49], were excluded due to calibration curves that exceeded our quality standards. Of note, we did not analyze for organic compounds, which have previously been shown to induce oxidative potential [41,42,50]. Organic compounds including polycyclic aromatic hydrocarbons and alkanes have been observed to be elevated at some locations in the summer however this is variable based on the locations sampled [51]. Thus, we cannot definitively determine if the organic compounds played a role in the increased oxidative potential observed but it is one likely rationale for these findings. Inclusion of these constituents may have identified positive correlations with DTT however that does not impact the observed findings which demonstrate the inability to consistently attribute oxidative potential to some elements, including Fe and Pb. Additionally, only a subset of days in two seasons were studied, so little is known about the chemical composition and oxidative potential during the fall and spring which is of importance since oxidative potential has previously been observed to be highest in the fall [23]. Finally, while the DTT assay is one method for the measurement of oxidative potential, there are alternative methods that can be utilized [22]. The various methods of oxidative potential assessment have limitations, including elevated affinity for transition metals compared to other components of $PM_{2.5}$ for the DTT assay [42]. These factors should be considered when interpreting results. Future studies investigating daily differences in these factors across all seasons in addition to paired in vitro and in vivo research would greatly support our findings and identify trends beyond those observed in our selective sampling periods.

4. Conclusions

We observed significant differences in elemental composition between sampling locations along highways in central Oregon, USA with trends suggesting increased concentrations during the winter but increased contributions to $PM_{2.5}$ mass from elements in the summer. These trends were driven by increased $PM_{2.5}$ mass concentrations and the variety of quantifiable elements in the winter. Interestingly, we observed increases in oxidative potential of the $PM_{2.5}$ samples in the summer suggesting that despite the elevated $PM_{2.5}$ mass in the winter, the components of $PM_{2.5}$ in the summer resulted in increased oxidative potential. This study demonstrates the influence of $PM_{2.5}$ composition in oxidative potential responses and highlights the importance of comparisons between $PM_{2.5}$ mass normalized and non-normalized data. Further research is needed to understand the variation within seasons of $PM_{2.5}$ composition and oxidative potential to support policies that protect human health.

Supplementary Materials: The following are available online at http://www.mdpi.com/2073-4433/11/10/1086/s1, Figure S1: DTT consumed (nM DTT/min) per Filter, Table S1: ICP-MS Instrumental Parameters, Table S2: Element Concentration (ppb), Table S3: Element Concentrations (ppb/µg), Table S4: Element Concentrations (µg/m^3), Table S5: $PM_{2.5}$ sample mass, Table S6: p-values for Pearson correlations, Table S7: p-values for Pearson correlations.

Author Contributions: Conceptualization, A.V. and C.R.; methodology, A.V., O.B.; validation, A.V. and O.B.; formal analysis, A.S.; investigation, A.V.; resources, C.R.; writing—original draft preparation, A.V. and A.S.; writing—review and editing, C.R. and O.B.; visualization, A.S.; supervision, C.R. All authors have read and agreed to the published version of the manuscript.

Funding: This research was funded by the University of Mississippi (UM), UM Department of BioMolecular Sciences, and UM School of Pharmacy.

Acknowledgments: All $PM_{2.5}$ samples were donated by the Lane Regional Air Protection Agency (LRAPA) in Lane County Oregon, USA. The ICP-MS facility at the University of Mississippi, University, MS, USA was utilized for all chemical analyses. Graphical abstract created through biorender.com.

Conflicts of Interest: The authors declare no conflict of interest. The funders had no role in the design of the study; in the collection, analyses, or interpretation of data; in the writing of the manuscript, or in the decision to publish the results.

References

1. Steenhof, M.; Gosens, I.; Strak, M.; Godri, K.J.; Hoek, G.; Cassee, F.R.; Mudway, I.S.; Kelly, F.J.; Harrison, R.M.; Lebret, E.; et al. In vitro toxicity of particulate matter (PM) collected at different sites in the Netherlands is associated with PM composition, size fraction and oxidative potential—The RAPTES project. *Part Fibre Toxicol.* **2011**, *8*, 26. [PubMed]
2. Pope, C.A.; Burnett, R.T.; Thurston, G.D.; Thun, M.J.; Calle, E.E.; Krewski, D.; Godleski, J.J. Cardiovascular mortality and long-term exposure to particulate air pollution: Epidemiological evidence of general pathophysiological pathways of disease. *Circulation* **2004**, *109*, 71–77. [CrossRef] [PubMed]
3. WHO. Ambient (Outdoor) Air Pollution. Available online: https://www.who.int/news-room/fact-sheets/detail/ambient-(outdoor)-air-quality-and-health (accessed on 30 June 2020).
4. European Commission. Standards—Air Quality—Environment—European Commission. Available online: https://ec.europa.eu/environment/air/quality/standards.htm (accessed on 9 July 2020).
5. US EPA O. National Ambient Air Quality Standards (NAAQS) for PM. US EPA, 2020. Available online: https://www.epa.gov/pm-pollution/national-ambient-air-quality-standards-naaqs-pm (accessed on 9 July 2020).
6. Feng, S.; Gao, D.; Liao, F.; Zhou, F.; Wang, X. The health effects of ambient PM2.5 and potential mechanisms. *Ecotoxicol. Environ. Saf.* **2016**, *128*, 67–74. [PubMed]
7. Lu, F.; Xu, D.; Cheng, Y.; Dong, S.; Guo, C.; Jiang, X.; Zhen, X. Systematic review and meta-analysis of the adverse health effects of ambient $PM_{2.5}$ and PM_{10} pollution in the Chinese population. *Environ. Res.* **2015**, *136*, 196–204. [CrossRef]
8. Peixoto, M.S.; de Oliveira Galvão, M.F.; de Medeiros, S.R.B. Cell death pathways of particulate matter toxicity. *Chemosphere* **2017**, *188*, 32–48. [CrossRef]
9. Cheung, K.; Daher, N.; Kam, W.; Shafer, M.M.; Ning, Z.; Schauer, J.J.; Sioutas, C. Spatial and temporal variation of chemical composition and mass closure of ambient coarse particulate matter ($PM_{10-2.5}$) in the Los Angeles area. *Atmos. Environ.* **2011**, *45*, 2651–2662. [CrossRef]
10. Demerjian, K.L.; Mohnen, V.A. Synopsis of the Temporal Variation of Particulate Matter Composition and Size. *J. Air Waste Manag. Assoc.* **2008**, *58*, 216–233.
11. Röösli, M.; Theis, G.; Künzli, N.; Staehelin, J.; Mathys, P.; Oglesby, L.; Camenzind, M.; Braun-Fahrländer, C. Temporal and spatial variation of the chemical composition of PM10 at urban and rural sites in the Basel area, Switzerland. *Atmos. Environ.* **2001**, *35*, 3701–3713.
12. Bell, M.L.; Ebisu, K.; Peng, R.D.; Samet, J.M.; Dominici, F. Hospital Admissions and Chemical Composition of Fine Particle Air Pollution. *Am. J. Respir. Crit. Care Med.* **2009**, *179*, 1115–1120. [CrossRef]
13. Bell, M.L. HEI Health Review Committee. Assessment of the health impacts of particulate matter characteristics. *Res. Rep. Health Eff. Inst.* **2012**, *161*, 5–38.
14. Kelly, F.J.; Fussell, J.C. Size, source and chemical composition as determinants of toxicity attributable to ambient particulate matter. *Atmos. Environ.* **2012**, *60*, 504–526. [CrossRef]
15. Boogaard, H.; Janssen, N.A.H.; Fischer, P.H.; Kos, G.P.A.; Weijers, E.P.; Cassee, F.R.; van der Zee, S.C.; de Hartog, J.J.; Brunekreef, B.; Hoek, G. Contrasts in Oxidative Potential and Other Particulate Matter Characteristics Collected Near Major Streets and Background Locations. *Environ. Health Perspect.* **2012**, *120*, 185–191. [PubMed]
16. Dagher, Z.; Garçon, G.; Billet, S.; Gosset, P.; Ledoux, F.; Courcot, D.; Aboukais, A.; Shirali, P. Activation of different pathways of apoptosis by air pollution particulate matter (PM2.5) in human epithelial lung cells (L132) in culture. *Toxicology* **2006**, *225*, 12–24. [CrossRef] [PubMed]
17. Pizzino, G.; Irrera, N.; Cucinotta, M.; Pallio, G.; Mannino, F.; Arcoraci, V.; Squadrito, F.; Altavilla, D.; Bitto, A. Oxidative Stress: Harms and Benefits for Human Health. *Oxid. Med. Cell Longev.* **2017**, *2017*, 1–13.
18. Shi, T.; Schins, R.P.F.; Knaapen, A.M.; Kuhlbusch, T.; Pitz, M.; Heinrich, J.; Borm, P.J.A. Hydroxyl radical generation by electron paramagnetic resonance as a new method to monitor ambient particulate matter composition. *J. Environ. Monit.* **2003**, *5*, 550–556. [CrossRef] [PubMed]
19. Borm, P.J.A.; Kelly, F.; Künzli, N.; Schins, R.P.F.; Donaldson, K. Oxidant generation by particulate matter: From biologically effective dose to a promising, novel metric. *Occup. Environ. Med.* **2007**, *64*, 73–74. [CrossRef]

20. Hedayat, F.; Stevanovic, S.; Miljevic, B.; Bottle, S.; Ristovski, Z.D. Review-evaluating the molecular assays for measuring the oxidative potential of particulate matter. *Chem. Ind. Chem. Eng. Q.* **2015**, *21*, 201–210. [CrossRef]
21. Roper, C.; Perez, A.; Barrett, D.; Hystad, P.; Massey Simonich, S.L.; Tanguay, R.L. Workflow for comparison of chemical and biological metrics of filter collected PM$_{2.5}$. *Atmos. Environ.* **2020**, *226*, 117379. [CrossRef]
22. Janssen, N.A.H.; Yang, A.; Strak, M.; Steenhof, M.; Hellack, B.; Gerlofs-Nijland, M.E.; Kuhlbusch, T.; Kelly, F.; Harrison, R.; Brunekreef, B.; et al. Oxidative potential of particulate matter collected at sites with different source characteristics. *Sci. Total Environ.* **2014**, *472*, 572–581. [CrossRef]
23. Wang, J.; Lin, X.; Lu, L.; Wu, Y.; Zhang, H.; Lv, Q.; Liu, W.; Zhang, Y.; Zhuang, S. Temporal variation of oxidative potential of water soluble components of ambient PM2.5 measured by dithiothreitol (DTT) assay. *Sci. Total Environ.* **2019**, *649*, 969–978. [CrossRef]
24. Shao, L.; Hu, Y.; Shen, R.; Schäfer, K.; Wang, J.; Wang, J.; Schnelle-Kreis, J.; Zimmermann, R.; BéruBé, K.; Suppan, P. Seasonal variation of particle-induced oxidative potential of airborne particulate matter in Beijing. *Sci. Total Environ.* **2017**, *579*, 1152–1160. [PubMed]
25. Cheung, K.; Shafer, M.M.; Schauer, J.J.; Sioutas, C. Diurnal Trends in Oxidative Potential of Coarse Particulate Matter in the Los Angeles Basin and Their Relation to Sources and Chemical Composition. *Environ. Sci. Technol.* **2012**, *46*, 3779–3787. [PubMed]
26. Szigeti, T.; Óvári, M.; Dunster, C.; Kelly, F.J.; Lucarelli, F.; Záray, G. Changes in chemical composition and oxidative potential of urban PM2.5 between 2010 and 2013 in Hungary. *Sci. Total Environ.* **2015**, *518–519*, 534–544. [CrossRef] [PubMed]
27. Visentin, M.; Pagnoni, A.; Sarti, E.; Pietrogrande, M.C. Urban PM$_{2.5}$ oxidative potential: Importance of chemical species and comparison of two spectrophotometric cell-free assays. *Environ. Pollut.* **2016**, *219*, 72–79. [CrossRef]
28. LRAPA. Air Quality Sensors | Lane Regional Air Protection Agency, OR. Available online: http://www.lrapa.org/307/Air-Quality-Sensors (accessed on 7 September 2020).
29. US EPA. List of Designated Reference and Equivalent Methods. 2020. Available online: www.epa.gov/ttn/amtic/criteria.html (accessed on 12 October 2020).
30. Gorai, A.K.; Tchounwou, P.B.; Biswal, S.; Tuluri, F. Spatio-Temporal Variation of Particulate Matter (PM2.5) Concentrations and Its Health Impacts in a Mega City, Delhi in India. *Environ. Health Insights* **2018**, *12*. [CrossRef]
31. Ho, K.; Lee, S.; Cao, J.; Chow, J.; Watson, J.; Chan, C. Seasonal variations and mass closure analysis of particulate matter in Hong Kong. *Sci. Total Environ.* **2006**, *355*, 276–287. [CrossRef]
32. Ledoux, F.; Courcot, L.; Courcot, D.; Aboukaïs, A.; Puskaric, E. A summer and winter apportionment of particulate matter at urban and rural areas in northern France. *Atmos. Res.* **2006**, *82*, 633–642. [CrossRef]
33. Lyman, S.; Tran, T. Inversion structure and winter ozone distribution in the Uintah Basin, Utah, USA. *Atmos. Environ.* **2015**, *123*, 156–165. [CrossRef]
34. Vitasse, Y.; Klein, G.; Kirchner, J.W.; Rebetez, M. Intensity, frequency and spatial configuration of winter temperature inversions in the closed La Brevine valley, Switzerland. *Theor. Appl. Climatol.* **2017**, *130*, 1073–1083. [CrossRef]
35. USDA. USDA—National Agricultural Statistics Service—Mississippi—2017–2020 County Estimates. Available online: https://www.nass.usda.gov/Statistics_by_State/Mississippi/Publications/County_Estimates/index.php (accessed on 30 June 2020).
36. Dall'Osto, M.; Querol, X.; Amato, F.; Karanasiou, A.; Lucarelli, F.; Nava, S.; Calzolai, G.; Chiari, M. Hourly elemental concentrations in PM2.5 aerosols sampled simultaneously at urban background and road site. *Atmos. Chem. Phys. Discuss* **2012**, *12*, 20135–20180. [CrossRef]
37. Sanders, P.G.; Xu, N.; Dalka, T.M.; Maricq, M.M. Airborne Brake Wear Debris: Size Distributions, Composition, and a Comparison of Dynamometer and Vehicle Tests. *Environ. Sci. Technol.* **2003**, *37*, 4060–4069. [CrossRef] [PubMed]
38. O'Sullivan, M. Iron metabolism of grasses: I. Effect of iron supply on some inorganic and organic constituents. *Plant Soil.* **1969**, *31*, 451–462. [CrossRef]
39. Bozlaker, A.; Peccia, J.; Chellam, S. Indoor/Outdoor Relationships and Anthropogenic Elemental Signatures in Airborne PM2.5 at a High School: Impacts of Petroleum Refining Emissions on Lanthanoid Enrichment. *Environ. Sci. Technol.* **2017**, *51*, 4851–4859. [CrossRef] [PubMed]

40. Titler, R.V. Chemical Analysis of Major Constituents and Trace Contaminants of Rock Salt. 2011. Available online: http://files.dep.state.pa.us/Water/Wastewater%20Management/WastewaterPortalFiles/Rock%20Salt%20Paper%20final%20052711.pdf (accessed on 10 July 2020).
41. Chung, M.Y.; Lazaro, R.A.; Lim, D.; Jackson, J.; Lyon, J.; Rendulic, D.; Hasson, A.S. Aerosol-Borne Quinones and Reactive Oxygen Species Generation by Particulate Matter Extracts. *Environ. Sci. Technol.* **2006**, *40*, 4880–4886. [CrossRef]
42. Charrier, J.G.; Anastasio, C. On dithiothreitol (DTT) as a measure of oxidative potential for ambient particles: Evidence for the importance of soluble transition metals. *Atmos. Chem. Phys.* **2012**, *12*, 9321–9333. [CrossRef]
43. Charrier, J.G.; Richards-Henderson, N.K.; Bein, K.J.; McFall, A.S.; Wexler, A.S.; Anastasio, C. Oxidant production from source-oriented particulate matter—Part 1: Oxidative potential using the dithiothreitol (DTT) assay. *Atmos. Chem. Phys.* **2015**, *15*, 2327–2340.
44. Yatkin, S.; Bayram, A. Elemental composition and sources of particulate matter in the ambient air of a Metropolitan City. *Atmos. Res.* **2007**, *85*, 126–139. [CrossRef]
45. Asano, H.; Aoyama, T.; Mizuno, Y.; Shiraishi, Y. Highly Time-Resolved Atmospheric Observations Using a Continuous Fine Particulate Matter and Element Monitor. *ACS Earth Space Chem.* **2017**, *1*, 580–590. [CrossRef]
46. Fang, T.; Verma, V.; Bates, J.T.; Abrams, J.; Klein, M.; Strickland, M.J.; Sarnat, S.E.; Chang, H.H.; Mulholland, J.A.; Tolbert, P.E.; et al. Oxidative potential of ambient water-soluble PM2.5 in the southeastern United States: Contrasts in sources and health associations between ascorbic acid (AA) and dithiothreitol (DTT) assays. *Atmos. Chem. Phys.* **2016**, *16*, 3865–3879. [CrossRef]
47. Gao, D.; Mulholland, J.A.; Russell, A.G.; Weber, R.J. Characterization of water-insoluble oxidative potential of PM2.5 using the dithiothreitol assay. *Atmos. Environ.* **2020**, *224*, 117327. [CrossRef]
48. Chen, Q.; Wang, M.; Wang, Y.; Zhang, L.; Li, Y.; Han, Y. Oxidative Potential of Water-Soluble Matter Associated with Chromophoric Substances in PM2.5 over Xi'an, China. *Environ. Sci. Technol.* **2019**, *53*, 8574–8584. [CrossRef]
49. Wei, J.; Yu, H.; Wang, Y.; Verma, V. Complexation of Iron and Copper in Ambient Particulate Matter and Its Effect on the Oxidative Potential Measured in a Surrogate Lung Fluid. *Environ. Sci. Technol.* **2019**, *53*, 1661–1671. [CrossRef] [PubMed]
50. Ntziachristos, L.; Froines, J.R.; Cho, A.K.; Sioutas, C. Relationship between redox activity and chemical speciation of size-fractionated particulate matter. *Part. Fibre Toxicol.* **2007**, *4*, 5. [CrossRef] [PubMed]
51. Cheung, K.; Olson, M.R.; Shelton, B.; Schauer, J.J.; Sioutas, C. Seasonal and spatial variations of individual organic compounds of coarse particulate matter in the Los Angeles Basin. *Atmos. Environ.* **2012**, *59*, 1–10. [CrossRef]

© 2020 by the authors. Licensee MDPI, Basel, Switzerland. This article is an open access article distributed under the terms and conditions of the Creative Commons Attribution (CC BY) license (http://creativecommons.org/licenses/by/4.0/).

Article

Differentiation of the Athens Fine PM Profile during Economic Recession (March of 2008 Versus March of 2013): Impact of Changes in Anthropogenic Emissions and the Associated Health Effect

Styliani Pateraki [1,*], Kyriaki-Maria Fameli [1], Vasiliki Assimakopoulos [1], Kyriaki Bairachtari [2], Alexandros Zagkos [3], Theodora Stavraka [4], Aikaterini Bougiatioti [1], Thomas Maggos [2] and Nikolaos Mihalopoulos [1,5]

[1] Institute for Environmental Research and Sustainable Development, National Observatory of Athens, 152 36 Athens, Greece; kmfameli@noa.gr (K.-M.F.); vasiliki@noa.gr (V.A.); abougiat@noa.gr (A.B.); nmihalo@noa.gr (N.M.)
[2] Environmental Research Laboratory/ INT-RP, National Centre for Scientific Research Demokritos, 153 10 Athens, Greece; kyriaki@ipta.demokritos.gr (K.B.); tmaggos@ipta.demokritos.gr (T.M.)
[3] Department of Applied Physics, Faculty of Physics, University of Athens, 157 84 Athens, Greece; azagkos@phys.uoa.gr
[4] Center for Environmental Effects on Health, Biomedical Research Foundation of Academy of Athens, 106 80 Athens, Greece; dorastauraka@gmail.com
[5] Environmental Chemical Processes Laboratory, Chemistry Department, University of Crete, 700 13 Heraklion, Greece
* Correspondence: paterakist@noa.gr

Received: 31 August 2020; Accepted: 14 October 2020; Published: 19 October 2020

Abstract: Despite the various reduction policies that have been implemented across Europe in the past few years, Particulate Matter (PM) exceedances continue to be recorded. Therefore, with the principal aim to clarify the complex association between emissions and fine particles levels, this work evaluates the impact of the anthropogenic contribution to the fine PM chemical profile. The fieldwork was conducted during March in 2008 and 2013 and covers the periods before and during the economic recession. The experimental data were analyzed in parallel with the emissions from the Flexible Emission Inventory for Greece and the Greater Athens Area (FEI-GREGAA). The differentiation of the mass closure results' and the aerosols' character is also discussed in combination with the calculated $PM_{2.5}$-Air Quality Indexes. The peak in the PM load and the Particulate Organic Matter (POM) component was recorded in 2013, corresponding to the enhancement of the anthropogenic input. Although the monitoring location is traffic-impacted, the sector of heating, from both wood burning and fossil fuel, proved to be the driving force for the configuration of the obtained PM picture. Especially in 2013, its contribution was two times that of traffic. Finally, the low wind speed values led to the deterioration of the air quality, especially for the sensitive groups.

Keywords: fine particles; carbonaceous and ionic constituents; health impact; field campaign; FEI-GREGAA emission inventory; sources; urban area

1. Introduction

In Europe, atmospheric pollution is responsible for more than 400,000 premature deaths a year, with the largest share to the exposure of fine particulate matter ($PM_{2.5}$, PM_1) [1]. Fine particles have increased potential health risk being compared to the one of coarse particles, for many reasons: (i) they penetrate deeper in the lung (ii) they can penetrate more readily into indoor environments (iii) they have longer periods of suspension (iv) they may be transported over long distances (v) they carry higher concentrations of toxic compounds and (vi) they can absorb larger amounts of semi-volatile compounds [2].

The Particulate Matter (PM) control is a challenging problem, especially in urban areas where large populations are exposed to increased concentration levels [3]. They have simultaneous primary and secondary sources [4] and their toxicity is highly depended on their chemical composition which in turn is linked to the emission sources, the atmospheric chemical processes, and the long range transport effects [5]. Therefore, the parallel knowledge of the temporal evolution of (i) their profile, (ii) their emission sources and (iii) their potential health risk is very important for the health protection and the set-up of effective policy strategies. In the case of fine particles, various experimental and modeling studies have been conducted all over the world investigating the change, for at least a four years' period, of the $PM_{2.5}$ mass [6–10] and/or its chemical composition [11–13]. Jiang et al. [14] and Zhang et al. [15], using a combined field and numerical modeling approach, gave insight on the differentiation of the $PM_{2.5}$ mass in connection with its health consequences for a 6 and 4 years' period, respectively. However, the works assessing the effect of the emissions to the obtained chemical PM profile, during several years' period, by combining experimental and emissions data are scarce [16,17]. To the best of our knowledge, information regarding the evolution of the PM_1 fraction over time in conjunction with emission inventories does not exist at all even if this fraction is considered as a better indicator of the anthropogenic sources [2,4,18].

Based on the above, the principal aim of this study is to assess the changes that have occurred, in the Athenian $PM_{2.5}$ and PM_1 levels and their chemical profile due to the different apportionment of the anthropogenic sources. The information concerns the changes that have taken place within an urban location of the Athens Basin between the years 2008 and 2013. The specific years were selected owing to their distinct characteristics. 2008 marks the beginning of the Greek economic crisis. Consequently, the effects of the decline of the economy on the everyday life had not yet been recorded. On the other hand, 2013 is within the economic recession with well-defined impacts of the crisis on emission inventories [19]. The fieldwork took place in March, an intermediate month of the year with contributions from the two significant anthropogenic contributors of the PM levels, traffic and residential heating [19–21]. The specific sectors are strongly associated with the changes that occur in everyday life. Therefore, with the principal aim of this work to uncover the impact of the changes of the local input to the obtained PM chemical, the data analysis is focused on the days that do not favor the dispersion of the air pollutants (ws < 2.5 m/s; [22]). Besides the analysis of the carbonaceous and ionic $PM_{2.5}$ and PM_1 contents (mass closure, ionic balance), the results are compared with the emissions for the two mentioned sectors from the FEI-GREGAA emission inventory, the most updated emission inventory for the Greater Athens Area (GAA; [19]). Additionally, the differentiation of the potential health impact of the $PM_{2.5}$ levels and its dependence on the prevailing (compositional, emissions, wind) conditions is also discussed.

2. Materials and Methods

2.1. Field Campaign

The monitoring was performed at a suburban site, very close to the Athens center and five of the city's main traffic avenues (Figure 1).

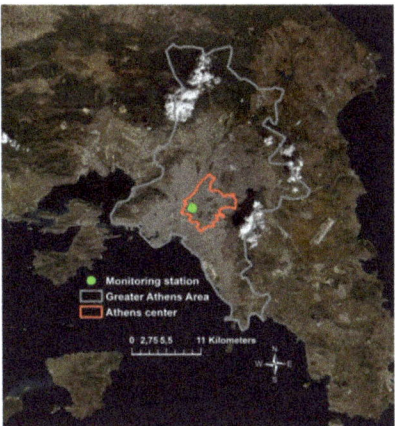

Figure 1. The location of the sampling station.

As a part of this study, a total of 40 and 50 PM samples were collected during March 2008 and 2013, respectively, by following the same methodology [23–25]. Two low-volume, controlled flow rate (2.3 m^3/h) samplers were used simultaneously, during each sampling period, collecting particles on quartz filters with diameter of 47 and 50 mm. The filters were pre-baked before the use at 550 °C in order to reduce residual carbon levels associated with them while they were conditioned, in a weighing room under controlled temperature (20 ± 1 °C) and relative humidity (50 ± 5%) conditions. Daily sampling periods lasted 24 h (08:00 a.m.–08:00 p.m.) covering both the weekdays and the weekends. The determination of the concentration of the particle mass was conducted gravimetrically using an electronic microbalance with a resolution of 1 µg according to the European Standard EN 12341 [26]. The collected PM samples were chemically analyzed for their carbonaceous (Organic Carbon (OC), Elemental Carbon (EC)) and ionic components (Cl$^-$, NO$_3^-$, SO$_4^{2-}$, NH$_4^+$, K$^+$, Mg^{2+}, Ca^{2+}). The chemical analysis for the period of 2008 was conducted at the Environmental Chemical Processes Laboratory of the University of Crete while for 2013, the Environmental Research Laboratory of the National Center of Scientific Research Demokritos has the responsibility. The water-soluble ions were detected using suppressed ion chromatography while the carbonaceous constituents were determined with the use of a carbon analyzer (Sunset Lab, Oregon, United States of America (USA); [18,26]). Note that at least for carbonaceous material (OC, EC) both laboratories participate to frequent inter-comparisons organized by the European Research Infrastructure for the observation of Aerosol, Clouds and Trace Gases (ACTRIS) network by using same procedure and with similar results.

With the principal aim to investigate the impact of the change of the share of the anthropogenic activities to the obtained fine PM chemical profile, the specific work focused on the two main sources of particulate pollution across the Greater Athens Area, the residential heating and the traffic. March was selected as a month with an intermediate contribution from both sectors. As it is well known, March belongs to the transition period and is characterized by intense dust intrusion from the Sahara Desert, a usual phenomenon for the Mediterranean basin [27,28]. Therefore, in order to focus on the impact of the local anthropogenic emissions and limit as much as possible the long range transport effects, only the days that were characterized by low wind speed conditions (ws ≤ 2.5 m/s) were taken into consideration. As it has been proven by Fourtziou et al. [22] and Liakakou et al. [29], wind values lower than 3 m/s favor the accumulation of the air pollutants within the mixing layer. Moreover, low winds trap the pollutants near their source and they provide sufficient time for the chemical reaction to form the secondary aerosols/constituents [12]. In March 2008, in 78.6% of the sampling days, the wind speed ranged between 1.40 m/s and 2.50 m/s while in the same month of 2013 it varied from 0.36 m/s to 2.38 m/s (90.3% of the sampling days).

2.2. Data Analysis

2.2.1. Mass Closure

For the purpose of mass closure, the detected chemical components were divided into five classes as follows: Elemental Carbon (EC), Particulate Organic Matter (POM; OC*1.6 (conversion factor for urban aerosols [30]), Secondary Inorganic aerosols (SIAs; sum of NO_3^-, NH_4^+ and non-sea salt sulfates ($nssSO_4^{2-}$)), Sea Salt (SS; sum of Na^+, Cl^-, ssK^+, $ssMg^{2+}$, $ssCa^{2+}$ and $ssSO_4^{2-}$) and Unidentified Material (UM; the difference between the gravimetrically measured aerosol mass and the reconstructed one (the sum of quantified chemical components)). More details about the procedure that was followed are given in Pateraki et al. [31].

As it has previously been mentioned (Section 2.1), the days with wind speed values higher than 2.5 m/s have been excluded from the data analysis. During the March of 2008, the chemical compounds measured during this work explained in a range between 61% to 94% and 65% and 84% for $PM_{2.5}$ and PM_1 mass, respectively. As far as the March of 2013 is concerned, the rates of the reconstructed mass varied from 66% to 88% and from 65% to 95% for $PM_{2.5}$ and PM_1, respectively. The aerosol-bound water, the undetected chemical components like oxygen, minerals and trace elements, the volatilization losses, the errors that affect the mass and chemical measurements as well as the uncertainty of the conversion of OC to POM could account for the UM Pateraki et al. [31].

2.2.2. Emissions

Concerning FEI-GREGAA [19], the methodology proposed by the Environmental Monitoring, Evaluation and Protection/Environmental Protection Agency (EMEP/EAA) Emission Inventory Guidebook 2016 was followed. More specifically, for the road transport and small combustion sectors the Tier 3 and Tier 2 approaches were used for the calculation of the annual emissions for the Attica region. For the road transport sector $PM_{2.5}$, PM_{10}, OM and EC emissions were calculated while for the small combustion sector $PM_{2.5}$, PM_{10} and BC emissions were produced. The respective energy consumption per fuel type for space heating was provided by the Odyssee-Mure project [32]. The annual values were afterwards spatially and temporarily disaggregated into gridded form and daily. The detailed methodology is described for both sectors in Fameli and Assimakopoulos [19] and [33]. It should be mentioned that for the road transport emissions local temporal coefficients, representative of the traffic counts profiles for the years 2008 and 2013, were used. As a result, different road emissions were calculated for each day of March 2008 and 2013. However, for the temporal allocation of residential heating emissions monthly and daily coefficients were used, provided by the TNO database [34], which were updated in order to correspond to the Greek temporal activity profiles. Moreover, for the needs of comparison with the measurements, the daily gridded emissions were supposed to represent a 1 m cell height.

2.2.3. Health Risk

The Air Quality Index (AQI) of the US Environmental Protection Agency (US EPA) is a widely used index which gives details about the daily air quality status [35]. In the present work, with the principal aim to investigate the possible relationship between the configured PM burden (mass, chemical composition) and the corresponding health effects, the USEPA Air Quality Index (AQI) for $PM_{2.5}$ was applied. Details about the specific methodology, the breakpoints for the $PM_{2.5}$ concentrations, the ranges of the $PM_{2.5}$-AQI categories and the possible health consequences are given in EPA 1999, [36] and Gorai et al. [35]. Dimitriou et al. [37] found that AQI in Athens is mainly affected by PM (by 72%) and only 28% by ozone with the contribution of other pollutants being negligible.

Finally, taking into consideration the extra adverse health effects of the acidic nature of fine particles [26,38] the differentiation of the aerosols' character between March of 2008 and 2013 was investigated performing the ionic balance. The ionic balance is expressed by the ratio of the equivalent cation sum (sum of NH_4^+, K^+, Na^+, Mg^{2+}, Ca^{2+} in neq/m^3) to the equivalent anion sum (sum of Cl^-, NO_3^-, SO_4^{2-}, in neq/m^3). Details about the procedure are given in Pateraki et al. [26] and [31].

3. Results and Discussion

3.1. Data Overview

A statistical summary of the primary PM$_{2.5}$ and PM$_1$ data (mass concentrations, chemical composition) is given on the Table S1.

The differentiation of the average obtained PM mass closure profile between the periods 2008 and 2013 is presented in Figure 2. Probably reflecting the changes in the prevailing emission sources, the PM burden appeared to be increased during March of 2013 compared to 2008. PM$_{2.5}$ was higher by 85% while in the case of PM$_1$ the rise was up to 29% compared to 2008. On a daily basis, the PM$_{2.5}$ concentrations reached 37.7 and 80.4 µg/m^3 in 2008 and 2013, respectively while the corresponding PM$_1$ peaks were 29.7 and 54.9 µg/m^3. Interestingly enough, the PM$_1$/PM$_{2.5}$ ratio in 2008 ranged between 0.73 and 0.97 while in 2013 it varied from 0.33 to 0.76. According to the findings, PM$_{2.5}$ in 2008 was mainly composed of PM$_1$ and influenced mainly by combustion and/or secondary sources. On the contrary, in 2013 the participation rate of PM$_1$ in the PM$_{2.5}$ mass was decreased indicating that the input from the natural sources and mechanical processes (e.g dust resuspension) was significant as well [39].

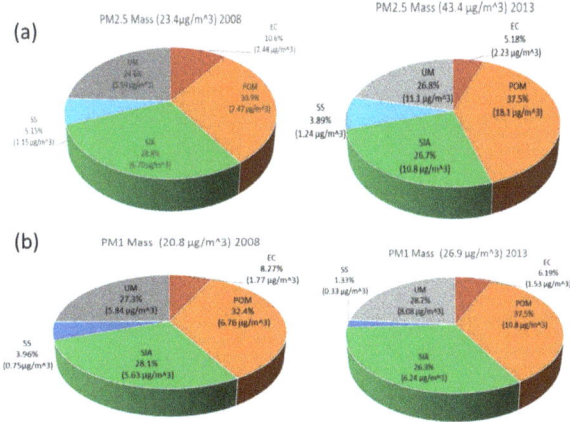

Figure 2. Differentiation of (**a**) the PM$_{2.5}$ and (**b**) the PM$_1$ mass closure profile in March 2008 and 2013.

In full agreement with the literature [2,31,40–42], the particles were mainly composed by carbonaceous material signifying the importance of the anthropogenic activities. Therefore, POM was the main constituent of the particles' mass being followed by SIA (PM$_{2.5}$: POM > SIA > UM > EC > SS, PM$_1$: POM > SIA ≈ UM > EC > SS). The more pronounced change between the two periods occurred for POM; its contribution was 7% and 5% higher in 2013 compared to 2008, for PM$_{2.5}$ and PM$_1$, respectively (in terms of mass is almost 59% and 37%, respectively). It is worth noting the different behavior of the other carbonaceous component, the EC. Its contribution to the PM$_{2.5}$ and PM$_1$ mass appeared to be decreased in 2013, at about 5% and 2%, respectively (in terms of mass is almost 11% and 15%, respectively). The different nature of the two carbonaceous constituents might be the explanation. EC is a primary pollutant formed during combustion of various fuels (coal, wood, fuel oil and motor fuel, especially diesel) whereas POM is a complex mixture composed of a primary (combustion derived) and a secondary material [43–46]. Therefore, the higher values of the POM/EC ratio in 2013, taking into consideration the previously mentioned lower EC levels, denote the decrease of the traffic input and the enhancement of the contribution from the heating sector.

This temporal change was more evident for PM$_{2.5}$ (2008: 3.02 and 3.82 as well as 2013: 8.12 and 7.02 for PM$_{2.5}$ and PM$_1$, respectively). The hypothesis of the enhancement of the emissions from the residential heating due to the intense use of biomass burning during 2013 is further supported by the calculation of the OC/EC and K$^+$/EC ratios. The average OC/EC ratio was 1.98 and 2.56 in 2008 while in 2013 it was 4.68 and 4.27 for PM$_{2.5}$ and PM$_1$, respectively ([47]; coal combustion (0.3–7.6), vehicle emission (0.7–2.4), biomass burning (4.1–14.5)). In the case of the K$^+$/EC ratio, its average values were 0.15 and 0.16 in 2008 as well as 0.24 and 0.24 in 2013 for PM$_{2.5}$ and PM$_1$, respectively ([47]; biomass burning (0.2–0.5) and fossil fuel combustion (0.03–0.09)).

3.2. Influence of the Emissions

According to the data produced by FEI-GREGAA [19], the total PM$_{2.5}$ emissions, from both the traffic and the heating sector, were higher in March 2013 (Figure 3a). In general, the increasing trend of the produced emissions, is in compliance with the experimental data, highlighting the enhancement of the anthropogenic contribution to the configured PM burden. The diverse rates of differentiation between the experimental and the emissions data are due to the consideration of the input only from the domestic and the traffic sector in the emission inventory.

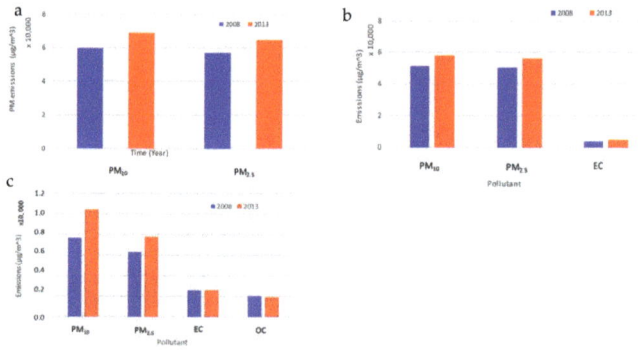

Figure 3. Differentiation of the average (**a**) total, (**b**) residential heating and (**c**) traffic PM$_{10}$ and PM$_{2.5}$ emissions between the March of 2008 and 2013.

Analyzing further the emissions' data, oil consumption for space heating decreased from 2008 to 2013 by 62.3% while wood consumption increased by 23%. As for the energy consumption by road transport, in Greece the majority of passenger cars use gasoline as fuel type. The gasoline consumption for the year 2008 was 4,046,516 metric tonnes and 1,453,481 metric tonnes in Greece and Attica respectively while for the year 2013 the respective consumption was 2,669,964 metric tonnes and 1,014,515 metric tonnes. The percentage decrease was 34.02% for Greece and 30.20% for Attica reflecting the impact of the economic crisis. In an attempt to analyze better the change on the share of the sources' contribution, the PM$_{10}$, EC and OM emissions are used where possible, from the two sectors. PM$_{10}$, even with a less evident differentiation (8%), presented the same behavior. As far as the traffic related carbonaceous input is concerned, EC was almost similar between the two years (+0.5% higher in 2013) while OM was lower in 2013 by about 7%.

Since the emissions of the residential heating were higher than those of traffic, it seemed that the specific sector drove the changes to the configuration of the PM burden (Figure 3b,c). The corresponding input was almost 7 and 8 times higher for PM$_{10}$ and PM$_{2.5}$ in 2008, and at about 6 and 7 times for PM$_{10}$ and PM$_{2.5}$ respectively in 2013. This is due to the fact that particle emissions from residential heating are highly dependent on the fuel type and the technology of the combustion installation. They are related to the biomass burning from fireplaces and stoves to which higher emission factors are attributed compared to the oil used by boilers [19]. Moreover, the decrease of the residential heating/traffic ratio from 2008 to 2013 is associated with the increase of the percentage contribution of the Heavy Duty Trucks (HDT) to traffic PM emissions. Based on the emissions data, HDT are the main contributor to the Athenian PM emissions. Approximately 36% and 42% of PM_{10} and $PM_{2.5}$ emissions originated from HDT in 2008 while the relevant percentages in 2013, were 39% and 46%. Even though the engine technology of passenger cars has improved and the withdrawal ratio of older passenger cars increased from 2008 to 2013 this was not the case for trucks the fleet of which remained rather stable during the above period (almost 53,000 HDT in 2008 and 52,000 in 2013). In 2013, it is worthy to note, the almost double EC input from the domestic heating than the one from traffic as well as the small decrease of the POM contribution from the traffic sector (7%).

3.3. Health Impact of the Aerosol Levels

The calculated $PM_{2.5}$ EPA-indexes, the frequency of occurrence of the corresponding $PM_{2.5}$-AQI health categories as well as the frequency of prevalence of the acidic PM nature during the two sampling periods are summarized in Figure 4.

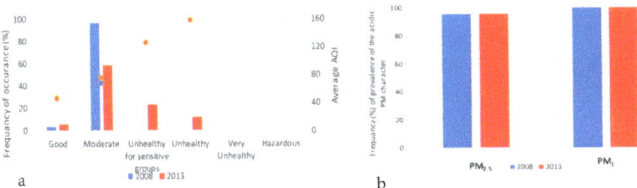

Figure 4. Frequency of occurrence (%) of (**a**) the health categories and the average AQI in connection with $PM_{2.5}$ and (**b**) the prevalence of the acidic PM character during March of 2008 and 2013.

The increasing trend of the AQIs from March, 2008, to March, 2013 Figure 4a (the blue and red dots are the values of the average AQI on March 2008 and March 2013, respectively), indicates the deterioration of the air quality within the GAA with time. For both periods, the "moderate pollution" days (People who are unusually sensitive to ozone or particle pollution may experience respiratory symptoms; [36]) was the category more often met with its frequency of occurrence reaching the 96% in 2008 and the 59% in 2013 (Figure 4a; the blue and red columns respectively).

In connection with the results of the ionic balance, for both fine PM fractions, the average cation/anion equivalent ratios were lower than unity indicating that the particles were acidic (2008: 0.74 and 0.63, 2013: 0.79 and 0.49 for $PM_{2.5}$ and PM_1, respectively). The frequency of prevalence of the acidic PM character is depicted in Figure 4b. The finding is consistent with other published works for the Mediterranean region [13,31]. Moreover, when the total cation equivalents were plotted against the total anion equivalents, the slope of the regression line was, in all the cases, lower than unity (Figure S1). The cation deficit (excess of anions) is commonly attributed to H^+ which was not measured using the ion chromatography and is mainly associated with $nssSO_4^{2-}$ (Figure S2). In general, high levels of secondary inorganic ions are indicated to enhance the acidity of particles [48,49].

In an attempt to determine the key parameters for the potential health risk of the airborne particles, the average compositional characteristics of the configured PM profile were associated with the obtained $PM_{2.5}$-AQI categories of the EPA and the prevailing wind speed/ temperature values. The results are summarized in Table S2. According to the data analysis, the lower the wind speed value, the more 'polluted' the investigated atmosphere is. The monitoring location was less polluted ('good days' category; Air quality is satisfactory and poses little or no health risk; [36]) when the average values of the wind speed and the temperature were 2.00 m/s and 15.2 °C, respectively while compositionally, the rates of the contribution of the total carbonaceous component to the $PM_{2.5}$ mass varied from 26% to 30%. On the contrary, when the wind velocity and the temperature varied between lower values (0.36–0.39 m/s and 12.6–13.5 °C) and the contribution of the total carbonaceous content was enhanced

(42–62%) the $PM_{2.5}$ status led to more serious health problems ('unhealthy days' category; Members of sensitive groups may experience more serious health effects; [36]).

4. Conclusions

The specific work makes an attempt to elucidate the temporal evolution of the PM chemical profile owing to the change of the share of the anthropogenic input during the economic recession period. Measurements took place both in 2008; a year landmark for the global economic crisis and in 2013; five years within the economic recession.

The main conclusions from the parallel analysis of the experimental and FEI-GREGAA emissions data can be summarized as follows:

- An enhanced input from the anthropogenic emissions and in full compliance with the emission data, was observed in 2013 for both the PM load and the POM component.
- The obtained PM chemical profile scheme was the result of both local and regional sources with POM and SIA, being the main constituents of the particles' mass.
- When moving from 2008 to 2013, the sector of residential heating seemed to drive the changes to the configuration of the PM burden.
- In 2013, the EC input from the domestic heating was almost double of the one from traffic while the POM contribution from the traffic sector was decreased at about 7%.
- The changes to the air quality were mainly driven by $PM_{2.5}$.
- From the health perspective, the considerable deterioration of the air quality in 2013, appeared to coincide with low average wind values (avg ws ≤ 0.69 m/s).
- Despite the position of the monitoring location next to the most trafficked avenues of the capital, the EC input from the heating sector was double compared to the traffic in 2013.

The parallel analysis of the profile of the airborne particles with experimental and emissions data is of great importance. Various (pollution, meteorological, health) parameters should be investigated simultaneously, to clarify the degree of the impact of the differentiation of the share of the anthropogenic sources to the PM chemical PM profile. An extensive field campaign including different diameter particles within different types of environments, with a more detailed chemical characterization and emissions from other sectors, could give a better understanding of the complex PM nature.

Supplementary Materials: The following are available online at http://www.mdpi.com/2073-4433/11/10/1121/s1, Table S1: A statistical summary of the primary $PM_{2.5}$ and PM_1 data (mass concentration, chemical composition), during the different sampling periods, Table S2: A summary of the temporal differentiation of the average compositional $PM_{2.5}$ and PM_1 characteristics, the wind speed and the temperature values, during the different EPA-Health categories, Figure S1: Sum of the anions vs sum of the cations regression analysis in (a) $PM_{2.5}$ and (b) PM_1 samples of 2008 and 2013, Figure S2: Differentiation of cation deficit versus the $nssSO_4$ concentrations in (a) $PM_{2.5}$ and (b) PM_1 samples between 2008 and 2013.

Author Contributions: The study was completed with cooperation between all authors. Conceptualization, S.P.; Data curation, S.P., K.-M.F., T.S., A.Z., A.B.; Methodology, S.P., K.-M.F., T.S., A.Z., K.B., A.B. and T.M.; Software, K.-M.F. and V.A.; Validation, S.P., Supervision, V.A., and N.M.; Writing—original draft, S.P.and K.-M.F.; Writing—review & editing, S.P., K.-M.F. and N.M. All authors have read and agreed to the published version of the manuscript.

Funding: The study was partly supported by the project THESPIA II (MIS 5002517), EU NSRF 2014-2020 and partly by the Post-Doctoral Research project 'Observatory of Air and Particulate Pollution over Greece (AirPaP)'. AirPaP has received funding from the Hellenic Foundation for Research and Innovation (HFRI) and the General Secretariat for Research and Technology (GSRT), under grant agreement No 409.

Acknowledgments: The authors would like to thank the National Observatory of Athens and the Hellenic National Meteorological Service for the provision of the meteorological data.

Conflicts of Interest: The authors declare no conflict of interest. The funder did not participate in any phase of the study, including the publication of data.

References

1. Paglione, M.; Gilardoni, S.; Rinaldi, M.; Decesari, S.; Zanca, N.; Sandrini, S.; Giulianelli, L.; Bacco, D.; Ferrari, S.; Poluzzi, V.; et al. The impact of biomass burning and aqueous-phase processing on air quality: A multi-year source apportionment study in the Po Valley, Italy. *Atmos. Chem. Phys.* **2020**, *20*, 1233–1254. [CrossRef]
2. Squizzato, S.; Masiol, M.; Agostini, C.; Visin, F.; Formenton, G.; Harrison, R.M.; Rampazzo, G. Factors, origin and sources affecting PM_1 concentrations and composition at an urban background site. *Atmos. Res.* **2016**, *180*, 262–273. [CrossRef]
3. Vecchi, R.; Chiari, M.; D'Alessandro, A.; Fermo, P.; Lucarelli, F.; Mazzei, F.; Navab, S.; Piazzalungac, A.; Pratie, P.; Silvania, F.; et al. A mass closure and PMF source apportionment study on the sub-micron sized aerosol fraction at urban sites in Italy. *Atmos. Environ.* **2008**, *42*, 2240–2253. [CrossRef]
4. Galindo, N.; Gil-Molto, J.; Varea, M.; Chofre, C.; Yubero, E. Seasonal and interanual trends in PM levels and associated inorganic ions in southeastern Spain. *Microchem. J.* **2013**, *110*, 81–88. [CrossRef]
5. Galindo, N.; Yubero, E.; Clemente, A.; Nicolas, J.F.; Varea, M.; Crespo, J. PM events and changes in the chemical composition of urban aerosols: A case study in the western Mediterranean. *Chemosphere* **2020**, *244*, 125520. [CrossRef]
6. Barzeghar, V.; Sarbakhsh, P.; Hassanvand, M.S.; Faridi, S.; Gholampour, A. Long-term trend of ambient air PM_{10}, $PM_{2.5}$, and O_3 and their health effects in Tabriz city, Iran, during 2006–2017. *Sustain. Cities Soc.* **2020**, *54*, 101988. [CrossRef]
7. Dong, Z.; Wang, S.; Xing, J.; Chang, X.; Ding, D.; Zheng, H. Regional transport in Beijing-Tianjin-Hebei region and its changes during 2014–2017: The impacts of meteorology and emission reduction. *Sci. Total Environ.* **2020**, *737*, 139792. [CrossRef]
8. Fan, Y.; Ding, X.; Hang, J.; Ge, J. Characteristics of urban air pollution in different regions of China between 2015 and 2019. *Build. Environ.* **2020**, *180*, 107048. [CrossRef]
9. Zhang, X.; Gu, X.; Cheng, C.; Yang, D. Spatiotemporal heterogeneity of $PM_{2.5}$ and its relationship with urbanization in North China from 2000 to 2017. *Sci. Total Environ.* **2020**, *744*, 140925. [CrossRef]
10. Chen, L.; Zhu, J.; Liao, H.; Yang, Y.; Yue, X. Meteorological influences on $PM_{2.5}$ and O_3 trends and associated health burden since China's clean air actions. *Sci. Total Environ.* **2020**, *744*, 140837. [CrossRef]
11. Gao, M.; Carmichael, G.R.; Saide, P.E.; Lu, Z.; Yu, M.; Streets, D.G.; Wang, Z. Response of winter fine particulate matter concentrations to emission and meteorology changes in North China. *Atmos. Chem. Phys.* **2016**, *16*, 11837–11851. [CrossRef]
12. Kim, Y.; Yi, S.-M.; Heo, J. Fifteen-year trends in carbon species and $PM_{2.5}$ in Seoul, South Korea (2003–2017). *Chemosphere* **2020**, *261*, 127750. [CrossRef]
13. Paraskevopoulou, D.; Liakakou, E.; Gerasopoulos, E.; Mihalopoulos, N. Sources of atmospheric aerosol from long-term measurements (5 years) of chemical composition in Athens, Greece. *Sci. Total Environ.* **2015**, *527*, 165–178. [CrossRef] [PubMed]
14. Jiang, Z.; Jolleys, M.D.; Fu, T.-M.; Palmer, P.I.; Ma, Y.; Tian, H.; Li, J.; Yang, X. Spatiotemporal and probability variations of surface $PM_{2.5}$ over China between 2013 and 2019 and the associated changes in health risks: An integrative observation and model analysis. *Sci. Total Environ.* **2020**, *723*, 137896. [CrossRef] [PubMed]
15. Zhang, Q.; Zheng, Y.; Tong, D.; Shao, M.; Wang, S.; Zhang, Y.; Xu, X.; Wang, J.; He, H.; Liu, W.; et al. Drivers of improved $PM_{2.5}$ air quality in China from 2013 to 2017. *Proc. Natl. Acad. Sci. USA* **2019**, *116*, 24463–24469. [CrossRef] [PubMed]
16. Gao, M.; Liu, Z.; Zheng, B.; Ji, D.; Sherman, P.; Song, S.; Xin, J.; Liu, C.; Wang, Y.; Zhang, Q.; et al. China's emission control strategies have suppressed unfavorable influences of climate on wintertime $PM_{2.5}$ concentrations in Beijing since 2002. *Atmos. Chem. Phys.* **2020**, *20*, 1497–1505. [CrossRef]
17. Querol, X.; Alastuey, A.; Pandolfi, M.; Reche, C.; Pérez, N.; Minguillón, M.C.; Moreno, T.; Viana, M.; Escudero, M.; Orio, A.; et al. 2001–2012 trends on air quality in Spain. *Sci. Total Environ.* **2014**, *490*, 957–969. [CrossRef]
18. Pateraki, S.; Asimakopoulos, D.N.; Maggos, T.; Assimakopoulos, V.D.; Bougiatioti, A.; Bairachtari, K.; Vasilakos, C.; Mihalopoulos, N. Chemical characterization, sources and potential health risk of $PM_{2.5}$ and PM_1 pollution across the Greater Athens Area. *Chemosphere* **2020**, *241*, 125026. [CrossRef]
19. Fameli, K.M.; Assimakopoulos, V.D. The new open Flexible Emission Inventory for Greece and the Greater Athens Area (FEI-GREGAA): Account of pollutant sources and their importance from 2006 to 2012. *Atmos. Environ.* **2016**, *137*, 17–37. [CrossRef]

20. Crilley, L.R.; Bloss, W.J.; Yin, J.; Beddows, D.C.S.; Harrison, R.M.; Allan, J.D.; Young, D.E.; Flynn, M.; Williams, P.; Zotter, P.; et al. Sources and contributions of wood smoke during winter in London: Assessing local and regional influences. *Atmos. Chem. Phys.* **2015**, *15*, 3149–3171. [CrossRef]
21. Mueller, W.; Loh, M.; Vardoulakis, S.; Johnston, H.J.; Steinle, S.; Precha, N.; Kliengchuay, W.; Tantrakarnapa, K.; Cherrie, J.W. Ambient particulate matter and biomass burning: An ecological time series study of respiratory and cardiovascular hospital visits in northern Thailand. *Environ. Health* **2020**, *19*, 77. [CrossRef]
22. Fourtziou, L.; Liakakou, E.; Stavroulas, I.; Theodosi, C.; Zarmpas, P.; Psiloglou, B.; Sciare, J.; Maggos, T.; Bairachtari, K.; Bougiatioti, A.; et al. Multi-tracer approach to characterize domestic wood burning in Athens (Greece) during wintertime. *Atmos. Environ.* **2017**, *148*, 89–101. [CrossRef]
23. Pateraki, S. Experimental and Arithmetic Study of the Particulate Matter (PM_{10}, $PM_{2.5}$, PM_1) Concentrations and Their Chemical Composition (PAHs, Ions, Organic/Elemental Carbon) over the Greater Athens Area. Ph.D. Thesis, Department of Physics, National Kapodistrian of Athens, Athens, Greece, 2012.
24. Stavraka, T. A Study of $PM_{2.5}$ and PM_1 in the Atmosphere of a Suburban Area within the Athens Basin. Bachelor's Thesis, Department of Physics, National Kapodistrian of Athens, Athens, Greece, 2013.
25. Zagkos, A. A Study of the Temporal Evolution of Particulate Pollution in Different Types of Environment across the Greater Athens Area: Concentrations, Chemical Composition and the Role of the Meteorology. Master's Thesis, Department of Physics, National Kapodistrian of Athens, Athens, Greece, 2014.
26. Pateraki, S.; Assimakopoulos, V.D.; Bougiatioti, A.; Kouvarakis, G.; Mihalopoulos, N.; Vasilakos, C. Carbonaceous and ionic compositional patterns of fine particles over an urban Mediterranean area. *Sci. Total Environ.* **2012**, *424*, 251–263. [CrossRef] [PubMed]
27. Kocak, M.; Mihalopoulos, N.; Kubilay, N. Chemical composition of the fine and coarse fraction of aerosols in the northeastern Mediterranean. *Atmos. Environ.* **2007**, *41*, 7351–7368. [CrossRef]
28. Theodosi, C.; Grivas, G.; Zarmpas, P.; Chaloulakou, A.; Mihalopoulos, N. Mass and chemical composition of size-segregated aerosols (PM_1, $PM_{2.5}$, PM_{10}) over Athens, Greece: Local versus regional sources. *Atmos. Chem. Phys.* **2011**, *11*, 11895–11911. [CrossRef]
29. Liakakou, E.; Stavroulas, I.; Kaskaoutis, D.G.; Grivas, G.; Paraskevopoulou, D.; Dumka, U.C.; Tsagkaraki, M.; Bougiatioti, A.; Oikonomou, K.; Sciare, J.; et al. Long-term variability, source apportionment and spectral properties of black carbon at an urban background site in Athens, Greece. *Atmos. Environ.* **2020**, *222*, 117137. [CrossRef]
30. Terzi, E.; Argyropoulos, G.; Bougiatioti, A.; Mihalopoulos, N.; Nikolaou, K.; Samara, C. Chemical composition and mass closure of ambient PM_{10} at urban sites. *Atmos. Environ.* **2010**, *44*, 2231–2239. [CrossRef]
31. Pateraki, S.; Asimakopoulos, D.N.; Bougiatioti, A.; Maggos, T.; Vasilakos, C.; Mihalopoulos, N. Assessment of $PM_{2.5}$ and PM_1 Chemical Profile in a Multiple-Impacted Mediterranean Urban Area: Origin, Sources and Meteorological Dependence. *Sci. Total Environ.* **2014**, *479*, 210–220. [CrossRef]
32. Odysseee Mure Project. Available online: https://www.odyssee-mure.eu/ (accessed on 15 March 2019).
33. Fameli, K.M.; Assimakopoulos, V.D. Development of a road transport emission inventory for Greece and the Greater Athens Area: Effects of important parameters. *Sci. Total Environ.* **2015**, *505*, 770–786. [CrossRef]
34. Schaap, M.; Roemer, M.; Sauter, F.; Boersen, G.; Timmermans, R.; Builtjes, P.J.H.; Vermeulen, A.T. *Lotos-Euros: Documentation*; TNO-Report B&O-A R 2005/297; TNO: Apeldoorn, The Netherlands, 2005.
35. Gorai, A.K.; Tchounwou, P.B.; Biswal, S.S.; Francis Tuluri, F. Spatio-Temporal Variation of Particulate Matter ($PM_{2.5}$) Concentrations and Its Health Impacts in a Mega City, Delhi in India. *Environ. Health Insights* **2018**, *12*, 1–9. [CrossRef]
36. EPA. *Guideline for Reporting of Daily Air Quality—Air Quality Index (AQI)*; EPA: Washington, DC, USA, 1999.
37. Dimitriou, K.; Liakakou, E.; Lianou, M.; Psiloglou, B.; Kassomenos, P.; Mihalopoulos, N.; Gerasopoulos, E. Implementation of an aggregate index to elucidate the influence of atmospheric synoptic conditions on air quality in Athens, Greece. *Air Qual. Atmos. Health* **2020**, *13*, 447–458. [CrossRef]
38. Wang, H.; Shooter, D. Coarse–fine and day–night differences of water-soluble ions in atmospheric aerosols collected in Christchurch and Auckland, New Zealand. *Atmos. Environ.* **2002**, *36*, 3519–3529. [CrossRef]
39. Khan, J.Z.; Sun, L.; Tian, Y.; Shi, G.; Feng, Y. Chemical characterization and source apportionment of PM_1 and $PM_{2.5}$ in Tianjin, China: Impacts of biomass burning and primary biogenic sources. *J. Environ. Sci.* **2021**, *99*, 196–209. [CrossRef]

40. Phairuang, W.; Inerb, M.; Furuuchi, M.; Hata, M.; Tekasakul, S.; Tekasakul, P. Size-fractionated carbonaceous aerosols down to PM$_{0.1}$ in southern Thailand: Local and long-range transport effects. *Environ. Pollut.* **2020**, *260*, 114031. [CrossRef]
41. Zhang, Y.-L.; Huang, R.-J.; El Haddad, I.; Ho, K.-F.; Cao, J.-J.; Han, Y.; Zotter, P.; Bozzetti, C.; Daellenbach, K.R.; Canonaco, F.; et al. Fossil vs. non-fossil sources of fine carbonaceous aerosols in four Chinese cities during the extreme winter haze episode of 2013. *Atmos. Chem. Phys.* **2015**, *15*, 1299–1312. [CrossRef]
42. Zotter, P.; Herich, H.; Gysei, M.; El-Haddad, I.; Zhang, Y.; Mocnik, G.; Hüglin, C.; Baltensperger, U.; Szidat, S.; Prévôt, A.S.H. Evaluation of the absorption Ångström exponents for traffic and wood burning in the Aethalometer-based source apportionment using radiocarbon measurements of ambient aerosol. *Atmos. Chem. Phys.* **2017**, *17*, 4229–4249. [CrossRef]
43. Lonati, G.; Giugliano, M.; Butelli, P.; Romele, L.; Tardivo, R. Major chemical components of PM$_{2.5}$ in Milan (Italy). *Atmos. Environ.* **2005**, *39*, 1925–1934. [CrossRef]
44. Sillanpaa, M.; Hillamoa, R.; Saarikoskia, S.; Frey, A.; Pennanen, A.; Makkonen, U.; Spolnik, Z.; Van Grieken, R.; Branis, M.; Brunekreef, B.; et al. Chemical composition and mass closure of particulate matter at six urban sites in Europe. *Atmos. Environ.* **2006**, *40*, S212–S223. [CrossRef]
45. Rattigan, O.V.; Felton, H.D.; Bae, M.-S.; Schwab, J.J.; Demerjian, K.L. Multi-year hourly PM$_{2.5}$ carbon measurements in New York: Diurnal, day of week and seasonal patterns. *Atmos. Environ.* **2010**, *44*, 2043–2053. [CrossRef]
46. Hai, C.D.; Oanh, N.T.K. Effects of local, regional meteorology and emission sources on mass and compositions of particulate matter in Hanoi. *Atmos. Environ.* **2013**, *78*, 105–112. [CrossRef]
47. Wang, Y.; Jia, C.; Tao, J.; Zhang, L.; Liang, X.; Ma, J.; Gao, H.; Huang, T.; Zhang, K. Chemical characterization and source apportionment of PM$_{2.5}$ in a semi-arid and petrochemical-industrialized city, Northwest China. *Sci. Total Environ.* **2016**, *573*, 1031–1040. [CrossRef] [PubMed]
48. Gao, J.; Tian, H.; Cheng, K.; Lu, L.; Zheng, M.; Wang, S.; Hao, J.; Wang, K.; Hua, S.; Zhu, C.; et al. The variation of chemical characteristics of PM$_{2.5}$ and PM$_{10}$ and formation causes during two haze pollution events in urban Beijing, China. *Atmos. Environ.* **2015**, *107*, 1–8. [CrossRef]
49. Hassan, S.K.; Khoder, M.I. Chemical characteristics of atmospheric PM$_{2.5}$ loads during air pollution episodes in Giza, Egypt. *Atmos. Environ.* **2017**, *150*, 346–355. [CrossRef]

Publisher's Note: MDPI stays neutral with regard to jurisdictional claims in published maps and institutional affiliations.

© 2020 by the authors. Licensee MDPI, Basel, Switzerland. This article is an open access article distributed under the terms and conditions of the Creative Commons Attribution (CC BY) license (http://creativecommons.org/licenses/by/4.0/).

Article

Source Apportionment and Assessment of Air Quality Index of PM$_{2.5-10}$ and PM$_{2.5}$ in at Two Different Sites in Urban Background Area in Senegal

Moustapha Kebe [1], Alassane Traore [1,2,*], Manousos Ioannis Manousakas [3], Vasiliki Vasilatou [3], Ababacar Sadikhe Ndao [1,2], Ahmadou Wague [2] and Konstantinos Eleftheriadis [3]

1. Institut de Technologie Nucléaire Appliquée, Université Cheikh Anta Diop Dakar, Dakar 10700, Senegal; moustapha5.kebe@ucad.edu.sn (M.K.); ababacar.ndao@ucad.edu.sn (A.S.N.)
2. Faculté des Sciences et Techniques, Université Cheikh Anta Diop Dakar, Dakar 10700, Senegal; ahmadou.wague@ucad.edu.sn
3. E.R.L., Institute of Nuclear & Radiological Sciences & Technology, Energy & Safety, N.C.S.R. Demokritos, 15310 Agia Paraskevi, Attiki, Greece; manosman@ipta.demokritos.gr (M.I.M.); vassiliki@ipta.demokritos.gr (V.V.); elefther@ipta.demokritos.gr (K.E.)
* Correspondence: alassane2.traore@ucad.edu.sn; Tel.: +221-77-605-08-45

Abstract: Identifying the particulate matter (PM) sources is an essential step to assess PM effects on human health and understand PM's behavior in a specific environment. Information about the composition of the organic or/and inorganic fraction of PM is usually used for source apportionment studies. In this study that took place in Dakar, Senegal, the identification of the sources of two PM fractions was performed by utilizing data on the elemental composition and elemental carbon content. Four PM sources were identified using positive matrix factorization (PMF): Industrial emissions, mineral dust, traffic emissions, and sea salt/secondary sulfates. To assess the effect of PM on human health the air quality index (AQI) was estimated. The highest values of AQI are approximately 497 and 488, in Yoff and Hlm, respectively. The spatial location of the sources was investigated using potential source contribution function (PSCF). PSCF plots revealed the high effect of transported dust from the desert regions to PM concentration in the sampling site. To the best of our knowledge, this is the first source apportionment study on PM fractions published for Dakar, Senegal.

Keywords: aerosol; PMF; elemental composition; PSCF

1. Introduction

As many studies have shown in the past, air pollution is a significant problem due to the multiple effects it has on human health, the environment [1–3], and climate [4]. Many resources and effort have been spent to reduce the particulate matter (PM) concentration levels and, consequently, their effects on human health and understand their behavior and properties in different environments [5,6]. Human health effects associated with PM pollution are premature death, hospital admissions, emergency room visit, asthma attack, chronic bronchitis, cancer, cardiovascular disease, diabetes, and restricted activity [7–10]. Identifying the (PM) sources and quantifying their contributions are essential steps towards effectively reducing ambient (PM) concentrations [11]. (PM) can be separated based on their size to coarse and fine. PM$_{10}$ are defined as particles with a diameter equal or smaller to 10 µm, while PM$_{2.5}$, which are often referred to as fine particles, have a diameter equal or smaller than 2.5 µm. Epidemiological studies have shown that PM$_{2.5}$, due to their smaller size, can penetrate the lower tissues of the respiratory tract and induce more severe health-related issues [12]. The lifetime of the PM$_{2.5}$ in the atmosphere is days to weeks, while for the PM$_{2.5-10}$ is minutes to hours [13]. The fine particles originate mainly from anthropogenic and combustion related activities, as well as other urban and industrial activities [14,15].

The first step to all source apportionment studies is sampling. For more comprehensive results it is preferable if more than one PM size fractions are sampled. The Dichotomous [16], Gent Stacked Filter Unit sampler manufactured by the International Atomic Energy Agency (IAEA) contracted with the University of Gent [17], offers the advantage of simultaneous collection of fine ($PM_{2.5}$) and coarse ($PM_{2.5-10}$) fractions. Several techniques, both destructive and non-destructive, are used to analyze the chemical composition of PM samples. Some commonly used techniques are: energy dispersive X-ray fluorescence (EDXRF); synchrotron induced X-ray fluorescence (S-XRF); proton induced X-ray emission (PIXE); inductively coupled plasma with atomic emission spectroscopy (ICP-AES); and inductively coupled plasma with mass spectroscopy (ICP-MS). These methods differ with respect to detection limits, sample preparation, and cost [18,19]. The advantage of XRF over other commonly used analytical techniques is that it is a non-destructive technique and enables direct analysis of a sample without pre-treatment, and with minimum damage to the sample itself [20]. It has often been the method of choice for analysis of trace elements on filters and allows fast and simultaneous analysis over the total spectrum, thus enabling the analysis of numerous elements simultaneously [21].

To effectively understand particulate-related pollution in an area, it is crucial to identify PM sources by doing source apportionment (SA). SA models aim to reconstruct the impacts of emissions from different sources of PM, based on ambient data registered at monitoring sites [22]. Positive matrix factorization (PMF) is the most commonly used SA model, and it has been successfully implemented in many areas around the world with different characteristics [23,24]. The model is described for the first time by Paatero and Trapper [25]. The big advantage of PMF is that it can be used with little to no external information about the sources in an area, making it a perfect choice for studies conducted in less studied receptors. In order for PMF to find the solution (source profiles and contributions) that best fits the studied site, a relatively large amount of data consisting of chemical constituents gathered from a number of samples is required. According to [26], at least 50 chemically characterized samples are required for running multivariate models. To increase the total number of samples, datasets simultaneously collected at different sites within the same region or different size fractions can be pooled together in one matrix [27]. Additional tools such as hybrid trajectory-based methods that provide information about the geographical origin of pollutants can be used to assist the user in assigning the identified source profiles to actual sources or/and to verify the results of the SA analysis.

Dissemination of information on air quality in urban areas can lead to raised awareness and increased citizens' involvement in those measures aiming at containing and reducing PM emissions. One of the first synthetic indices used for reporting air quality in an area adopted by the United States Environmental Protection Agency (US-EPA), was the Pollution Standard Index (PSI). In 1999, the EPA replaced the PSI index with Air Quality Index (AQI), which includes two new sub-indices, the ozone at ground level and fine particulate. (AQI) is an index to effectively communicate to the public the health risk due to the ambient concentrations of pollutants [28].

Dakar is the capital of Senegal, and the most developed city of the country, having an estimated population of five million inhabitants. The explosion of the population and the rapid increase of industrial and agricultural activities combined with a significant rise of the vehicular fleet resulted in a change of the city's environment and a deterioration of the area's air quality. Although the air-quality of the city is monitored by the management center of air quality, from the Environment Direction and Reserved Buildings [29], the number of studies in the area are very limited, and none of them is referring to source apportionment of $PM_{2.5}$ and/or $PM_{2.5\text{-coarse}}$ [30,31]. Since Dakar is in close proximity to the many arid areas of the African region, coarse particles are expected to have a high impact on PM concentration levels in the area, and thus the study of this particular size fraction is critical. Overall, Dakar is an area in the world for which results regarding the composition and sources of PM are still scarce.

To address the aforementioned points, the main objective of this study was the identification of PM sources in Dakar, as well as to assess the AQI. The work focused on $PM_{2.5}$ and $PM_{2.5-10}$ fractions. The PM samples were simultaneously collected at two sites, Hlm (industrial area), and Yoff (near road area) in Dakar. The elemental composition of the samples was identified using a commercially available XRF instrument (EPSILON 5 by PANalytical, The Netherlands).

2. Experiments

2.1. Sampling

The sampling was performed in two sites, Hlm and Yoff. The sampler at Hlm site (latitude 14°42′53″ N, longitude 17°26′41″ W) was installed in an urban/industrial area where many small factories as gases company, metal, food, and chemical industries are located. Hlm is considered an industrial district with 39,100 inhabitants in an area of approximately 2 km². At Hlm, the sampler was installed above 2.5 m height, from the ground, to reduce the high impact of local dust resuspension. Yoff (latitude 14°45′14″ N, longitude 17°28′04″ W) is an area with 895 inhabitants and surface space of approximately 15 km². The sampler was installed on the roadside across paint and printing factories, and having a distance of 2 km from the coast. The duration of each daily collection was about 24 h. Figure 1 shows the locations of the sampling sites.

Air masses could reach the sampling points undisturbed from all directions, and therefore the collected PM samples can be considered representative of the wider urban area. A low volume sampler from the University of GENT was used for sampling. The sampler is described in detail elsewhere [32]. Briefly, the sampler operates at a flow rate of 16 L min^{-1}. It collects particulates that have an equivalent aerodynamic diameter (EAD) of less than 10 μm in separate "coarse" (2.5–10 μm EAD) and "fine" (<2.5 μm EAD) size fractions on two sequential 47 mm diameter filters. The discrimination against the >10 μm EAD particles is accomplished by a PM_{10} pre-impaction stage upstream of the stacked filter cassette. The air is drawn through the sampler by means of a diaphragm vacuum pump, which is enclosed in a special housing together with a needle valve, vacuum gauge, flow meter, volume meter, time switch (for interrupted sampling), and hour meter.

$PM_{2.5}$ and $PM_{2.5-10}$ were collected onto Nucleopore polycarbonate filters 47 mm diameter with 0.4 μm and 8 μm pore size, respectively [32]. After collection, the filters were kept in a desiccator to stabilize without hydration or contamination. The gravimetric quantification was performed with a Sartorius microbalance Secura and Quintix model 26 with an accuracy of 10^{-5} g. Filter weight was obtained from the average of three measurements when the variations were less than 0.5%.

During the years 2018–2019, a yearlong measurement campaign was performed to collect the $PM_{2.5-10}$ and $PM_{2.5}$ samples. The samples were collected twice per week; one sample during working days, and one during the weekend. This sampling protocol was selected to achieve the highest possible variation of source contributions, which is crucial for optimum performance of the source apportionment models. Several samples were collected throughout the period of 2018 and 2019, and 71 of each size fraction were selected to be analyzed by XRF. The selection was made to achieve maximum time coverage.

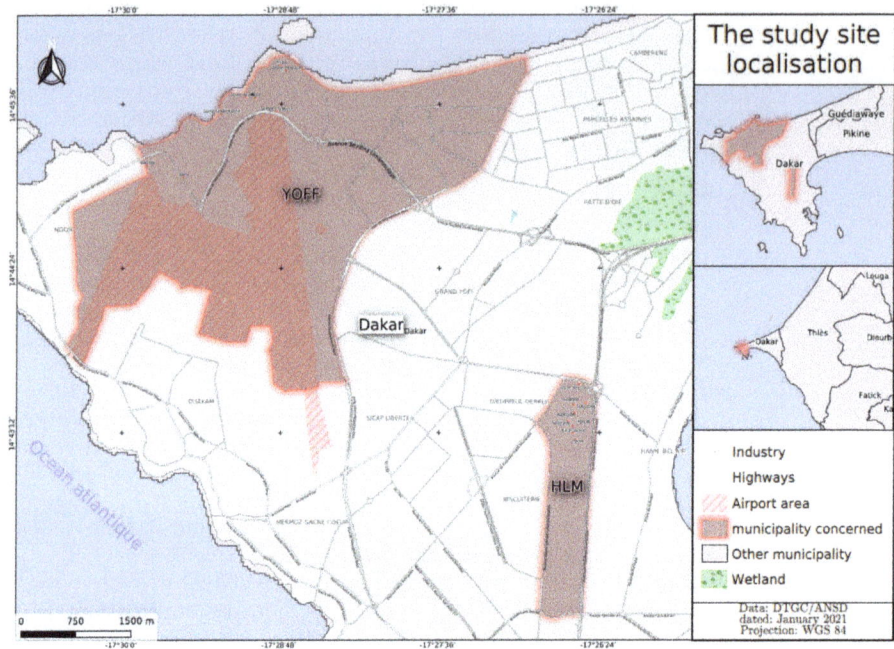

Figure 1. Location of the Hlm and Yoff sampling sites.

2.2. Elemental Analysis

The elemental composition of the samples was determined by Energy dispersive X-ray Fluorescence Spectroscopy (ED-XRF) using a commercially available system (Epsilon 5 by PANalytical, The Netherlands) [33]. Epsilon 5 is manufactured with optimized Cartesian-triaxial geometry and an extended K line excitation. A Ge X-ray detector is used to detect the characteristic X-rays emitted from the sample. The XRF system that was used is a secondary target system (ST-XRF). The system is equipped with nine secondary targets, and the operating conditions can be set for each target to achieve optimum analysis and performance. The total analysis time per sample was about 1.5 hour, and the operating conditions of the instrument are described in detail elsewhere [33].

Out of the 35 elements determined by the ED-XRF method, the ones used in the source apportionment analysis were Na, Mg, Al, Si, S, Cl, K, Ca, Ti, V, Cr, Mn, Fe, Co, Ni, Cu, Zn, Br, Sr, and Pb. For an element to be included in the source apportionment analysis, a threshold of at least 50% of points higher than the detection limit (DL) was set. The DLs were calculated following the approach described in [34]. The DLs were in the range of 10 to 24 ng m^{-3}, and the uncertainties in the range of 1 to 16%. Since the uncertainties are a very important input in the PMF analysis, they were realistically estimated taking into account the individual uncertainties of every step of the analytical and sampling process [34].

2.3. Black Carbon Analysis

Black carbon (BC) or light absorbing carbon (LAC) is one of the most significant constituents of PM. A precise and accurate determination of both the concentration and source contribution of BC in air particulate samples can provide key information for both environmental management regulators and researchers. To measure (LAC) on the samples the Multi-wavelength absorption black carbon instrument (MABI) was used. The MABI (Figure 2), which is developed at ANSTO, measures the (Io) (measured light transmission through blank/unexposed filter) and (I) (measured light transmission through filter after particle sampling) values for seven different wavelengths which are used to determine the mass absorption coefficient and black carbon concentrations at each of these wavelengths. These wavelengths are 405, 465, 525, 639, 870, 940, and 1050 nm. MABI unit does not automatically calculate (BC) values in concentration units ($\mu g\ m^{-3}$).

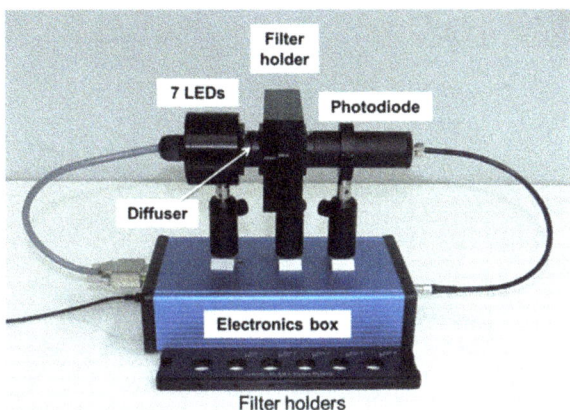

Figure 2. ANSTO Multi-wavelength absorption black carbon instrument (MABI) Unit.

The first step for the calculation of BC values in concentration units ($\mu g\ m^{-3}$) is the calculation b_{abs} (black carbon light absorption coefficient) using Equation (1). For that, it is necessary to know the exposed area of the filters (A) and sampled air volume (V).

$$b_{abs} = 100 \frac{A}{V} \ln\left[\frac{Io}{I}\right] = BC * \varepsilon / 1000 \tag{1}$$

ε = Mass absorption coefficient in $m^2\ g^{-1}$
A = Filter collection area in cm^2
V = Volume of air sampled through the filter in m^3
I_o, R_o = Measured light transmission and reflection through blank (unexposed) filter
I, R = Measured light transmission and reflection through filter after particle sampling.

The calculation of (BC) values in concentration units ($\mu g\ m^{-3}$) is done by using Equation (3). The accuracy of these equations relies on the mass absorption coefficient (ε), which is a strong function of the size and density of the light absorbing particles. As black carbon can be emitted from a range of different sources with a range of different densities and sizes, ε is wavelength and density dependent. Ignoring this can result in an inaccurate estimation of black carbon concentrations depending on the dominant sources present. In the absence of any detailed information on aerosol state of mixing we use the default value 5.39 $m^2\ g^{-1}$ at 870 nm for ε as provided by the manufacturer [34]. However, some indication on certain sources of PM is derived from the Angstrom exponent calculated from the MABI measurements on the 7 wavelengths. An indication for the Angstrom exponent is given when we plot the mass absorption coefficient for each wavelength (Figure 3). The value of −1267 indicated that part of the BC in our measurement sites originates from biomass burning [35].

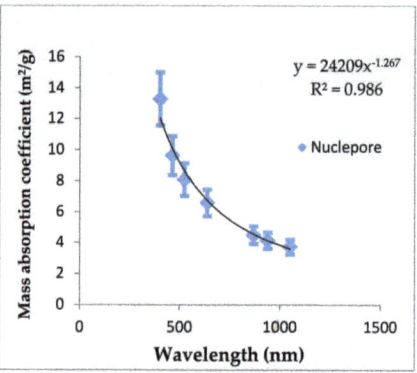

Figure 3. Mass absorption coefficient vs wavelength for nucleopore filters.

2.4. PMF

The basic mass balance equation that is used by PMF can be written as

$$X = G \times F + E \tag{2}$$

where X is the PM chemical composition matrix, G is the source contribution matrix, F the factor profile matrix, and E the residuals [36]. In PMF, G and F are constrained not to be negative (more accurately, only very slightly negative values are allowed). These constraints help the model achieve environmentally reasonable solutions, as it is impossible to have sources with negative contributions to PM mass.

The approach PMF follows to solve Equation (2), is to minimize the sum of the squares of the residuals scaled by the uncertainties of the data points. The task of PMF analysis can thus be described as to minimize Q, which is defined as:

$$Q = \sum_{j=1}^{m}\sum_{i=1}^{n} \frac{e_{ij}^2}{s_{ij}^2} \tag{3}$$

where s_{ij} is the uncertainty of the j_{th} species concentration in sample i, n is the number of samples [37].

As mentioned earlier, US-EPA PMF 5.0 version was used in this study. Twenty five variables were used as input in the model, and were namely PM, BC, Br, Rb, Sr, Ba, Pb, V, Cr, Mn, Fe, Co, Ni, Cu, Zn, Na, Mg, Al, Si, P, S, Cl, K, Ca, Ti. The variables with signal to noise (S/N) between 0.2 and 2 were defined as "Weak" in the model. PM mass was used as "total variable". Species with S/N < 0.2 are assigned as "Bad" variables [38] and were excluded from the analysis. The elements defined as "Weak" variables were Na, P, V, Cr, Co, Ni, Br, Zn, Rb and Ba, while Mg was defined as "Bad" variable.

When the decrease in Q becomes small with an increasing number of factors, it is an indication that too many factors are being included in the fit [39]. In this study, a large number of factors (3 to 10) were tested and 4 were found to yield the optimum results. A < 5% difference was estimated between the Q robust and true values, as well as between the Q true and theoretical values. The total number of base runs was set to 100. All runs provided very similar results indicated by the very low difference between the scaled residuals of the different runs. To evaluate the rotational ambiguity of the solution, the uncertainty estimation tools provided by EPA PMF 5.0 were utilized. BS (bootstrap) analysis showed that the factors were stable, and were successfully reproduced at a level of at least 85%. displacement (DIS) and bootstrap-displacement (BS-DIS) analysis did not show any factor swaps for the lowest relevant Q change level. The correlation between modeled and true PM mass was estimated to be high (R^2 = 0.75). All statistical indicators and uncertainty estimation tools show that the solution was well resolved, and the results had low uncertainty.

Because the number of species used for the source apportionment analysis represented a small fraction of PM mass, the contribution of the sources cannot be considered representative of the PM mass. Additionally, in such cases, the estimations have high relative uncertainty. For that reason, the source contributions are not reported.

2.5. Trajectory Analysis and Long-Range Transport

To assess the potential influence of long range transport events to PM concentrations, statistical trajectory methods (STMs) were used. The analysis was performed by using the OPENAIR software [40]. The HYSPLIT 4.0 model [41,42] developed by the NOAA (National Oceanic and Atmospheric Administration) was utilized to produce the back trajectory files. For the calculation of the trajectories, the GDAS meteorological database was used, and the calculations were done every 3 h for 120 h back, at a height level of 500 m above ground level (AGL). The STM that was used was the Potential Source Contribution Function (PSCF) [36,43]. The PSCF is estimated as:

$$PSCF = \frac{m_{ij}}{n_{ij}} \tag{4}$$

where n_{ij} is the number of times that the trajectories passed through the cell (i, j) and m_{ij} is the number of times that a source concentration at the receptor was higher than a certain threshold (90th percentile) when the trajectories passed through the cell (i, j).

2.6. Air Quality Index (AQI)

AQI is used to identify the poor air quality zones in urban or industrial zones. The AQI of each pollutant is calculated by Equation (5):

$$I_p = \frac{I_{Hi} - I_{Lo}}{BP_{Hi} - BP_{LO}} (C_P - BP_{LO}) + I_{LO} \tag{5}$$

where I_p is the index for pollutant p

C_P is the truncated concentration of pollutant p
BP_{Hi} is the concentration breakpoint that is greater than or equal to C_P
BP_{LO} is the concentration breakpoint that is less than or equal to C_P
I_{Hi} is the AQI value corresponding to BP_{Hi}
I_{Lo} is the AQI value corresponding to BP_{LO}

3. Results

3.1. Concentration Level of Particulate Matter

The concentration levels of the collected PM size fractions in the two study sites are presented in Table 1.

Table 1. Particulate matter (PM)$_{2.5-10}$ and PM$_{2.5}$ mass concentrations at Hlm and Yoff sites during the study period in µg m^{-3}.

Sites	Hlm		Yoff	
Particulate Matter	PM$_{2.5-10}$	PM$_{2.5}$	PM$_{2.5-10}$	PM$_{2.5}$
Mean	246.16	280.58	240.03	302.73
Median	207.80	259.55	203.81	290.13
Maximum	538.47	482.29	501.89	494.69
Minimum	123.55	184.80	22.24	148.27

The mean values of PM$_{2.5-10}$ concentrations at Hlm and Yoff are 246.16 and 240.03 µg m^{-3}, respectively. The average PM$_{2.5-10}$ concentration at the two sites is very similar, even though they have different characteristics (industrial and urban). This fact is an indication that anthropogenic emissions are not the dominant factor that affects PM mass in the area, which is most likely affected mainly by natural sources. The 24 hours limit value for the coarse particles in Senegal is 150 µg m^{-3} [44].

The average $PM_{2.5}$ concentration at Hlm and Yoff is 280.58 and 302.73 µg m^{-3}, respectively. $PM_{2.5}$ concentration is higher in Yoff than in Hlm, indicating that urban activities in Dakar have a higher contribution in $PM_{2.5}$ concentration levels than industrial activities. This might be related to the characteristics of the car fleet used in the city, which might lead to increased traffic-related emissions. Due to the extremely high $PM_{2.5}$ concentrations, it can be assumed that dust also contributes to this fraction. Although dust particles are mostly related to coarse fractions of PM, it can also contribute significantly to $PM_{2.5}$, as according to the size distributions, the lower tail of the coarse mode particles is found in $PM_{2.5}$.

3.2. Elemental Concentration

The mean values and standard deviations of the elemental concentrations of the $PM_{2.5}$ and $PM_{2.5-10}$ samples in the two sites are presented in Table 2.

Table 2. The average concentration and standard deviation (Stdev) of the elemental concentrations in sites during the studied period in µg m^{-3}.

	BC	Na	Mg	Al	Si	P	S	Cl	K	Ca	Ti	Pb
Mean	3.6	0.57	0.11	1.26	3.10	0.13	0.50	2.19	0.37	5.71	0.2	0.13
Stdev	0.47	0.18	0.03	0.21	0.51	0.03	0.05	0.36	0.03	0.61	0.02	0.03
	V	Cr	Mn	Fe	Co	Ni	Cu	Zn	Br	Rb	Sr	Ba
Mean	0.01	0.01	0.01	2.17	0.002	0.01	0.03	0.07	0.01	0.013	0.03	0.02
Stdev	0.001	0.001	0.004	0.233	0.001	0.001	0.003	0.009	0.002	0.003	0.003	0.004

The elements that present the highest concentrations are Ca, Si, Cl, Al, Fe, Na, S, and K. All these elements originate mainly from natural sources, with the exception of K that can be a soil component but can also be emitted from biomass burning. Ca, Si, Al, and Fe are well-known soil components [45]. Na and Cl usually originate from sea salt, especially in coastal areas. S is found in the particulate phase mainly in the form of SO_4^{2-}, which is formed by the oxidation of its precursor gas SO_2. The elemental concentrations also indicate the importance of natural sources in the area, and especially of soil resuspension source.

3.3. Source Apportionment

Four factors with physical meaning were obtained using the EPA PMF model. The identified sources were namely mineral dust, sea salt/secondary sulfates, traffic emissions, and industrial emissions. The chemical composition of the sources is represented by the four factors from the PMF analysis presented in Figure 4.

The first factor has high loadings of Al, Si, and K that clearly indicates mineral dust as the origin source. This source might be somewhat mixed with biomass burning emissions, as K is a tracer of that source as well [46]. The low concentration of BC in the factor indicates that biomass burning contribution to the factor is low.

The second factor contains high concentration of Cl and Br corresponding to sea salt source. The factor also contains a very high percentage of S, which indicates that a factor is mixed with secondary sulfates. The fact that the two sources are mixed might be attributed to synchronous transportation from the sea or coastal region to the sampling point.

Factor 3 is identified as traffic emissions, and it contains a high concentration of BC, Cu, Ni, and Pb (>60%). In this factor we found also Rb and Ba. According to previous studies, Ni, Cu, Zn, Br, and Pb are related to exhaust traffic emissions [47]. Cu originates from break abrasion and Zn from the combustion of lubricating oil and tire wear [48].

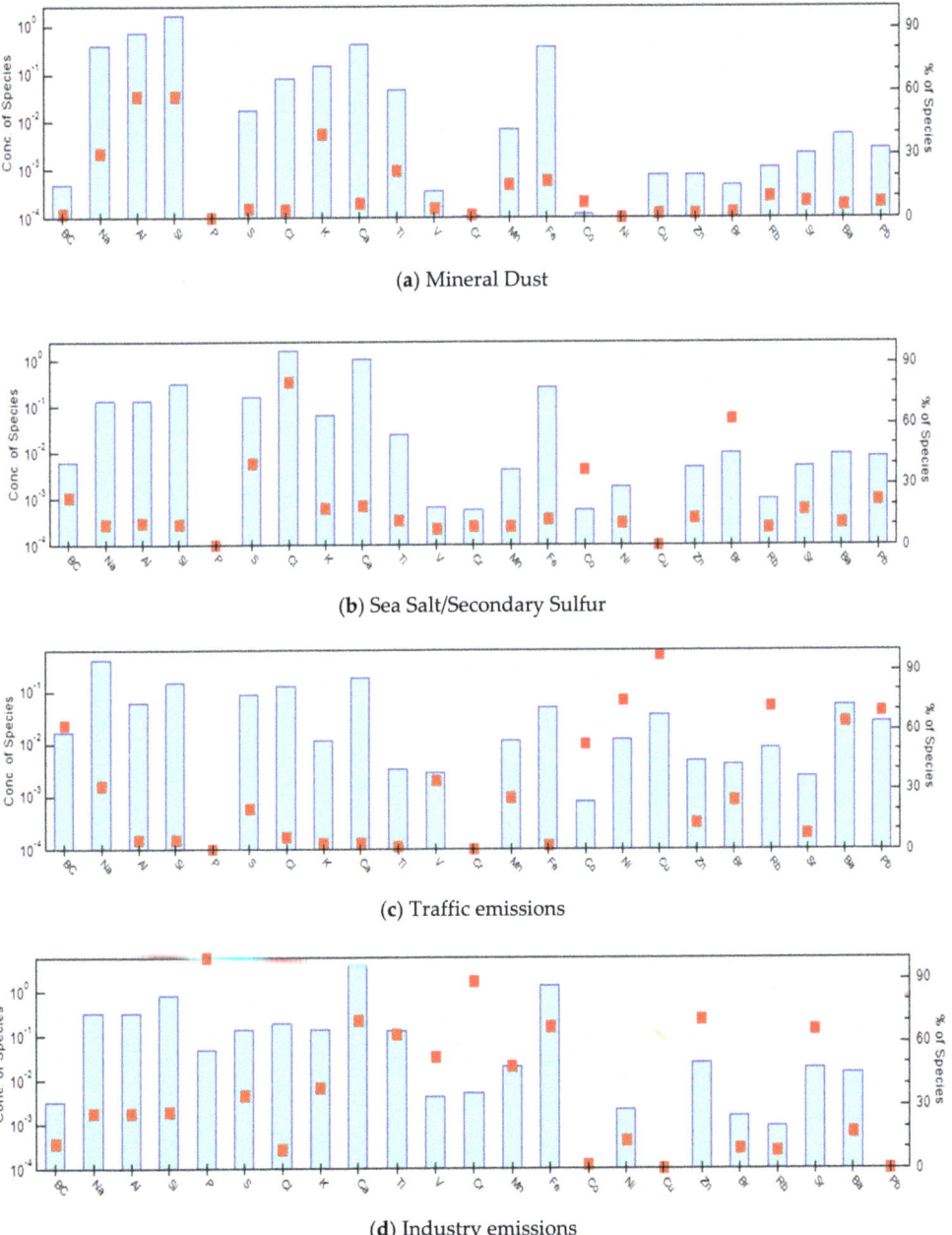

Figure 4. PM sources profiles. The bars represent the normalized concentration of the species in the profile, and the squares show the contribution of the source to the average species concentration with (**a**) Mineral Dust, (**b**) Sea Salt/Secondary Sulfur, (**c**) Traffic and (**d**) Industry emissions.

Factor 4 is characterized by a high concentration (>50%) of P, Cr, Zn, Fe, Ti, Ca, and Mn that clearly indicates heavy industry, primary refinery, and/or coal mines, as well as the use of lubricant oils [49,50]. Around Hlm site, we found metallurgy industry, food factory, chemical industrial, and gas company.

Probably because of the very high dust load in the region (as indicated by the high concentrations of earth components/elements, >12 µg m^{-3} as sum of the elements in pure form), the presence of crustal elements/tracers is apparent in every factor. It has been shown in previous publications that if PM of certain origin have really high concentrations in a region (>50 µg m^{-3}), their effect can be identified in other factors [51].

3.4. The Role of Long-Range Transport

The highest concentrations of coarse particulate matter (PM$_{2.5-10}$) appear to be related to long-range transport from the Sahara desert. The highest concentrations of fine particulate matter (PM$_{2.5}$) are transported from Mali. PSCF analysis (Figure 5) indicates source areas that affect the receptor during the days that the observed pollutant concentrations at the receptor are high. The fact that long-range transport affects the local PM concentrations, does not mean that the local contributions are low. PSCF analysis also indicates the strong influence of natural sources to PM levels in Dakar, and especially the influence of transported dust from the nearby arid regions. The fact that mineral dust particles affect the concentrations of PM in region, can also be confirmed by the high contributions of soil-related elements (Table 3).

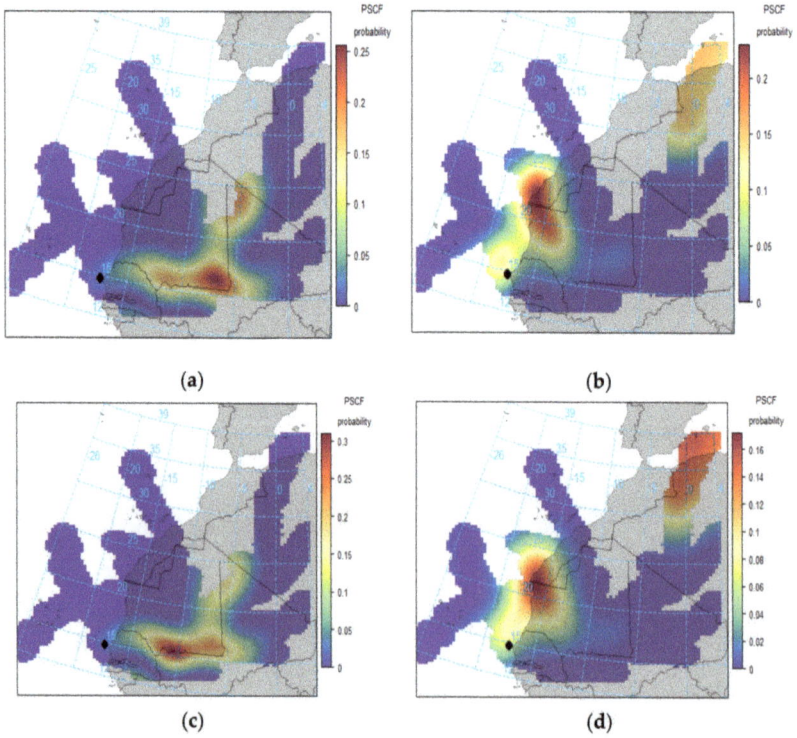

Figure 5. Potential Source Contribution Functions (PSCFs) for the 85th percentile of (a) PM$_{2.5}$ to Hlm, (b) PM$_{2.5-10}$ to Hlm, (c) PM$_{2.5}$ to Yoff and (d) PM$_{2.5-10}$ to Yoff.

3.5. Air Quality Index (AQI)

The breakpoint concentrations have been defined by the EPA on the basis of National Ambient Air Quality Standards (NAAQS) as shown in Table 3 [52].

Table 3. Breakpoint Concentration of air pollutants defined by U.S. EPA.

		These Breakpoints					Equal This AQI	This Category
O_3 (ppm) 8-h	O_3 (ppm) 1-h	$PM_{2.5}$ (µg m^{-3}) 24-h	$PM_{2.5-10}$ (µg m^{-3}) 24-h	CO (ppm) 8-h	SO_2 (ppb) 1-h	NO_2 (ppb) 1-h	AQI	
0.000–0.054	-	0.0–12.0	0–54	0.0–4.4	0–35	0–53	0–50	Good
0.055–0.070	-	12.1–35.4	55–154	4.5–9.4	36–75	54–100	51–100	Moderate
0.071–0.085	0.125–0.164	35.5–55.4	155–254	9.5–12.4	76–185	101–360	101–150	Unhealthy for Sensitive Groups
0.086–0.105	0.165–0.204	55.5–150.4	255–354	12.5–15.4	(186–304)	361–649	151–200	Unhealthy
0.106–0.200	0.205–0.404	150.5–250.4	355–424	15.5–30.4	(305–604)	650–1249	201–300	Very unhealthy
	0.405–0.504	250.5–350.4	425–504	30.5–40.4	(605–804)	1250–1649	301–400	Hazardous
	0.505–0.604	350.5–500.4	505–604	40.5–50.4	(805–1004)	1650–2049	401–500	Hazardous

The highest individual pollutant index, Ip, represents the (AQI) of the location. The pollution level and status will be highlighted according to the (AQI) number, as shown in Table 3. The situation in an area can be classified from good to hazardous [53]. The conditions for ideal air quality was defined by [28].

Figures 6 and 7 show the distribution of the air quality index (AQI) levels for the particulate matter $PM_{2.5-10}$ and $PM_{2.5}$ in Dakar. $PM_{2.5-10}$ and $PM_{2.5}$ represent in this study the "key pollutants". The four major dynamic variables to interpreting the (AQI) time series were energy consumption structure, pollutant emissions, weather and topography of the city [54].

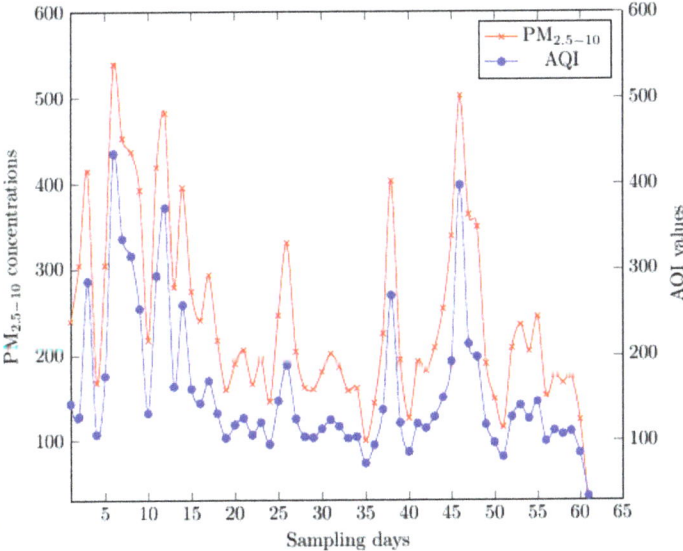

Figure 6. Evolution of $PM_{2.5-10}$ concentration and air quality index (AQI) values with time.

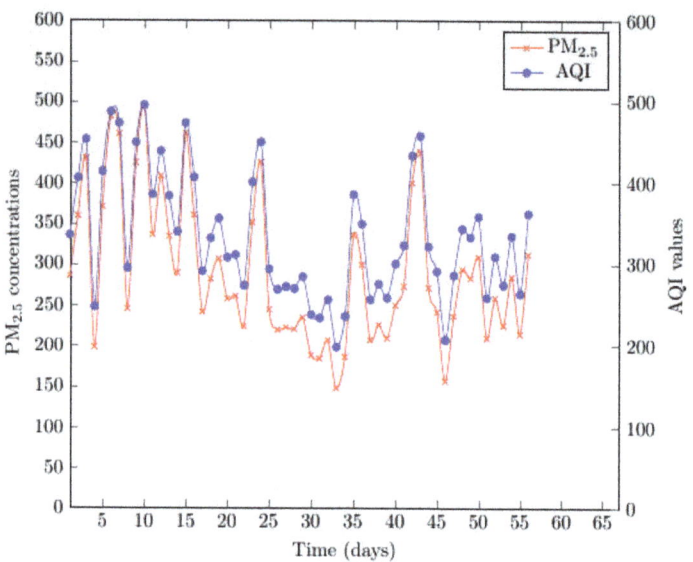

Figure 7. Evolution of PM$_{2.5}$ concentration and AQI values with time.

Figure 6 shows that (AQI) for the coarse particles ranges between 35 and 435. The figure includes the measurements conducted at both sites (Yoff and Hlm). The values of (AQI) are high in both areas. Furthermore Yoff site is close to the sea meaning that the air there is generally cleaner. The PM$_{2.5-10}$ concentration is proportional to values (AQI) of 1.55. According to the findings of the study, the highest (AQI) value recorded in Hlm is approximately 488. This value is very high, even for an industrial area. Regarding Yoff the highest value of (AQI) was found to be approximately 497. We can conclude that Hlm and Yoff cities are severely polluted in PM$_{2.5-10}$.

In Figure 7, the AQI for PM$_{2.5}$ is presented. The (AQI) ranged between 199 and 497, which classifies the situation in the city from very unhealthy to hazardous. The (AQI) values for PM$_{2.5}$ are more important than those of PM$_{2.5-10}$ because the potential toxicology impact PM$_{2.5}$ is higher. The highest values of (AQI) are approximately 497 and 488, in Yoff and Hlm, respectively. In contrast to what was found for the coarse particles, the (AQI) for PM$_{2.5}$ is higher in the urban zone.

4. Conclusions

This study took place in the capital city of Senegal, Dakar; a coastal area of 5 million inhabitants. The measurement campaign took place during the years 2018 and 2019. PM$_{2.5-10}$ and PM$_{2.5}$ samples were collected twice per week (one on a working day and one during weekend). The average concentration of PM$_{2.5-10}$ at Hlm and Yoff were found equal to 246.16 and 240.03 µg m^{-3}, respectively, while the average concentration of PM$_{2.5}$ was 280.58 and 302.73 µg m^{-3}, respectively. The concentration of the coarse particles was higher in Hlm, whereas the fine particles are more important in Yoff site. The elemental composition of the PM samples was determined using XRF. The elements that present the highest concentrations are Ca, Si, Cl, Al, Fe, Na, S, and K. All these elements originate mainly from natural sources, with the exception of K that can be a soil component but can also be emitted from biomass burning. According to PSCF, the high PM$_{2.5}$ and PM$_{2.5-10}$ concentration levels were related to long-range transportation events from Mali and Sahara desert, respectively. The sources of PM$_{2.5-10}$ and PM$_{2.5}$ were identified using EPA PMF 5.0. Four PM sources were identified: industrial emissions, mineral dust, traffic emissions, and sea salt/secondary sulfates. (AQI) for the coarse particles ranged between 35 and 435, and based on those findings, it can be said that Hlm and Yoff cities are severely polluted in $PM_{2.5-10}$. (AQI) for $PM_{2.5}$ ranged between 199 and 497, which classifies the situation in the city from very unhealthy to hazardous.

This study is the first study that investigates the sources of $PM_{2.5-10}$ and $PM_{2.5}$ in Dakar. The information summarized are important as it can provide information to the people and the government of Senegal about the particulate related pollution in the area. It must be notted here that the extremely high particulate levels found in Dakar are mainly due to mineral particles (desert dust and locally resuspended mineral particles). It is not easy to aseess the pollution levels based only on an AQI calculation that takes into account the concentrations of suspended particles, even though the concentrations of fine particles are indeed very high. The relatively limited available data concerning the chemical composition of PM (and the lack of data about gas concentrations) do not allow to conclude definitively about the seriousness of the air pollution in Dakar, since as stated before the estimates presented in this study are based mainly on the overall concentration, and not on PM chemical composition. Further studies will be necessary to arrive at a more realistic state of air quality in this city, that include the measurement of additional PM chemical components and better quantification of source contributions.

Author Contributions: Conceptualization, A.T. and M.I.M.; methodology, A.T. and M.K.; validation, M.K., A.T., and M.I.M.; formal analysis, M.K., M.I.M., and V.V.; writing—original draft preparation, A.T. and M.K.; writing—review and editing, A.T. and M.I.M.; supervision, A.T.; Administrative issue, A.S.N., A.W., and K.E. All authors have read and agreed to the published version of the manuscript.

Funding: This research was funded by the International Atomic Energy Agency in Vienna under the framework of the Regional technical project RAF7016.

Institutional Review Board Statement: Not applicable.

Informed Consent Statement: Not applicable.

Data Availability Statement: Not applicable.

Acknowledgments: We are very grateful to Roman Padilla Alvarez for his support during the entire duration of the project, the sampling protocol and advices.

Conflicts of Interest: The authors declare no conflict of interest.

References

1. Elichegaray, C. Département Air à l'Agence de l'environnement et de la maîtrise de l'énergie (ADEME). *Pollut. Atmosphérique* **2001**, 8.
2. Li, W.; Bai, Z.; Liu, A.; Chen, J.; Chen, l. Characteristics for major PM2.5 components during winter in Tianjin, China. *Aerosol Air Qual. Res.* **2009**, *9*, 105–119. [CrossRef]
3. Katsouyanni, K.; Touloumi, G.; Samoli, E.; Gryparis, A.; Le Tertre, A.; Monopolis, Y.; Rossi, G.; Zmirou, D.; Ballester, F.; Boumghar, A.; et al. Confounding and effect modification in the short-term effects of ambient particles on total mortality: Results from 29 European cities within the APHEA2 Project. *Epidemiology* **2001**, *12*, 521–531. [CrossRef] [PubMed]
4. Liu, Y.; Daum, P.H. Anthropogenic aerosols: Indirect warming effect from dispersion forcing. *Nature* **2002**, *419*, 580–581. [CrossRef]
5. Pandolfi, M.; Alastuey, A.; Pérez, N.; Reche, C.; Castro, I.; Shatalov, V.; Querol, X. Trends analysis of PM source contributions and chemical tracers in NE Spain during 2004–2014: A multi-exponential approach. *Atmospheric Chem. Phys.* **2016**, *16*, 11787–11805. [CrossRef]
6. Manousakas, M.; Popovicheva, O.; Evangeliou, N.; Diapouli, E.; Sitnikov, N.; Shonija, N.; Eleftheriadis, K. Aerosol carbonaceous, elemental and ionic composition variability and origin at the Siberian High Arctic, Cape Baranova. *Tellus B Chem. Phys. Meteorol.* **2020**, *72*, 1–14. [CrossRef]
7. Guaita, R.; Pichiule, M.; Mate, T.; Linares, C.; Diaz, J. Short-term impact of particulate matter (PM2.5) on respiratory mortality in Madrid. *Int. J. Environ. Health Res.* **2011**, *21*, 260–274. [CrossRef]
8. Halonen, J.I.; Lanki, T.; Tuomi, T.Y.; Tiittanen, P.; Kulmala, M.; Pekkanen, J. Particulate air pollution and acute cardiorespiratory hospital admissions and mortality among the elderly. *Epidemiology* **2009**, *20*, 143–153. [CrossRef]
9. Perez, L.; Tobías, A.; Querol, X.; Pey, J.; Alastuey, A. Saharan dust, particulate matter and cause-specific mortality: A case-crossover study in Barcelona (Spain). *Environ. Int.* **2012**, *48*, 150–155. [CrossRef]

10. Samoli, E.; Peng, R.; Ramsay, T.; Pipikou, M.; Touloumi, G.; Dominici, F.; Burnett, R.; Cohen, A.; Krewski, D.; Samet, J.; et al. Acute effects of ambient particulate matter on mortality in Europe and North America: Results from the APHENA Study. *Environ. Health Perspect.* **2008**, *116*, 1480–1486. [CrossRef]
11. Cesari, D.; Donateo, A.; Conte, M.; Contini, D. Inter-comparison of source apportionment of PM10 using PMF and CMB in three sites nearby an industrial area in central Italy. *Atmos. Res.* **2016**, *182*, 282–293. [CrossRef]
12. Freer-Smith, P.H.; Beckett, K.P.; Taylor, G. Deposition velocities to Sorbus aria, Acer campestre, Populus deltoides, Pinus nigra and Cupressocyparis leylandii for coarse, fine and ultra-fine particles in the urban environment. *Environ. Pollut.* **2005**, *133*, 157–167. [CrossRef] [PubMed]
13. Cheung, K.; Daher, N.; Kam, W.; Shafer, M.M.; Ning, Z.; Schauer, J.J.; Sioutas, C. Spatial and temporal variation of chemical composition and mass closure of ambient coarse particulate matter (PM10–2.5) in the Los Angeles area. *Atmos Environ.* **2011**, *45*, 2651–2662.
14. Psanis, C.; Triantafyllou, E.; Giamarelou, M.; Manousakas, M.; Eleftheriadis, K.; Biskos, G. Particulate matter pollution from aviation-related activity at a small airport of the Aegean Sea Insular Region. *Sci. Total Environ.* **2017**, *596–597*, 187–193. [CrossRef] [PubMed]
15. Dall'Osto, M.; Querol, X.; Amato, F.; Karanasiou, A.; Lucarelli, F.; Nava, S.; Calzolai, G.; Chiari, M. Hourly elemental concentrations in PM2.5 aerosols sampled simultaneously at urban background and road site during SAPUSS—Diurnal variations and PMF receptor modelling. *Atmos. Chem. Phys.* **2013**, *13*, 4375–4392. [CrossRef]
16. Shaka', H.; Saliba, N.A. Concentration measurements and chemical composition of PM10–2.5 and PM2.5 at a coastal site in Beirut, Lebanon. *Atmos. Environ.* **2004**, *38*, 523–531. [CrossRef]
17. Maenhaut, W.; Francois, F.; Cafmeyer, J. *The Gent Stacked Filter Unit (Sfuj Sampler for the Collection of Atmospheric Aerosols in Two Size Fractions: Description and Instructions for Installation and Use)*; Report N°. NAHRES-19; IAEA: Vienna, Austria, 1993; pp. 249–263.
18. Chow, J.C. Critical review: Measurement methods to determine compliance with ambient air quality standards for suspended particles. *J. Air Waste Manag. Assoc.* **1995**, *45*, 320–382. [CrossRef]
19. Wilson, W.E.; Chow, J.C.; Claiborn, C.; Fusheng, W.; Engelbrecht, J.; Watson, J.G. Monitoring of particulate matter outdoors. *Chemosphere* **2002**, *49*, 1009–1043. [CrossRef]
20. Van Griekan, R.; Markowicz, A.; Veny, P. Current trends in the literature on x-ray emission spectrometry. *X-ray Spectr.* **1991**, *20*, 271–276.
21. Dzubay, T.G.; Hines, L.E.; Stevens, R.K. Particle bounce errors in cascade impactors. *Atmos. Environ.* **1976**, *10*, 229–234. [CrossRef]
22. Viana, M.; Kuhlbusch, T.A.J.; Querol, X.; Alastuey, A.; Harrison, R.M.; Hopke, P.K.; Winiwarter, W.; Vallius, M.; Szidat, S.; Prévôt, A.S.H.; et al. Source apportionment of particulate matter in Europe: A review of methods and results. *J. Aerosol Sci.* **2008**, *39*, 827–849. [CrossRef]
23. Belis, C.A.; Pernigotti, D.; Pirovano, G.; Favez, O.; Jaffrezo, J.L.; Kuenen, J.; van Der Gon, H.G.; Reizer, M.; Riffault, V.; Alleman, L.Y.; et al. Evaluation of receptor and chemical transport models for PM10 source apportionment. *Atmos. Environ. X* **2020**, *5*, 100053. [CrossRef]
24. Manousakas, M.; Diapouli, E.; Belis, C.A.; Vasilatou, V.; Gini, M.; Lucarelli, F.; Querol, X.; Eleftheriadis, K. Quantitative assessment of the variability in chemical profiles from source apportionment analysis of PM10 and PM2.5 at different sites within a large metropolitan area. *Environ. Res.* **2021**, *192*, 110257. [CrossRef] [PubMed]
25. Paatero, P.; Tapper, U. Positive matrix factorization: A non-negative factor model with optimal utilization of error estimates of data values. *Environmetrics* **1994**, *5*, 111–126. [CrossRef]
26. Johnson, T.M.; Guttikunda, S.; Wells, G.J.; Artaxo, P.; Bond, T.C.; Russell, A.G.; Watson, J.G.; West, J. *Tools for Improving Air Quality Management, A Review of Top-Down Source Apportionment Techniques and Their Application in Developing Countries*; ESMAP: Washington, DC, USA, 2011; p. 220.
27. Amato, F.; Alastuey, A.; Karanasiou, A.; Lucarelli, F.; Nava, S.; Calzolai, G.; Severi, M.; Becagli, S.; Gianelle, V.L.; Colombi, C.; et al. AIRUSE-LIFEC: A harmonized PM speciation and source apportionment in five southern European cities. *Atmos. Chem. Phys.* **2016**, *16*, 3289–3309.
28. Kanchan, K.; Gorail, A.K.; Pramila, G. A Review on Air Quality Indexing System. *Asian J. Atmos. Environ.* **2015**, *9*, 101–113. [CrossRef]
29. Available online: http://www.denv.gouv.sn/index.php/air-et-climat/centre-de-gestion-de-la-qualite-de-l-air-cgqa/pollution (accessed on 7 August 2019).
30. Rivellini, L.H.; Chiapello, I.; Tison, E.; Fourmentin, M.; Féron, A.; Diallo, A.; N'Diaye, T.; Goloub, P.; Canonaco, F.; Prévôt, A.S.; et al. Chemical characterization and source apportionment of submicron aerosols measured in Senegal during the 2015 SHADOW campaign. *Atmos. Chem. Phys.* **2017**, *17*, 10291–10314. [CrossRef]
31. Ba, A.N.; Verdin, A.; Cazier, F.; Garcon, G.; Thomas, J.; Cabral, M.; Dewaele, D.; Genevray, P.; Garat, A.; Allorge, D.; et al. Individual exposure level following indoor and outdoor air pollution exposure in Dakar (Senegal). *Environ. Pollut.* **2019**, *248*, 397–407.
32. Hopke, P.K.; Xie, Y.; Raunemaa, T.; Biegalski, S.; Landsberger, S.; Maenhaut, W.; Artaxo, P.; Cohen, D. Characterization of the Gent Stacked Filter Unit PM10 Sampler. *Aerosol Sci. Technol.* **1997**, *27*, 726–735. [CrossRef]
33. Manousakas, M.; Diapouli, E.; Papaefthymiou, H.; Kantarelou, V.; Zarkadas, C.; Kalogridis, A.-C.; Karydas, A.G.; Eleftheriadis, K. XRF characterization and source apportionment of PM10 samples collected in a coastal city. *X-ray Spectr.* **2018**, *47*, 190–200.

34. Cohen, D.D. Summary of Light Absorbing Carbon and Visibility Measurements and Terms. In *ANSTO External Report ER-790*; ANSTO: Sydney, Australia, 2020; ISBN 1 921268 32 8.
35. Diapouli, E.; Kalogridis, A.; Markantonaki, C.; Vratolis, S.; Fetfatzis, P.; Colombi, C.; Eleftheriadis, K. Annual variability of black carbon concentrations originating from biomass and fossil fuel combustion for the suburban aerosol in Athens, Greece. *Atmosphere* **2017**, *8*, 234. [CrossRef]
36. Manousakas, M.I.; Florou, K.; Pandis, S.N. Source Apportionment of Fine Organic and Inorganic Atmospheric Aerosol in an Urban Background Area in Greece. *Atmosphere* **2020**, *11*, 330. [CrossRef]
37. Chueinta, W.; Hopke, P.K.; Paatero, P. Investigation of sources of atmospheric aerosol at urban and suburban residential areas in Thailand by positive matrix factorization. *Atmos. Environ.* **2000**, *34*, 3319–3329. [CrossRef]
38. Paatero, P.; Hopke, P.K. Discarding or down weighting high-noise variables in factor analytic models. *Anal. Chim. Acta.* **2003**, *490*, 277–289. [CrossRef]
39. Xie, Y.L.; Hopke, P.K.; Paatero, P.; Barrie, L.A.; Li, S.-M. Identification of source nature and seasonal variations of Arctic aerosol by positive matrix factorization. *J. Atmos. Sci.* **1999**, *56*, 249–260. [CrossRef]
40. Carslaw, D.C.; Ropkins, K. Openair—An R package for air quality data analysis. *Environ. Model. Softw.* **2012**, *27*, 52–61. [CrossRef]
41. Stein, A.F.; Draxler, R.R.; Rolph, G.D.; Stunder, B.J.B.; Cohen, M.D.; Ngan, F. Noaa's hysplit atmospheric transport and dispersion modeling system. *Bull. Am. Meteorol. Soc.* **2015**, *96*, 2059–2077. [CrossRef]
42. Rolph, G.; Stein, A.; Stunder, B. Real-time environmental applications and display system: READY. *Environ. Model. Softw.* **2017**, *95*, 210–228. [CrossRef]
43. Stohl, A. Trajectory statistics—A new method to establish source-receptor relationships of air pollutants and its application to the transport of particulate sulfate in Europe. *Atmos. Environ.* **1996**, *30*, 579–587. [CrossRef]
44. Senegalese Norm. 2018; ISO NS 05-062.
45. Vasilatou, V.; Manousakas, M.; Gini, M.; Diapouli, E.; Scoullos, M.; Eleftheriadis, K. Long Term Flux of Saharan Dust to the Aegean Sea around the Attica Region, Greece. *Front. Mar. Sci.* **2017**, *4*, 42. [CrossRef]
46. Diapouli, E.; Popovicheva, O.; Kistler, M.; Vratolis, S.; Persiantseva, N.; Timofeev, M.; Kasper-Giebl, A.; Eleftheriadis, K. Physicochemical characterization of aged biomass burning aerosol after long-range transport to Greece from large scale wildfires in Russia and surrounding regions, Summer 2010. *Atmos. Environ.* **2014**, *96*, 393–404. [CrossRef]
47. Zhang, N.; Cao, J.; Ho, K.; He, Y. Chemical characterization of aerosol collected at Mt. Yulong in wintertime on the southeastern Tibetan Plateau. *Atmos. Res.* **2012**, *107*, 76–85. [CrossRef]
48. Sternbeck, J.; Sjodin, A.; Andreasson, K. Metal emissions from road traffic and the influence of resuspension—results from two tunnel studies. *Atmos. Environ.* **2002**, *36*, 4735. [CrossRef]
49. Manousakas, M.; Papaefthymiou, H.; Diapouli, E.; Migliori, A.; Karydas, A.G.; Bogdanovic-Radovic, I.; Eleftheriadis, K. Assessment of PM2.5 sources and their corresponding level of uncertainty in a coastal urban area using EPA PMF 5.0 enhanced diagnostics. *Sci. Total Environ.* **2017**, *574*, 155–164. [CrossRef]
50. Pateraki, S.; Manousakas, M.; Bairachtari, K.; Kantarelou, V.; Eleftheriadis, K.; Vasilakos, C.; Assimakopoulos, V.D.; Maggos, T. The traffic signature on the vertical PM profile: Environmental and health risks within an urban roadside environment. *Sci. Total Environ.* **2019**, *646*, 448–459. [CrossRef]
51. Gunchin, G.; Manousakas, M.; Osan, J.; Karydas, A.G.; Eleftheriadis, K.; Lodoysamba, S.; Shagjjamba, D.; Migliori, A.; Padilla-Alvarez, R.; Streli, C.; et al. Three-year Long Source Apportionment Study of Airborne Particles in Ulaanbaatar Using X-ray Fluorescence and Positive Matrix Factorization. *Aerosol Air Qual. Res.* **2019**, *19*, 1056–1067. [CrossRef]
52. Mintz, D. *Technical Assistance Document for the Reporting of Daily Air Quality—The Air Quality Index (AQI)*; US-EPA: Washington, DC, USA, 2018; 454/B-18-007.
53. Wang, L.; Zhang, P.; Tan, S.; Zhao, X.; Cheng, D.; Wei, W.; Su, J.; Pan, X. Assessment of urban air quality in China using air pollution indices (APIs). *J. Air Waste Manag.* **2013**, *63*, 170–178. [CrossRef]
54. Yu, B.; Huang, C.M.; Liu, Z.H.; Wang, H.P.; Wang, L.L. A chaotic analysis on air pollution index change over past 10 years in Lanzhou, northwest China. *Stochastic Environ. Res.* **2011**, *25*, 643–653. [CrossRef]

Article

First-Time Source Apportionment Analysis of Deposited Particulate Matter from a Moss Biomonitoring Study in Northern Greece

Chrysoula Betsou [1], Evangelia Diapouli [2,*], Evdoxia Tsakiri [3], Lambrini Papadopoulou [4], Marina Frontasyeva [5], Konstantinos Eleftheriadis [2] and Alexandra Ioannidou [1]

1. Nuclear Physics Laboratory, Department of Physics, Aristotle University of Thessaloniki, 54 124 Thessaloniki, Greece; chbetsou@physics.auth.gr (C.B.); anta@physics.auth.gr (A.I.)
2. Institute of Nuclear & Radiological Sciences & Technology, Energy & Safety, NCSR "Demokritos", 15 341 Athens, Greece; elefther@ipta.demokritos.gr
3. Laboratory of Systematic Botany and Phytogeography, Department of Botany, School of Biology, Aristotle University of Thessaloniki, 54 124 Thessaloniki, Greece; tsakiri@bio.auth.gr
4. Division of Mineralogy-Petrology-Economic Geology, Geology Department, Aristotle University of Thessaloniki, 54 124 Thessaloniki, Greece; lambrini@geo.auth.gr
5. Frank Laboratory of Neutron Physics, Joint Institute for Nuclear Research, Dubna, 141980 Moscow, Russia; marina@nf.jinr.ru
* Correspondence: ldiapouli@ipta.demokritos.gr

Citation: Betsou, C.; Diapouli, E.; Tsakiri, E.; Papadopoulou, L.; Frontasyeva, M.; Eleftheriadis, K.; Ioannidou, A. First-Time Source Apportionment Analysis of Deposited Particulate Matter from a Moss Biomonitoring Study in Northern Greece. *Atmosphere* **2021**, *12*, 208. https://doi.org/10.3390/atmos12020208

Academic Editor: Antoaneta Ene
Received: 30 December 2020
Accepted: 30 January 2021
Published: 4 February 2021

Publisher's Note: MDPI stays neutral with regard to jurisdictional claims in published maps and institutional affiliations.

Copyright: © 2021 by the authors. Licensee MDPI, Basel, Switzerland. This article is an open access article distributed under the terms and conditions of the Creative Commons Attribution (CC BY) license (https://creativecommons.org/licenses/by/4.0/).

Abstract: Moss biomonitoring is a widely used technique for monitoring the accumulation of trace elements in airborne pollution. A total of one hundred and five samples, mainly of the *Hypnum cupressiforme* Hedw. moss species, were collected from the Northern Greece during the 2015/2016 European ICP Vegetation (International Cooperative Program on Effects of Air Pollution on Natural Vegetation and Crops) moss survey, which also included samples from the metaliferous area of Skouries. They were analyzed by means of neutron activation analysis, and the elemental concentrations were determined. A positive matrix factorization (PMF) model was applied to the results obtained for source apportionment. According to the PMF model, five sources were identified: soil dust, aged sea salt, road dust, lignite power plants, and a Mn-rich source. The soil dust source contributed the most to almost all samples (46% of elemental concentrations, on average). Two areas with significant impact from anthropogenic activities were identified. In West Macedonia, the emissions from a lignite power plant complex located in the area have caused high concentrations of Ni, V, Cr, and Co. The second most impacted area was Skouries, where mining activities and vehicular traffic (probably related to the mining operations) led to high concentrations of Mn, Ni, V, Co, Sb, and Cr.

Keywords: positive matrix factorization model; trace elements; moss; biomonitoring; neutron activation analysis

1. Introduction

Trace elements are dispersed widely in all ecosystems. They interact with different natural components of the environment and they have a significant impact on the biosphere [1–3]. The majority of trace elements are deposited on the terrestrial environment from the atmosphere through either wet or dry deposition. The atmospheric deposition of trace elements is a matter of great concern worldwide, as, depending on their concentrations, trace elements may be hazardous both for humans and the environment [4–7].

In order to determine the atmospheric deposition of trace elements, complicated and expensive classical analytical methods are usually required. However, there is another reliable technique that uses living organisms (grass, leaves, pine needles, mosses, and lichens) in order to study the environmental quality and the effects of trace elements on the biosphere [8–13]. The biomonitoring technique is a rapid, effective, and low-cost technique that can provide data for the atmospheric deposition of trace elements in even remote areas. Over the last few decades, among the different living organisms, mosses are the bioaccumulators that have been widely used for this purpose, as they display very unique characteristics, both morphologically and physiologically [6,14,15].

Mosses were first used as biomonitors of the atmospheric fallout in the Scandinavian countries [16–19]. They have no rooting system; all the necessary elements and water are obtained directly through wet and dry deposition, while some trace amounts of nutrients may also come from the soil [11,20–25]. They grow up slowly and they can provide information about the atmospheric fallout over a long time of exposure. Each annual accumulation of trace elements can be distinguished with the use of the annual growth segments of mosses [26]. Mosses are suitable for monitoring elemental concentrations on a large geographic scale [6,8,23,27,28] while also achieving high-sampling-density networks [3,15,29].

The moss biomonitoring technique is used as part of a European monitoring program for air pollutants that has been performed since the 1990s on a regular basis (every five years) [30]. The "European moss survey" is performed under the auspices of the ICP Vegetation (International Cooperative Program on Effects of Air Pollution on Natural Vegetation and Crops), aiming at the atmospheric deposition of trace elements on mosses across Europe [13,31]. Greece contributed for the first time to such an extent to the European moss survey by providing data on the elemental concentrations in moss samples collected in all the northern part of the country.

Different statistical tools are usually applied in moss studies for the identification of the emission sources of the trace elements, such as the factor analysis (FA) and principle component analysis (PCA) methods [15,32]. In the present study, a more robust source apportionment technique, positive matrix factorization (PMF), was applied for the first time in mosses in the Greek territory. PMF displays enhanced features with respect to the FA and PCA methods, namely the use of the experimental uncertainty in order to weigh the data values and non-negativity constraints for the obtained factors [33]. It is one of the most popular source apportionment techniques for atmospheric aerosol and is extensively used in the framework of air quality management [34–36]. PMF is a receptor modeling tool that attributes the observed elemental concentrations to their major emission sources without "a priori" information on source chemical profiles [36–38]. To our knowledge, this source apportionment technique has been applied for the identification of natural and anthropogenic sources of trace elements in mosses in only very few recent research works [39,40].

The overall objective of this research was to study the bioaccumulation of trace elements in mosses of Northern Greece, to identify their sources by applying the positive matrix factorization model, and to establish a Greek database for future studies.

2. Materials and Methods

2.1. Study Area and Sampling

The "Moss survey" sampling protocol followed the instructions of the ICP Vegetation Monitoring manual; thus, only samples of natural growing pleurocarp mosses were collected. In drier areas—as Greece is—the taxa *Hypnum cupressiforme* Hedw., *Pseudoscleropodium purum* (Hedw.) M. Fleisch, *Camptothecium lutescens* (Hedw.) Schimp., and *Homalothecium sericeum* (Hedw.) Schimp. were among the recommended taxa to be collected [23,30,31]. All samples of fresh moss material were collected from open areas, at least 300 m from main roads and 200 m from populated areas, and from different substrate types (rocks, surface soil, branches, and near roots). It was recommended to make one composite sample consisting of five sub-samples from each sampling point, if possible, and to collect them within an area of about 50 × 50 m. Sub-samples were placed side by side in large plastic bags, which were tightly closed to prevent contamination during transportation.

For each sample, the amount of fresh moss collected for elemental analysis was about one liter (ICP Vegetation Protocol) [32].

The moss sampling took place in the area of Northern Greece during the end of the summer of 2016. There was no rain during the sample collection and no contact with surface water. Any extraneous materials that might have been found on the moss surface were removed manually from the samples. Mosses were not washed, and after sampling, they were frozen until further treatment according to the ICP Vegetation Protocol [31].

In total, 105 moss samples were collected, with the sampling sites' altitudes ranging between 30 and 1450 m above sea level. More specifically, 95 moss samples were collected from 39.97° North to 41.65° North and from 20.97° East to 26.26° East, in a grid of 30 × 30 km (Figure 1), while the remaining 10 moss samples were collected from the area of Skouries (40°28′ N, 23°41′ E), close to active mining facilities. The area of Skouries is located almost in the middle of Northern Greece, in the Northeastern Chalkidiki peninsula (Central Macedonia Region, Chalkidiki Prefecture), and has had a long mining history since ancient times (6th century B.C.). In recent years, many studies have been performed on the environmental pollution of the area due to the mining activities [41–46]. The present study is the first one to use moss samples as monitors of trace elements originating from the mining activities in the Skouries area.

Among the moss taxa recommended for collection, in Northern Greece, the most commonly found taxon was the *Hypnum cupressiforme* Hedw., and it was collected in the majority of the sampling sites (in 92 sites), with the other moss samples being *Pseudoscleropodium purum* (Hedw.) M. Fleisch, *Camptothecium lutescens* (Hedw.) Schimp., and *Scleropodium touretii* (Brid.) L.F. Koch. From the Skouries area, the ten moss samples collected belonged to taxa found in great abundance in the specific area: *Hypnum cupressiforme* Hedw., *Homalothecium sericeum* (Hedw.) Schimp., *Isothecium alopecuroides* (Lam. ex Dubois) Isov., *Antitrichia curtipendula* (Hedw.) Brid., *Pterogoniumgracile* (Hedw.) Sm., and *Dicranum scoparium* Hedw. [47].

It must be noted that, according to the ICP Vegetation Protocol, when bryophyte species other than the main preferred ones (i.e., *Hylocomium splendens* (Hedw.) Schimp. or *Pleurozium schreberi* (Willd. ex Brid.) Mitt., which are species more commonly found in the northern European countries but are not so common in drier areas) are used for a survey, a comparison and calibration of their uptake of heavy metals relative to the main preferred species must be performed [31].

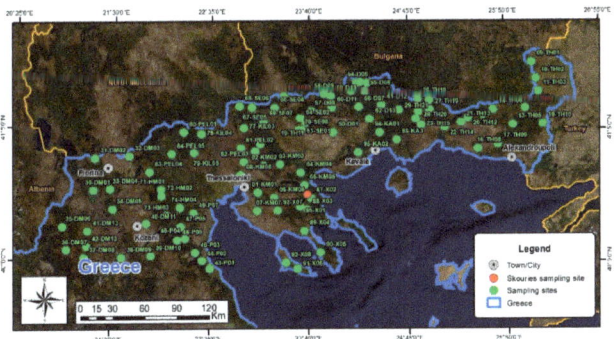

Figure 1. The moss sampling sites in Northern Greece.

2.2. Sample Preparation and NAA Analysis

After sampling, mosses were prepared and analyzed through neutron activation analysis (NAA) at the IBR-2 reactor of the Frank Laboratory of Neutron Physics, Joint Institute for Nuclear Research (JINR), in Dubna, Russia.

More specifically, samples were air-dried to a constant weight for 3 h at 40 °C, and they were then homogenized and pelletized using a press form. Each sample was packed into plastic bags for short-term irradiation, and then into aluminum foil cups for long-term irradiation. Through a pneumatic system, REGATA, the packed samples were irradiated in two different irradiation channels of the IBR-2 reactor for short- and long-term irradiation. For the determination of the short-lived isotopes (Al, Na, Mg, Ti, Cl, Mn, V, Ag, Ni), samples were irradiated for 180 s, while for the long-lived isotopes (Ca, K, Sc, Br, Rb, Sr, Cr, Fe, Co, Zn, Th, U, As, Ce, Cs, Mo, Zr, Sb, Ba, La, Sm, Gd, Tb, Hf, Au), they were irradiated for 4–5 days. In accordance with the quality assurance procedures of the Moss Monitoring Manual of the ICP Vegetation Protocol [8,31], simultaneously with the moss samples, certified reference materials (NIST SRM 1547—peach leaves, NIST SRM 1575a—pine needles, NIST SRM 1633b—coal fly ash, NIST SRM 1632c—coal (bituminous), NIST SRM 2709—San Joaquin soil, and IRMM SRM-667—estuarine sediment), which were provided by the National Institute of Standards and Technology (NIST) and the Institute for Reference Materials and Measurements (ERM), were also irradiated in the same channels of the reactor and under the same conditions. The standard reference materials (SRMs) varied between 1% and 14%, with the exception of Mn, Ti, Mg, Br, and Hf, for which the relative differences were 20%. There was no use of correction factors in calculating the final elemental content for the moss species approved by the program, and no additional variability was added to the data.

After irradiation, samples were measured by High Purity Germanium (HPGe) detectors of 40% relative efficiency and 1.74 keV FWHM (full-width at half-maximum) at the 1332 keV line of ^{60}Co. More specifically, for the determination of the short-lived isotopes, the samples that were irradiated for a short time period were measured with gamma spectrometry for 15 min immediately after their irradiation. Concerning the long-lived isotopes, samples were measured with gamma spectrometry once for 5 h after 3 days of their irradiation, and once more for 20 h at 20 days after their irradiation. The GENIE-2000 software was used for the analysis of the gamma spectrum. Based on the certified values of the reference materials and by using special software that was developed in the JINR [48], the concentrations of 34 elements in mosses were defined.

2.3. Positive Matrix Factorization (PMF) Analysis

The measured elemental concentrations were used for the application of receptor modeling for the identification of the sources of trace elements. More specifically, the positive matrix factorization (PMF) source apportionment method was applied by means of the EPA PMF 5.0 model [37].

The PMF model is a mathematical model based on the mass conservation between an emission source and a study site [49]. It apportions the concentration of each chemical element in each sample to the different sources impacting this element using the following mass balance equation

$$c_{ij} = \sum_{k=1}^{p} g_{ik} \times f_{kj} + e_{ij} \tag{1}$$

where

c_{ij} is the concentration of a chemical element j measured in sample i,
p is the number of factors (sources) that contribute to the measured concentrations,
g_{ik} is the contribution of the source k to the sample i,
f_{kj} is the concentration of the chemical element j in the source k, and
e_{ij} is the residual (the difference between the measured value and the value fitted by the model) for chemical element j in sample i.

The model solves Equation (1) by minimizing the following object function:

$$Q = \sum_{j=1}^{m}\sum_{i=1}^{n} \frac{e_{ij}^2}{s_{ij}^2} = \sum_{j=1}^{m}\sum_{i=1}^{n} \frac{\left(c_{ij} - \sum_{k=1}^{p} g_{ik} \times f_{kj}\right)^2}{s_{ij}^2} \tag{2}$$

where s_{ij} is the uncertainty of the measured concentration of chemical element j in sample i, n is the number of samples, and m is the number of chemical elements [50].

Thus, as shown by Equation (2), the PMF solution is a weighted least squares fit, where the known standard deviations of the input concentration values (s_{ij}) are used for determining the weights of the residuals, e_{ij}. In addition, the model implements non-negative constraints [51,52]. The final output consists of two matrices: Matrix G, including the contributions of each source to each sample (g_{ik}), and Matrix F, including the concentration of each chemical element in each source (f_{kj}). This output depends on the number of factors/sources (p), which should be defined by the user.

The input database for the source apportionment analysis included 30 chemical elements (Na, Mg, Al, Si, Cl, K, Ca, Cd, Ti, V, Cr, Mn, Fe, Co, Ni, As, Br, Rb, Sr, Mo, Sc, Sb, Cs, Ba, La, Ce, Tb, Hf, Ta, and Th) that were detected in the 105 samples. The uncertainties of the concentrations of all chemical elements were calculated based on the sampling and analytical uncertainties. Elements with high uncertainty values (low signal-to-noise (S/N) ratio) were excluded from the analysis. In addition, four chemical elements (Mg, Al, Cd, and Th) were set as "weak" chemical elements because, based on the initial model runs, they produced high residual values. In the case of "weak" elements, the model triples all uncertainties associated with these chemical elements [53,54]. Additionally, an extra 10% modeling uncertainty was added in the model to account for modeling errors.

Solutions with different numbers of sources (ranging from 4 to 10) were examined [36]. Based on different criteria, such as the distributions of the residuals, the G-space plots, and the Q values, the best solution of the model was extracted, including five sources with a physical meaning. The final solution was based on 100 runs, while the rotational ambiguity of the solution was also assessed through the use of different Fpeak values (in the range −1.0 to 1.5). The best solution was found for Fpeak = −0.5.

3. Results

3.1. Elemental Concentrations in Mosses

The descriptive analyses (max, min, mean, median, and standard deviation) of the elemental concentrations of the moss samples collected from the whole region of Northern Greece are presented in Table 1, while the concentrations of mosses collected specifically from the area of Skouries are presented in Table 2. This distinction was performed in order to separately study the case of Skouries and to understand the impact that the mining activities have on the area. The results of the current study are in agreement with the data obtained from a previous moss biomonitoring study held in Northeastern Greece, close to the Bulgarian boarders [55]. A more detailed discussion of the measured elemental concentrations may be found in [56].

3.2. Source Apportionment of Elemental Concentrations

According to the PMF model results and all the performance criteria described above, the best solution that was extracted included the following five sources: soil dust, aged sea salt, road dust, lignite power plants, and a Mn-rich source. The chemical profiles of all the identified sources and the percentage of each chemical element assigned to each source are presented in Figure 2.

The soil dust source was identified by the high contribution of crustal elements, such as Mg, Al, Ti, Si, Ca, Th, Ba, La, K, and Fe [57]. More than 40% of the mass of key crustal elements (such as Si, Ca, K, and Mg) was assigned to this factor by the model. This source may represent the local soil dust resuspension or soil dust transported from greater distances, such as during the long-range transport of African dust [34]. It should be mentioned that the dry climate of Greece favors soil resuspension, even over long distances [58].

Table 1. The elemental concentrations (in µg g^{-1}) in 95 moss samples collected in Northern Greece and determined by means of neutron activation analysis (NAA).

Element	Na	Mg	Al	Cl	K	Ca	Sc	Si	Ti	V	Cr	Mn	Fe	Co	Ni	Zn	As	Ce	Br	Rb	Sr	Zr	Mo	Ag	Sb	Cs	Ba	La	Tb	Hf	Ta	Th	U	Au
mean	1730	4432	7886	145	6360	8900	2.12	78,458	440	10.08	24.05	269	5974	3.02	12.83	56.56	2.45	11.07	6.48	21.59	52.13	35.18	0.28	0.061	0.27	0.88	104.7	5.42	0.13	0.9	0.20	2.07	0.55	0.00123
StDev	2266	2813	6812	68	2956	3236	1.93	64,318	327	6.12	35.58	200	5182	3.25	12.77	52.74	2.99	7.86	2.95	17.18	38.14	38.91	0.25	0.025	0.39	0.79	94.2	5.40	0.12	0.9	0.21	2.44	0.60	0.00216
median	751	3600	5840	130	5670	8170	1.44	62,200	327	8.17	11.50	219	3770	1.69	7.26	37.60	1.44	9.11	5.85	15.50	38.20	21.50	0.23	0.055	0.20	0.68	65.9	3.22	0.08	0.6	0.12	0.99	0.30	0.00073
min	184	705	1350	47	2160	3940	0.29	11,500	97	2.61	2.04	34	1010	0.43	1.72	14.60	0.52	1.87	1.69	5.11	12.70	3.29	0.02	0.018	0.02	0.17	15.9	0.50	0.02	0.1	0.03	0.28	0.07	0.00004
max	9210	17,300	46,100	380	17,200	23,400	8.92	340,000	1760	33.40	222	1090	28,700	20.3	90.20	282.0	17.90	46.10	15.00	82.9	197.0	219.0	2.31	0.16	3.23	5.08	519	35.2	0.57	4.7	1.09	13.6	3.38	0.0072

Table 2. The concentrations of trace elements (in µg g^{-1}) in 10 moss samples collected only from the area of Skouries and determined by means of neutron activation analysis (NAA).

Element	Na	Mg	Al	Cl	K	Ca	Sc	Si	Ti	V	Cr	Mn	Fe	Co	Ni	Zn	As	Ce	Br	Rb	Sr	Zr	Mo	Ag	Sb	Cs	Ba	La	Tb	Hf	Ta	Th	U	Au
mean	1079	10,094	11,595	599	7078	10,088	2.73	65,120	670	21	118	652	9866	6.8	86	48	11.4	11.14	18	26	57	28.88	0.22	0.09	0.69	1.01	151	5.53	0.13	0.91	0.14	1.91	0.41	0.0097
StDev	401	4353	3449	549	2827	3064	0.94	37,186	268	6	46	408	3256	2.4	30	12	3.1	3.65	16	5	24	11.70	0.08	0.03	0.24	0.18	87	1.80	0.04	0.38	0.06	0.59	0.16	0.0083
median	901	11,350	12,050	353	6195	9270	2.48	56,050	636	20	102	538	8955	6.4	77	47	10.7	10.34	14	26	49	25.05	0.24	0.09	0.65	1.04	123	5.19	0.13	0.82	0.14	1.76	0.40	0.0069
min	668	1880	7170	271	4910	6090	1.74	40,400	352	14	66	338	6350	4.5	52	35	7.5	7.39	5	19	35	16.60	0.05	0.06	0.32	0.74	75	3.65	0.08	0.54	0.08	1.27	0.16	0.0043
max	1710	16,400	17,300	2070	14,800	16,800	4.12	168,000	1230	30	200	1760	14,300	10.6	138	72	15.4	17.50	59	34	104	50.70	0.33	0.14	1.03	1.28	367	8.78	0.18	1.71	0.27	2.92	0.67	0.0319

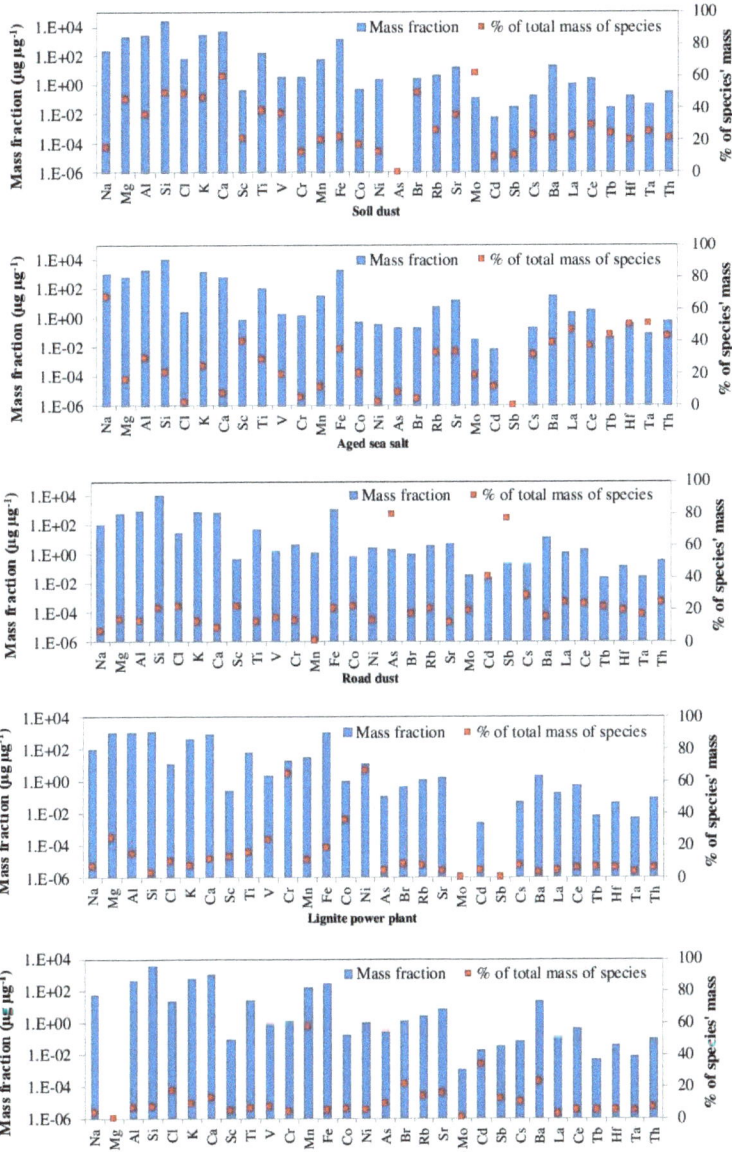

Figure 2. The chemical profiles of the soil dust, aged sea salt, road dust, lignite power plant, and Mn-rich source. The light blue bars (left axis) represent the concentrations as mass fractions, while the red dots (right axis) represent the percentage of each chemical element's mass that is assigned to each source.

The aged sea salt source was identified by the very high contribution of Na. Chlorine was also present in this factor, but in much lower concentrations than expected for the sea salt based on stoichiometry. The Cl/Na ratio for fresh sea salt is equal to 1.8 [59], while the respective ratio in this factor was below 0.01. This indicates the depletion of Cl from this source, suggesting that the sea salt particles deposited on mosses were not fresh and had previously interacted with anthropogenic pollutants [53]. A significant part of the Hf and

Ta mass (above 50%) was also apportioned to this factor. These two elements have been associated with heavy oil combustion [60] and may be indicative of shipping emissions arriving together with sea salt particles.

The dispersed road dust source included high loadings of Sb and As. This factor is related to vehicular traffic and reflects the road dust that has entered into the atmosphere through the wind action or through the turbulent action of motor vehicles passing across road surfaces [61]. More than 70% of the mass of Sb was assigned to this factor. Sb is a common traffic tracer, and is mainly associated with the brake pads in cars [22,37,62–65]. The dispersed road dust source also included high concentrations of crustal elements from the dust deposited in the road and resuspended due to vehicular traffic (such as Fe, Si, and Al), as well as Ba, an element that is also related to brake abrasion [65].

The lignite power plant source was identified by the presence of Ni, Cr, and Co. More than 60% of the Ni and Cr mass was assigned to this source. Ni is a common component of fly ash, while Cr and Co are linked to fossil fuel burning [37,66]. This factor is also influenced by high loadings of the V and Mg, which are also elements included in the fly ash produced by the burning process. A part of the fly ash may escape from the filters of the power plants and be emitted into the air. It should be noted that the contribution of this source is very high in the Skouries area and in West Macedonia, an area greatly impacted by the Ptolemaida lignite power plants. Regarding the area of Skouries, this source may be also related to local rocks, specifically amphibolites, which are considered to have derived from ophiolites, which are rich in Cr, Ni, and Co [67–69].

The Mn-rich source was characterized by high concentrations of Mn and was associated with mining activities. It should be noted that this source was more prevalent in the Skouries samples, further supporting its association with the mining activities in the area. Mixed sulphide and manganese deposits exist in the area north of Skouries, contributing to the observed Mn concentration [69].

The relative source contributions for each sample are presented in Figure 3. The average source contributions for the whole region of Northern Greece are presented in Figure 4. Among the five sources identified by source apportionment, the soil dust source was the one that contributed the most to the elemental concentrations found in mosses (46%), followed by the aged sea salt source and the road dust source.

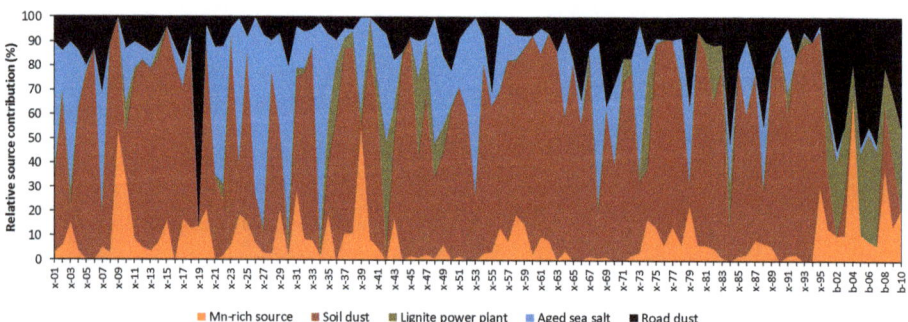

Figure 3. The relative contributions of the sources to each sample.

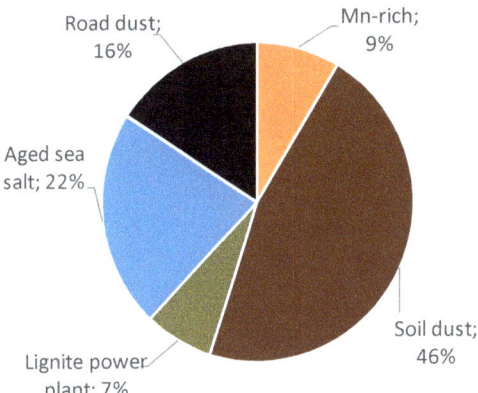

Figure 4. The average source contributions for the region of Northern Greece (based on all 105 moss samples).

4. Discussion

Mosses are ideal bioindicators of trace elements. Moss biomonitoring is an efficient, non-expensive, and widely used method for the determination of the atmospheric deposition of trace elements. In the present work, 105 moss samples were collected from the region of Northern Greece (10 from the area of Skouries, known for its mining activities, and 95 from the remaining part of Northern Greece). They were analyzed using the NAA, and the concentrations of 34 elements were determined. The elemental concentrations were further exploited for the determination of the trace elements' sources through the application of the positive matrix factorization (PMF) model. It should be noted that this is the first time that the PMF model has been used in conjunction with the moss biomonitoring technique for the identification of trace elements' sources in Greece. PMF has also been used for moss elemental concentration source apportionment in a few recent studies in Europe and Asia, with positive results with respect to source identification [39,40].

As seen in Tables 1 and 2, significant differences in the elemental concentrations in mosses of the area of Skouries and the rest of Northern Greece are observed. For example, the mean values of the elements (Fe, Al, Mg, Cl, K, Ca, Sc, Ti, V, Cr, Mn, Co, Ni, As, Br, Ag, Sb, Cs, Ba, La) are higher in the samples collected from the area of Skouries than the rest of Northern Greece. This is more pronounced for Mn and Ni, which were found to be 1.5 times higher in Skouries in comparison to the remaining samples, and for Au, which was 80% higher in Skouries. The above-mentioned elements may be associated with the extraction of gold and, generally, with the mining activities taking place in the Skouries area. The main reason for these observed elevated elemental concentrations in the moss samples of Skouries is the resuspension of the soil dust that is released during the mining activities. The dust, which contains different trace elements, can be easily transferred by air, be deposited in the surrounding area, and be absorbed by the biota [46].

Several other studies concerning the levels of trace elements in different matrices (water, seaweeds, fish, sediments, soil) have been conducted over the last decades in the area of Skouries. For instance, high concentrations of As, Zn, and Mn (320–4100 $\mu g g^{-1}$, 940–4100 $\mu g g^{-1}$, and 3840–26,000 $\mu g g^{-1}$, respectively) have been measured in sediment samples collected close to the coastal mining facilities (Ierissos Gulf) [70]. Furthermore, high concentrations of Cr and Ni have been also measured in surface sediment samples in the area near the load-out facility of the mining operations in Stratoni Bay, and they were considered as the result of the weathering action of the rivers that flow into the Gulf [71,72]. Additional studies that have included water samples from Ierissos Gulf and soil samples from the Stratoni area have shown exceedances of the World Health Organization (WHO) limit values and of the global soil mean values, respectively, mainly for the elements As, Sb, Cr, Ni, Cd, Pb, Mn, and Fe [43,45,73–75].

According to the PMF model results, five sources were identified: soil dust, aged sea salt, road dust, lignite power plants, and a Mn-rich source. Overall, the results indicate that the majority of trace elements found in mosses are related with soil dust particles that have entered the atmosphere through resuspension and transport mechanisms and are subsequently deposited on mosses. Relatively high contributions (on average 26%) from a geogenic source were also found in moss samples collected in Norway [39]. It should be noted that the arid climate of Greece favors the resuspension of the soil and road dust particles; with the aid of the local winds, the particles can easily travel over even further distances, influencing the elemental concentrations in mosses [40]. Nevertheless, anthropogenic sources have been also identified, with the most prevailing source being road dust from vehicular traffic. Xiao et al. performed a similar study in alpine ecosystems in China and also identified vehicular traffic as a major source contribution to moss elemental concentrations (average contribution of 15%) [40]. Sea salt particles were found to contribute significantly to the elemental concentrations as well; the respective source chemical profile displayed a low Cl/Na ratio, pointing towards aged sea spray produced through chemical reactions of fresh sea salt particles with anthropogenic emissions. Cl depletion in atmospheric aerosol has been observed to occur when urban pollutants, such as nitric and sulfuric acid, interact with sea salt particles to form sodium salts [53].

Based on the relative source contributions for each sampling site, the soil dust source and the aged sea salt contributed more than 60% to the majority of the moss samples, while the vehicular-related dust source was present in almost all samples (Figure 3).

The application of the source apportionment methodology led to the identification of two areas with significant impact from anthropogenic activities: the area of Ptolemaida in the region of Western Macedonia (sites x-30–x-42) and the area of Skouries (sites b-01–b-10). These two areas are both linked with intense anthropogenic activities, whose impact can be visualized through the high loadings of key anthropogenic elements in mosses and, correspondingly, high contributions from specific anthropogenic sources, as discussed below.

In the area of Ptolemaida, where there is a complex of lignite power plants, the concentrations of Ni, V, Cr, and Co were found in high levels, and these are directly related with the mining and burning processes of the lignite [37]. The lignite power plant source was found to contribute significantly to moss samples collected from this area (12% in comparison to 4%, which was the overall mean contribution for all samples). The second area, which corresponds to the area of Skouries, is characterized by intense mining activities and displayed high contributions of the Mn-rich source, the lignite power plant source, and the road dust source (21%, 30%, and 42% of the measured concentrations, respectively). The Mn-rich source has been directly associated with the mining activities in the area, while the high contribution of the road dust source may be indicative of the vehicular traffic associated with mining (e.g., transport of materials, etc.). The source identified as lignite power plants may, in this case, be mostly related to the resuspension of local soil dust, as discussed above. It should be noted that this is the first time that the impact of the mining activities in the area has been imprinted on moss biomonitors, and it was simultaneously verified by the source apportionment results. This discrimination of the area of Skouries according to the PMF results confirms the impact that the mining activities have on the environment in the most appropriate way, and it verifies the results of previous studies about the influence of the mining activities on the surrounding area.

Author Contributions: This work is the product of the collaboration of all authors. The specific contributions of each author are provided below: Conceptualization, A.I., K.E., E.D. and C.B.; Methodology, E.D., A.I. and C.B.; Validation, All; Formal analysis, C.B. and E.D.; Investigation, E.T., L.P. and M.F.; Resources, All; Data curation, C.B.; Writing—original draft preparation, C.B. and E.D.; Writing—review and editing, A.I., L.P., E.T., M.F. and K.E.; Visualization, C.B. and E.D.; Supervision,

A.I.; Project administration, A.I.; Funding acquisition, A.I. All authors have read and agreed to the published version of the manuscript."

Funding: This research was financially supported by the General Secretariat for Research and Technology (GSRT) and the Hellenic Foundation for Research and Innovation (HFRI) (Scholarship Code: Contr. No 131183/I2, Code 569, 1st action 2016).

Informed Consent Statement: Not applicable.

Data Availability Statement: Publicly available datasets were analyzed in this study. These data can be found here: https://thesis.ekt.gr/thesisBookReader/id/47451#page/12/mode/2up.

Conflicts of Interest: The authors declare no conflict of interest. The funder had no role in the design of the study; in the collection, analyses, or interpretation of data; in the writing of the manuscript, or in the decision to publish the results.

References

1. Ulrich, B.; Pankrath, J. *Effects of Accumulation of Air Pollutants in Forest Ecosystems*; D. Reidel Publishing Company: Dordrecht, The Netherlands, 1983.
2. Aničić, M.; Tomašević, M.; Tasić, M.; Rajšić, S.; Popović, A.; Frontasyeva, M.V.; Lierhagen, S.; Steinnes, S. Monitoring of trace element atmospheric deposition using dry and wet moss bags: Accumulation capacity versus exposure time. *J. Hazard. Mat.* **2009**, *171*, 182–188. [CrossRef]
3. Meyer, C.; Diaz-de-Quijano, M.; Monna, F.; Franchi, M.; Toussaint, M.; Gilbert, D.; Bernard, N. Characterization and distribution of deposited trace elements transported over long and intermediate distances in north-eastern France using *Sphagnum* peatlands as a sentinel ecosystem. *Atmos. Environ.* **2015**, *101*, 286–293. [CrossRef]
4. Eleftheriadis, K.; Colbeck, I. Coarse atmospheric aerosol: Size distributions of trace elements. *Atmos. Environ.* **2001**, *35*, 5321–5330. [CrossRef]
5. Basile, A.; Sorbo, S.; Pisani, T.; Paoli, L.; Munzi, S.; Loppi, S. Bioacumulation and ultrastructural effects of Cd, Cu, Pb and Zn in the moss *Scorpiurum circinatum* (Brid.) Fleisch. & Loeske. *Environ. Pollut.* **2012**, *166*, 208–211.
6. Gerdol, R.; Marchesini, R.; Iacumin, P.; Brancaleoni, L. Monitoring temporal trends of air pollution in an urban area using mosses and lichens as biomonitors. *Chemosphere* **2014**, *108*, 388–395. [CrossRef] [PubMed]
7. Diapouli, E.; Popovicheva, O.; Kistler, M.; Vratolis, S.; Persiantseva, N.; Timofeev, M.; Kasper-Giebl, A.; Eleftheriadis, K. Physicochemical characterization of aged biomass burning aerosol after long-range transport to Greece from large scale wildfires in Russia and surrounding regions, Summer 2010. *Atmos. Environ.* **2014**, *96*, 393–404. [CrossRef]
8. Frontasyeva, M.; Harmens, H.; Uzhinskiy, A.; Chaligava, O.; the participants of the moss survey. Mosses as biomonitors of air pollution: 2015/2016 survey on heavy metals, nitrogen and POPs in Europe and beyond. In *Report of ICP Vegetation Moss Survey Coordination Center*; Joint Institute for Nuclear Research: Dubna, Russia, 2020; p. 136. ISBN 978-5-9530-0508-1.
9. Adamo, P.; Giordano, S.; Vingiani, S.; Castaldo Cobianchi, R.; Violante, P. Trace element accumulation by moss and lichen exposed in bags in the city of Naples (Italy). *Environ. Pollut.* **2003**, *122*, 91–103. [CrossRef]
10. Szczepaniak, K.; Biziuk, M. Aspects of the biomonitoring studies using mosses and lichens as indicators of metal pollution. *Environ. Res.* **2003**, *93*, 221–230. [CrossRef]
11. Krmar, M.; Wattanavatee, K.; Radnović, D.; Olivka, J.; Bhongsuwan, T.; Frontasyeva, M.V.; Pavlov, S.S. Airborne radionuclides in mosses collected at different latitudes. *J. Environ. Radioact.* **2013**, *117*, 45–48. [CrossRef] [PubMed]
12. Krmar, M.; Radnović, D.; Hansman, J.; Mesaroš, M.; Betsou, C.H.; Jakšić, T.; Vasić, P. Spatial distribution of ^{7}Be and ^{137}Cs measured with the use of biomonitors. *J. Radioanal. Nucl. Chem.* **2018**, *318*, 1845–1854. [CrossRef]
13. Harmens, H.; Norris, D.; Mills, G. The participants of the moss survey. In *Heavy Metals and Nitrogen in Mosses: Spatial Patterns in 2010/2011 and Long-Term Temporal Trends in Europe*; ICP Vegetation Programme Coordination Centre, Centre for Ecology and Hydrology: Bangor, UK, 2013; p. 63.
14. Czarnowska, K.; Rejment-Grochowska, I. Concentration of heavy metals-iron, manganese, zinc and copper in mosses. *Acta Soc. Bot. Pol.* **1974**, *43*, 39–44. [CrossRef]
15. Hristozova, G.; Marinova, S.; Svozilík, E.; Nekhoroshkov, P.; Frontasyeva, M.V. Biomonitoring of elemental atmospheric deposition: Spatial distributions in the 2015/2015 moss survey in Bulgaria. *J. Radioanal. Nucl. Chem.* **2020**, *323*, 839–849. [CrossRef]
16. Rühling, E.; Tyler, G. An ecological approach to the lead problem. *Bot. Notiser* **1968**, *121*, 321–342.
17. Little, P.; Martin, M.H. Biological monitoring of heavy metal pollution. *Environ. Pollut.* **1974**, *6*, 1–19. [CrossRef]
18. Gjengedal, E.; Steinnes, E. Uptake of metal ions in moss from artificial precipitation. *Environ. Monit. Assess.* **1990**, *14*, 77–87. [CrossRef]
19. Steinnes, E.; Rambaek, J.; Hanssen, E. Large scale multi-element survey of atmospheric deposition using naturally growing moss as biomonitor. *Chemosphere* **1992**, *25*, 735–752. [CrossRef]
20. Krmar, M.; Radnović, D.; Hansman, J. Correlation of unsupported ^{210}Pb in soil and moss. *J. Environ. Radioact.* **2014**, *129*, 23–26. [CrossRef]

21. Krmar, M.; Radnović, D.; Hansman, J.; Repić, P. Influence of broadleaf forest vegetation on atmospheric deposition of airborne radionuclides. *J. Environ. Radioact.* **2017**, *177*, 32–36. [CrossRef]
22. Harmens, H.; Norris, D.A.; the participants of the Moss Survey. *Spatial and Temporal Trends in Heavy Metal Accumulation in Mosses in Europe (1990–2005)*; ICP Vegetation Programme Coordination Centre, Centre for Ecology and Hydrology: Bangor, UK, 2008.
23. Harmens, H.; Norris, D.A.; Steinnes, E.; Kubin, E.; Piispanen, J.; Alber, R.; Aleksiayenak, Y.; Blum, O.; Coşkun, M.; Dam, M.; et al. Mosses as biomonitors of atmospheric heavy metal deposition: Spatial patterns and temporal trends in Europe. *Environ. Pollut.* **2010**, *158*, 3144–3156. [CrossRef]
24. Glime, J.M. Bryophyte ecology. In *Physiological Ecology*; Michigan Technological University and the International Association of Bryologists: Houghton, MI, USA, 2017; Volume 1, Available online: http://digitalcommons.mtu.edu/bryophyte-ecology/ (accessed on 25 March 2017).
25. Betsou, C.H.; Tsakiri, E.; Kazakis, N.; Hansman, J.; Krmar, K.; Frontasyeva, M.; Ioannidou, A. Heavy metals and radioactive nuclide concentrations in mosses in Greece. *Radiat. Eff. Defects Solids* **2018**, *173*, 851–856. [CrossRef]
26. Chakrabortty, S.; Paratkar, G.T. Biomoitoring of trace element air pollution using mosses. *Aerosol Air Qual. Res.* **2006**, *6*, 247–258. [CrossRef]
27. Frontasyeva, M.V.; Steinnes, E.; Harmens, H. Monitoring long-term and large-scale deposition of air pollutants based on moss analysis. In *Biomonitoring of Air Pollution Using Mosses and Lichens: Passive and Active Approach—State of the Art and Perspectives*; Aničić Urošević, M., Vuković, G., Tomašević, M., Eds.; Nova Science Publishers: New York, NY, USA, 2016; ISBN 978-1-53610-051-8. Available online: https://novapublishers.com/shop/biomonitoring-of-air-pollution-using-mosses-and-lichens-a-passive-and-active-approach-%E2%80%92-state-of-the-art-research-and-perspectives/ (accessed on 29 December 2020).
28. Holy, M.; Pesch, R.; Schröder, W.; Harmens, H.; Ilyin, I.; Alber, R.; Aleksiayenak, Y.; Blum, O.; Coşkun, M.; Dam, M.; et al. First thorough identification of factors associated with Cd, Hg and Pb concentrations in mosses sampled in the European Surveys 1990, 1995, 2000 and 2005. *J. Atmos. Chem.* **2009**, *63*, 109–124. [CrossRef]
29. Gusev, A.; Iliyn, I.; Rozovskaya, O.; Shatalov, V.; Sokovych, V.; Travnikov, O. Modelling of heavy metals and persistent organic pollutants: New developments. In *EMEP/MSC-E Technical Report 1/2010*; Meteorological Synthesizing Centre e East: Moscow, Russia, 2010; Available online: http://en.msceast.org/reports/1_2010.pdf (accessed on 29 December 2020).
30. Harmens, H.; Mills, G.; Hayes, F.; Norris, D.A.; Sharps, K. Twenty-eight years of ICP Vegetation: An overview of its activities. *Ann. Bot.* **2015**, *5*, 31–43. [CrossRef]
31. Frontasyeva, M.; Harmens, H. *Heavy Metals, Nitrogen and POPs in European Mosses: 2015 Survey—Monitoring Manual, 2015*; ICP Vegetation: Bangor, UK, 2015; Available online: https://icpvegetation.ceh.ac.uk/sites/default/files/Moss%20protocol%20manual.pdf (accessed on 29 December 2020).
32. Qarri, F.; Lazo, P.; Stafilov, T.; Frontasyeva, M.; Harmens, H.; Bekteshi, L.; Baceva, K.; Goryainov, Z. Multi-elements atmospheric deposition in Albania. *Environ. Sci. Pollut. Res.* **2014**, *21*, 2506–2518. [CrossRef] [PubMed]
33. Belis, C.A.; Larsen, B.R.; Amato, F.; El Haddad, I.; Favez, O.; Harrison, R.M.; Hopke, P.K.; Nava, S.; Paatero, P.; Prévôt, A.; et al. European guide on air pollution source apportionment with receptor models—Revised version 2019. In *2019 JRC Technical Reports*; Publications Office of the European Union: Luxembourg, 2019. [CrossRef]
34. Diapouli, E.; Manousakas, M.I.; Vratolis, S.; Vasilatou, V.; Pateraki, S.; Bairachtari, K.A.; Querol, X. AIRUSE-LIFE+: Estimation of natural source contributions to urban ambient air PM10 and PM2.5 concentrations in southern Europe—Implications to compliance with limit values. *Atmos. Chem. Phys.* **2017**, *17*, 3673–3685. [CrossRef]
35. Almeida, S.M.; Manousakas, M.; Diapouli, E.; Kertesz, Z.; Samek, L.; Hristova, E.; Šega, K.; Padilla Alvarez, R.; Belis, C.A.; Eleftheriadis, K. The IAEA European Region Study GROUP. Ambient particulate matter source apportionment using receptor modeling in European and Central Asia urban areas. *Environ. Pollut.* **2020**, *266*, 115199. [CrossRef]
36. Manousakas, M.; Diapouli, E.; Bellis, C.A.; Vasilatou, V.; Gini, M.; Lucarelli, F.; Querol, X.; Eleftheriadis, K. Quantitative assessment of the variability in chemical profiles from source apportionment analysis of PM10 and PM2.5 at different sites within a large metropolitan area. *Environ. Res.* **2021**, *192*, 110257. [CrossRef]
37. Manousakas, M.; Diapouli, E.; Papaefthymiou, H.; Migliori, A.; Padilla-Alvarez, R.; Bogovac, M.; Kaiser, R.B.; Jaksic, M.; Bogdanovic-Radovic, I.; Eleftheriadis, K. Source apportionment by PMF on elemental concentrations obtained by PIXE analysis of PM10 samples collected at the vicinity of lignite power plants and mines in Megalopolis, Greece. *Nucl. Instrum. Methods Phys. Res. B* **2015**, *349*, 114–124. [CrossRef]
38. Belis, C.A.; Pernigotti, D.; Pirovano, G.; Favez, O.; Jaffrezo, J.L.; Kuenen, J.; van Der Gon, H.D.; Reizer, M. Evaluation of receptor and chemical transport models for PM10 source apportionment. *Atmos. Environ.* **2020**, *X5*, 100053. [CrossRef]
39. Christensen, E.R.; Steinnes, E.; Eggen, O.A. Anthropogenic and geogenic mass input of trace elements to moss and natural surface soil in Norway. *Sci. Total Environ.* **2018**, *613–614*, 371–378. [CrossRef]
40. Xiao, J.; Han, X.; Sun, S.; Wang, L.; Rinklebe, J. Heavy metals in different moss species in alpine ecosystems of Mountain Gongga, China: Geochemical characteristics and controlling factors. *Environ. Pollut.* **2020**, in press. [CrossRef]
41. Lazaridou-Dimitriadou, M.; Koukoumides, C.; Lekka, E.; Gaidagis, G. Integrative evaluation of the ecological quality of metalliferous streams (Chalkidiki, Macedonia, Hellas). *Environ. Monit. Assess.* **2004**, *90*, 59–86. [CrossRef]
42. Pappa, F.K. Radioactivity and metal concentrations in marine sediments associated with mining activities in Ierissos Gulf, North Aegean Sea, Greece. *Appl. Radiat. Isot.* **2016**. [CrossRef]

43. Chantzi, P.; Dotsika, E.; Raco, B. Isotope Geochemistry Survey in Ierissos Gulf Basin, North Greece, IOP Conf. Ser. *Earth Environ. Sci.* **2016**, *44*, 052036. [CrossRef]
44. Argyraki, A. Garden soil and house dust as exposure media for lead uptake in the mining village of Stratoni, Greece. *Environ. Geochem. Health* **2016**, *36*, 677–692. [CrossRef] [PubMed]
45. Argyraki, A.; Boutsi, Z.; Zotiadis, V. Towards sustainable remediation of contaminated soil by using diasporic bauxite: Laboratory experiments on soil from the sulfide mining village of Stratoni, Greece. *J. Geochem. Explor.* **2017**, *183*, 214–222. [CrossRef]
46. Hovardas, T. "Battlefields" of blue flags and seahorses: Acts of "fencing" and "defencing" place in a gold mining controversy. *J. Environ. Psychol.* **2017**, *53*, 100–111. [CrossRef]
47. Hill, M.O.; Bell, N.; Bruggeman-Nannenga, M.A.; Brugués, M.; Cano, M.J.; Enroth, J.; Flatberg, K.I.; Frahm, J.-P.; Gallego, M.T.; Garilleti, R.; et al. An annotated checklist of the mosses of Europe and Macaronesia. Bryological Monograph. *J. Bryol.* **2006**, *28*, 198–267. [CrossRef]
48. Pavlov, S.S.; Dmitriev, A.Y.; Chepurchenko, I.A.; Frontasyeva, M.V. Automation system for measurement of gamma-ray spectra of induced activity for neutron activation analysis at the reactor IBR-2 of Frank Laboratory of Neutron Physics at the Joint Institute for Nuclear Research. *Phys. Elem. Part. Nucl.* **2014**, *11*, 737–742. [CrossRef]
49. Belis, C.A.; Larsen, R.; Amato, F.; El Hadad, I.; Olivier, F.; Harisson, R.M.; Hopke, P.K.; Nava, S.; Paatero, P.; Prévot, A.; et al. *Reference Report of the European Guide on Air Pollution Source Apportionment with Receptor Models*; Joint Research Centre Institute for Environment and Sustainability: Ispra, Italy, 2014; pp. 1–92. [CrossRef]
50. *EPA Positive MatrixFactorization (PMF) 5.0 Fundamentals and User Guide*; US EPA: Washington, DC, USA, 2014.
51. Paatero, P.; Tapper, U. Analysis of different modes off actor analysis as least squares fit problems. *Chemom. Intell. Lab. Syst.* **1993**, *18*, 183–194. [CrossRef]
52. Vaccaro, S.; Sobiecka, E.; Contini, S.; Locoro, G.; Free, G.; Gawlik, B.M. The application of positive matrix factorization in the analysis, characterization and detection of contaminated soils. *Chemosphere* **2007**, *69*, 1055–1063. [CrossRef]
53. Diapouli, E.; Manousakas, M.; Vratolis, S.; Vasilatou, V.; Maggos, T.H.; Saraga, T.; Grigoratos, T.H.; Argyropoulos, G.; Voutsa, D.; Samara, C.; et al. Evolution of air pollution source contributions over one decade, derived by PM_{10} and $PM_{2.5}$ source apportionment in two metropolitan urban areas in Greece. *Atmos. Environ.* **2017**, *164*, 416–430. [CrossRef]
54. Dörter, M.; Karadeniz, H.; Saklangıç, U.; Yenisoy-Karakaş, S. The use of passive lichen biomonitoring in combination with positive matrix factor analysis and stable isotopic ratios to assess the metal pollution sources in throughfall deposition of Bolu plain, Turkey. *Ecol. Indic.* **2020**, *113*, 106212. [CrossRef]
55. Yurukova, L.; Tsakiri, E.; Çayir, A. Cross-Border Response of Moss, *Hypnum cupressiforme* Hedw. to Atmospheric Deposition in Southern Bulgaria and Northeastern Greece. *Bull. Environ. Contam. Toxicol.* **2009**, *83*, 174–179. [CrossRef]
56. Betsou, C.H.; Tsakiri, E.; Kazakis, N.; Vasilev, A.; Frontasyeva, M.; Ioannidou, A. Atmospheric deposition of trace elements in Greece using moss *Hypnum cupressiforme* Hedw as biomonitors. *J. Radioanal. Nucl. Chem.* **2019**, *320*, 597–608. [CrossRef]
57. Lage, J.; Wolterbeek, H.T.H.; Reis, M.A.; Chaves, P.C.; Garcia, S.; Almeida, S.M. Source apportionment by positive matrix factorization on elemental concentration obtained in PM10 and biomonitors collected in the vicinities of a steelworks. *J. Radioanal. Nucl. Chem.* **2016**, *309*, 397–404. [CrossRef]
58. Manousakas, M.; Diapouli, E.; Papaefthymiou, H.; Kantarelou, V.; Zarkadas, C.; Kalogridis, A.-C.; Karydas, A.-G.; Eleftheriadis, K. XRF characterization and source apportionment of PM_{10} samples collected in a coastal city. *X-Ray Spectrom.* **2017**, 1–11. [CrossRef]
59. Bowen, H.J.M. *Environmental Chemistry of the Elements*; Academic Press: London, UK, 1979.
60. Aničić, M.; Frontasyeva, M.V.; Tomašević, M.; Popović, A. Assessment of atmospheric deposition of heavy metalsand other elements in Belgrade using the moss biomonitoring technique and neutron activation analysis. *Environ. Monit. Assess.* **2007**, *129*, 207–219. [CrossRef]
61. Davy, P.K.; Trompetter, W.J.; Markwitz, A. *Source Apportionment of Airborne Particles in the Auckland Region: 2010 Analysis*; GNS Science: Wellington, New Zealand, 2011.
62. Amato, F.; Pandolfi, M.; Viana, M.; Querol, X.; Alastuey, A.; Moreno, T. Spatial and chemical patterns of PM10 in road dust deposited in urban environment. *Atmos. Environ.* **2009**, *43*, 1650–1659. [CrossRef]
63. Gianini, M.; Fischer, A.; Gehrig, R.; Ulrich, A.; Wichser, A.; Piot, C.; Hueglin, C. Comparative source apportionment of PM10 in Switzerland for 2008/2009 and 1998/1999 by positive matrix factorization. *Atmos Environ.* **2012**, *54*, 149–158. [CrossRef]
64. Boamponsem, L.K.; De Freitas, C.R.; Wiliams, D. Source apportionment of air pollutants in the Greater Auckland Region of New Zealand using receptor models and elemental levels in the lichen, Parmotrema reticulatum. *Atmos. Pollut. Res.* **2017**, *8*, 101–113. [CrossRef]
65. Sternbeck, J.; Sjodin, A.; Andreasson, K. Metal emissions from road traffic andthe influence of resuspension—Results from two tunnel studies. *Atmos. Environ.* **2002**, *36*, 4735–4744. [CrossRef]
66. Weinstein, J.P.; Hedges, S.R.; Kimbrough, S. Characterization and aerosol mass balance of PM2.5 and PM10 collected in Conakry, Guinea during the 2004 Harmattan period. *Chemosphere* **2010**, *78*, 980–988. [CrossRef]
67. Kockel, F.; Mollat, H.; Walther, H.W. *Erlauterungenzur Geologischen Karte der Chalkidiki und Angrenzender Gebiete 1:100000 (Nord-Griechenland)*; Bundesanstalt fur Geowissenschaften und Rohstoffe: Hannover, Germany, 1977; p. 119.
68. Kydonakis, K.; Brun, J.-P.; Poujol, M.; Monié, P.; Chatzitheodoridis, E. Inference on the Mesozoic evolution of the north Aegean from the isotopic record of the Chalkidiki block. *Tectonophysics* **2016**, *682*, 65–84. [CrossRef]

69. Siron, C.R.; Thompson, J.F.H.; Baker, T.; Friedman, R.; Tsitsanis, P.; Russell, S.; Randall, S.; Mortensen, J. Magmatic and metallogenic framework of Au-Cu porphyry and polymetallic carbonate-hosted replacement deposits of the Kassandra Mining District, Northern Greece. *Soc. Econ. Geol. Spec. Publ.* **2016**, *19*, 29–55.
70. Pappa, F.K.; Tsabaris, C.; Patiris, D.L.; Kokkoris, M.; Vlastou, R. Temporal investigation of radionuclides and heavy metals in a coastal mining area at Ierissos Gulf, Greece. *Environ. Sci. Pollut. Res.* **2019**, *26*, 27457–27469. [CrossRef]
71. Stamatis, N.; Ioannidou, D.; Christoforidis, A.; Koutrakis, E. Sediment pollution by heavy metals in the Strymonikos and Ierissos Gulf, North Aegean Sea, Greece. *Environ. Monit. Assess.* **2001**, *80*, 33–49. [CrossRef] [PubMed]
72. Stamatis, N.; Kamidis, N.; Pigada, P.; Sylaios, G.; Koutrakis, E. Quality Indicators and Possible Ecological Risks of Heavy Metals in the Sediments of three Semi-closed East Mediterranean Gulfs. *Toxics* **2019**, *7*, 30. [CrossRef]
73. Alloway, B.J. *Heavy Metals in Soils*; Blackie Academic and Professional: London, UK, 1995.
74. Kelepertsis, A.; Argyraki, A.; Alexakis, D. Multivariate statistics and spatial interpretation of geochemical data for assessing soil contamination by potentially toxic elements in the mining area of Stratoni, north Greece. *Geochem. Explor. Environ. Anal.* **2006**, *6*, 349–355. [CrossRef]
75. Kelepertzis, E.; Argyraki, A.; Daftsis, E. Geochemical signature of surface water and stream sediments of a mineralized drainage basin at NE Chalkidiki, Greece: A pre-mining survey. *J. Geochem. Explor.* **2012**, *114*, 70–81. [CrossRef]

MDPI
St. Alban-Anlage 66
4052 Basel
Switzerland
Tel. +41 61 683 77 34
Fax +41 61 302 89 18
www.mdpi.com

Atmosphere Editorial Office
E-mail: atmosphere@mdpi.com
www.mdpi.com/journal/atmosphere